D1061970

THE RECEPTION OF COPERNICUS' HELIOCENTRIC THEORY

THE RECEPTION OF COPERNICUS' HELIOCENTRIC THEORY

PROCEEDINGS OF A SYMPOSIUM ORGANIZED BY THE
NICOLAS COPERNICUS COMMITTEE
OF THE INTERNATIONAL UNION OF THE HISTORY AND PHILOSOPHY
OF SCIENCE

TORUŃ, POLAND 1973

Edited by

JERZY DOBRZYCKI

D. REIDEL PUBLISHING COMPANY
DORDRECHT-HOLLAND/BOSTON-U. S. A.

Published in co-edition with OSSOLINEUM,
the Polish Academy of Sciences Press, Wrocław/Warsaw.

The book has been prepared in the series Studia Copernicana, V: Colloquia Copernicana, I, by the Research Centre for the History of Science and Technology of the Polish Academy of Sciences in Warsaw.

Printed in Poland.

Published by D. Reidel Publishing Company,
P. O. Box 17, Dordrecht, Holland

Sold and distributed in the U.S.A., Canada, and Mexico
by D. Reidel Publishing Company, Inc.
306 Dartmouth Street, Boston,
Mass. 02116, U.S.A.

Library of Congress Catalog Card Number 72-95980

ISBN 90 277 0311 6

FOREWORD

In 1965 the International Union of the History and Philosophy of Science founded the Nicolas Copernicus Committee whose main task was to explore the means by which different nations could co-operate in celebrating the 5th centenary of the great scholar's birth.

The committee initiated the publication of a collection of studies dealing with the effect that Copernicus' theory has had on scientific developments in centres of learning all over the world.

An Editorial Board, consisting of J. Dobrzycki (Warsaw), J. R. Ravetz (Leeds), H. Sandblad (Göteborg) and B. Sticker (Hamburg), was nominated. We found that our initiative aroused a lively interest among Copernicus scholars; the present volume, with 11 articles by authors from nine American, Asian and European countries, contains the result of their research. It appears in the series 'Studia Copernicana' by agreement with the Polish Academy of Science, and we hope to publish a number of other contributions in a subsequent volume.

We are happy to say that our efforts have been fruitful and that this volume presents not only several in-depth studies, but also a more general survey of the rules governing the evolution of science, rules set within the framework of Copernicus' theory as it developed among various nations and in various scientific institutions over the centuries.

It has been shown once again that, 500 years after his birth, the work of Copernicus remains a source of scientific interest and continues to stimulate fresh study and research.

Warsaw, January 1972

Jerzy Bukowski
Chairman, Nicolas Copernicus Committee

TABLE OF CONTENTS

ROBERT S. WESTMAN
University of California at Los Angeles

THE COMET AND THE COSMOS: KEPLER, MÄSTLIN AND THE COPERNICAN HYPOTHESIS[1]

Of all the initial factors involved in Kepler's decision to become an advocate of the Copernican system, there is no doubt among historians that a paramount weight must be assigned to the role of his teacher of astronomy at the University of Tübingen, Michael Mästlin (1550–1631)[2]. There are no sceptics here simply because Kepler himself tells us, with typical candor, that it was Mästlin who first acquainted him with the difficulties inherent in the usual opinion of the world and that, subsequently, he was stimulated to pursue and defend the views of Copernicus on mathematical and "Physical or, if you prefer, Metaphysical" grounds[3]. But beyond a few well-known statements by Kepler to this effect, little has come to light on the precise nature of Mästlin's influence nor on his attitude toward the Copernican theory[4]. Did Mästlin then do no more than familiarize Kepler with the fundamentals of the new astronomy? Not at all. Although the evidence is incomplete, enough of it is available to enable us to reconstruct an important discovery by Mästlin that was to affect both his own and Kepler's acceptance of the Copernican hypothesis. That discovery concerned the orbit of the Comet of 1577.

---

[1] An earlier version of this essay was read at the Winter Meeting, 1969 of the British Society for the History of Science. I am immensely grateful to Dr. Chester Raymo (Stonehill College) for his constructive criticisms of that paper and for his valuable suggestions and generous encouragement which have helped to make possible the appearance of this article in its present form. I should also like to thank my colleague, Dr. John G. Burke, for reading and commenting on the final draft. Any errors are, of course, entirely my own.

[2] See, for example, Arthur Koestler, *The Sleepwalkers* (New York, 1963), 247–48; Max Caspar, *Kepler*, trans. C. Doris Hellman (London and New York, 1959), 46–47; William Whewell, *History of the Inductive Sciences*, 3rd ed., I (London, 1857), 293.

[3]. Johannes Kepler, *Gesammelte Werke*, ed. Max Caspar, I (Munich, 1938), 9: 11–21 (hereafter cited as G. W.)

[4] C. Doris Hellman has a lengthy description of Mästlin's work, but she has virtually no discussion of Mästlin's influence on Kepler (*The Comet of 1577: Its Place in the History of Astronomy* (New York, 1944), 137–159).

Our first, and most significant, clue regarding the influence of Mästlin comes from a passage in the *Cosmographic Mystery* which, hitherto, has been overlooked. Here Kepler informs us:

I did not follow this way of life rashly, nor without the profoundest authority of my teacher, the famous Mathematician Mästlin. For he was also my first leader and guide in that [he introduced me] to the art of philosophical argumentation and to other things, and therefore, he ought rightly to receive the highest mention. Indeed, he showed me a... reason [in support of Copernicus] by another special argument: for he found that the Comet of the year '77 moved constantly with respect to the motion of Venus professed by Copernicus; and he conjectured from its superlunary height that [the Comet] completed its orbit in the same orb as the Copernican Venus. But if one considers how easily falsehood is inconsistent with itself and, on the contrary, how the truth is always consistent with itself, then one may perhaps begin from this [argument] alone to correctly understand the most important argument for the arrangement of the Copernican orbs[5].

Kepler reaffirms this point in an early letter to Mästlin in which he summarizes the three principal arguments he intended to use in the *Cosmographic Mystery*: "The third [argument] is yours: the Comet of Venus"[6]. Yet, with the exception of a few scattered references in his later writings, this "most important argument" receives no further mention. The reason for its disappearance will become clear at the end of our account. But let us now turn our full attention to the comet of 1577 and Mästlin's decision to embrace the Copernican theory.

Like Kepler, Mästlin began his studies in pursuit of a career in theology; but, unlike the former, whose vocational plans were altered abruptly by his appointment as an instructor of mathematics at Graz in 1594, Mästlin became a deacon at Backnang in Wurttemberg in 1576. Four years later, we find him ensconced as a professor of mathematics at the University of Heidelberg and, in 1583, he succeeded his former

---

[5] *Neque tamen temere, et sine grauissima praeceptoris mei Maestlini clarissimi Mathematici authoritate, hanc sectam amplexus sum. Nam is etsi primus mihi dux et praemonstrator fuit, cum ad alia, tum praecipue ad haec philosophemata, atque ideo iure primo loco recenseri debuisset: tamen alia quadam peculiari ratione tertiam mihi causam praebuit ita sentiendi: dum Cometam anni 77 deprehendit, constantissime ad motum Veneris a Copernico proditum moueri, et capta ex altitudine superlunari coniectura, in ipso orbe Venerio Copernicano curriculum suum absoluere. Quod si quis secum perpendat, quam facile falsum a seipso dissentiat, et econtra, quam constanter verum vero consonet: non iniuria maximum argumentum dispositionis orbium Copernicanae vel ex hoc solo coeperit* (G. W. I, 16: 39–40 17: 1–10).

[6] The entire passage reads as follows: "First, I offer some theses with regard to Sacred Letters and I show how, in a like manner, its authority agrees with these [Copernican theses] and, if anyone should protest that Copernicus is not in agreement [with Scripture], then it will be impossible [for him] to refute [Copernicus] by such an appeal. Next, I shall put forth three reasons by which I have always adhered to the Copernican motions [quibus motus semper Copernico adhaeserim]. The first is astronomical, where I shall reply to those who believe that the true occasionally follows from the false and that, therefore, it is most likely that Copernicus assumed a false [premise] so that he might demonstrate that even from this [falsehood] the truth might follow most beautifully. I deny this possibility. The next premise is Physical, where I promise that I shall defend the entire mass of Copernican hypotheses much more correctly and uniquely from Nature [ex Natura fontibus]. The third [argument] is yours: the Comet of Venus. On the basis of these premises I agree with the substance itself". (Kepler to Mästlin, October 3, 1595, G. W. XIII (1945), 34–35: 43–52).

teacher, Philip Apian (d. 1589) at the University of Tübingen[7]. It was presumably while he was a student that he observed the supernova of 1572. We know of his findings from a reprint which Tycho Brahe published in his *Astronomiae Instauratae Progymnasmata*[8].

Three indications of Mästlin's early attitude toward the Copernican theory stand forth in this brief work. First, Mästlin claimed that the new celestial light had neither a "natural cause" nor did it resemble a comet — "unless we wish to say against Aristotle and all the Physicists and Astronomers that Comets can be generated not only in the Elementary Region but also in the Stellar Orb, which, according to Copernicus, is the outermost sky containing everything in itself, and, therefore, the Sky is not free from generation and corruption". Here he departs from Copernicus' belief that comets exist in "the highest regions of the air"[9]; instead, he maintained the possibility that imperfect, transient objects, such as comets, might flourish above the moon in the celestial region. Secondly, Mästlin attributed his inability to measure the star's magnitude and "Height from the Center of the World" to the "immense Height of the Stellar Orb" whose distance, according to Copernicus, was unknown. And finally, we are told that Copernicus, "Prince of Astronomers after Ptolemy", had demonstrated "the certain distances of all the Planetary Orbits from the Center of the World".

One is impressed as much by Mästlin's concern with celestial distances as by his conclusion that there must be change in the hitherto uncorruptible heavenly sphere. Not only did the size of the Copernican universe offer an explanation of why the parallax of the New Star could not be obtained, it also contained a persuasive assessment of the distances of the planetary orbs. This interest again surfaces in Mästlin's lectures at Tübingen where he used tables of planetary distances whose values, he told Kepler, "I copied straight down from Copernicus himself"[10]. And it is altogether likely that Mästlin's early concerns lay behind Kepler's great desire to verify his well-known polyhedral hypothesis, purporting to show the metaphysical causes of the mean distances among the six planets, using the most reliable data available. Indeed, it was Mästlin who constantly stressed the need to "reform the motions of the celestial bodies" according to more accurate observations[11], who

---

[7] For further biographical details, consult Hellman, *op. cit.*, 139 ff. and Caspar, *Kepler*, *op. cit.*, 46 ff.

[8] Tycho Brahe, *Opera Omnia*, ed. I. L. E. Dreyer, III (Copenhagen, 1916), 58—62 (hereafter cited as "Brahe").

[9] *De revolutionibus orbium caelestium*, Book I, Chapter 8.

[10] Mästlin to Kepler, Feb. 27, 1596, G. W. XIII, 55: 20—23. Excerpts from these tables appear in Kepler's correspondence until he acquires his own personal copy of the *De revolutionibus* (Nuremberg, 1543) sometime between September 14 and October 3, 1595. I should like to thank Dr. Martha List (Kepler-Kommission der Bayerischen Akademie der Wissenschaften) for providing information on the date when Kepler first obtained his volume of Copernicus' work.

[11] See, for example, Mästlin's letter to the Prorector of the University of Tübingen, End of May, 1596, G. W. XIII, 84: 24 ff.

expressed doubt about discrepancies between Kepler's beloved scheme of the regular solids and the relative planetary distances[12] and who provided supportive evidence for his pupil's theory by affixing an appendix to the *Cosmographic Mystery* entitled, "De Dimensionibus Orbium et Sphaerarum Coelestium Iuxta Tabulas Prutenicas, ex sententia Nicolai Copernici"[13]. If the much praised relationship between Tycho and Kepler was a vital one, then we have surely to thank Mästlin for setting Kepler on the road to their meeting. And yet, in spite of Mästlin's early attraction to the harmonious dimensions of the Copernican universe, there is so far no convincing testimony to cause us to believe that, in 1572, he accepted it as the true system of the world.

The appearance of a new and errant body in the heavens of November 1577 was a momentous event not only for Mästlin but also for a great many writers with an awe of its astrological consequences, yet no training in astronomy[14]. It was destined to become one of the greatest subjects of astronomical investigation in the last quarter of the sixteenth century. Of the many observers who witnessed this "fiery apparition" there were five who concluded that it was located in the area beyond the moon: Tycho Brahe[15], Helisaeus Roeslin (1544–1616), a physician and astrologer[16], William IV, Landgrave of Hesse Cassel (1532–92)[17] Cornelius Gemma (1535–79)[18] and Mästlin[19]. As a student and even as a child Kepler heard much about this comet[20].

What effect did the absence of measurable cometary parallax have in persuading these diverse observers to favorably consider the Copernican theory? Generally

---

[12] Ibid., 54: 10 ff.

[13] Ibid., 56: 63 ff.; I, 132–145.

[14] See Hellman, *op. cit.*, Chapters 6–8.

[15] Brahe, *De Mundi Aetherei... Phaenomenis* (1588), IV. On those who believed in the comet's supra-lunar position, see Hellman, Chapter 4.

[16] See Plate No. 1. Helisaeus Roeslin, *Theoria Nova Coelestium ΜΕΤΕΩΡΩΝ*, Argentorati, 1578. Roeslin later adopted a geoheliocentric universe which he portrayed in comparison with the other two cosmologies (*De Opere Dei Creationis*, Frankfort, 1597); cf. Brahe, 306–316.

[17] Brahe, 212–245.

[18] Ibid., 288–305.

[19] Ibid., 245–288. Michael Mästlin, *Observatio et demonstratio cometae aetherae qui anno 1577 et 1578 constitutus in sphaera Veneris apparuit cum admirandius eius passionibus varietate scilicet motus loco orbe distantia a terro centro etc. adhibitis demonstrationibus geometricis et calculo arithmetico cuius modi de alio quoquam cometa nunquam visa est*, Tübingen, 1578 (hereafter cited as "Mästlin").

[20] In 1603, Kepler reported to his friend David Fabricius concerning the comets which he had observed: "The problem is very obscure. I have not seen many; I saw the one in Nov. '96; I never was able to see the one of the year '93. I did not even hear about the comet of the year '90. Concerning the Comet of the year '77 I heard the most of all; I was even led up to a high place by my mother in order to view it, although I was but a boy [of six years]. I think that I saw one or the other of the years '80 or '82 for I saw the brightest star and the shortest tail; but I think that I was only a boy of ten years [at the time]" (Kepler to David Fabricius, July 4, 1603, XIV (1949), 416: 285–92).

speaking, historians have called attention to the comet's role in undermining some of the essential features of the accepted cosmology. The new evidence contradicted Aristotle's widelyheld theory of comets as atmospheric or meteorological phenomena — dense, dry exhalations from the earth which burned until consumed in the highest portions of the atmosphere[21]. It called into question the fundamental dichotomy between the perfect celestial region and the corruptible sublunary sphere by showing that the heavens were not immutable[22]. And, it has been argued, if indeed the comet was moving in the area above the moon, then its path would intersect the crystalline spheres which, for those astronomers who countenanced their reality, meant that they had to be abandoned[23]. Thomas Kuhn states this case succintly when he maintains that any break with the Aristotelian cosmological tradition worked for the Copernicans.

> Somehow, in the century after Copernicus' death, all novelties of astronomical observation and theory, whether or not provided by Copernicans, turned themselves into evidence for the Copernican theory. That theory, we should say, was proving its fruitfulness. But, at least in the case of comets and novas, the proof is very strange, for the observations of comets and novas have nothing whatsoever to do with the earth's motion. They could have been made and interpreted by a Ptolemaic astronomer just as readily as by a Copernican[24].

In a broad sense, I think, Kuhn's intuition is correct: the existence of an alternative cosmological scheme opened the way to a radically different interpretation of long-known phenomena. But while specific issues, such as the rejection of crystalline spheres and the imperfection of the celestial sky, might urge further examination of the new world view, they were not *sufficient* conditions for the overthrow of the ancient one. For such adverse testimony to the physical foundations of the old cosmos said nothing about its mathematical structure. In an interesting and unforeseen way, however, the comet of 1577 did force a reconsideration of the old order of the planets. And, at least in Mästlin's instance, this reordering of the planets followed directly from his attempts to compute the comet's orbit.

The desire to find an orbit for a purely transitory, ephemeral phenomenon marked an important shift in the theoretical interpretation of comets: an atmospheric effect now came to be treated as though it were a celestial body and, therefore, subject to all astronomical theories and techniques of mensuration. In principle, this was a fundamental step in the right direction; in practice, however, the techniques developed in planetary astronomy were to prove of limited value in determining the comet's orbit. The information obtained by Mästlin for the comet of 1577 differed from ordinary planetary data in two significant ways. In the first place, with only a small

---

[21] *Meteorologica* 344a 9–344b 18.

[22] Marie Boas, *The Scientific Renaissance: 1450–1630* (New York, 1962), 119; Thomas S. Kuhn, *The Copernican Revolution* (New York, 1959), 207.

[23] Ibid., Boas, 113–115.

[24] Kuhn, *op. cit.*, 208.

section of the orbit with which to work (about 60° of motion in longitude) and the absence of conjunctions and oppositions, Mästlin was forced to base his calculations of the comet's mean motion in its entire orbit upon its average motion over roughly one-sixth of the complete revolution. Secondly, the great inclination of the plane of the comet to the ecliptic prevented a separate treatment of the latitudes as was customary in planetary astronomy. This was the first time that any astronomer had attempted to reconstruct the orbit of a body moving out of the plane of the ecliptic by more than a few degrees; and many of Mästlin's difficulties and his greatest errors may be traced to this source[25].

Yet, if Mästlin's solutions to these new problems were frequently founded upon arbitrary assumptions and trial and error, they also reveal a measure of genuine innovation. Had his work not been so rapidly superseded and overshadowed by Tycho Brahe's comprehensive study of the comet, in which the latter put forth a geoheliocentric orbit for the new body, then perhaps Mästlin's contribution would be more widely recognized today. For it was not Tycho who directly influenced Kepler, but Mästlin. And this is significant because Kepler's cautious teacher believed that only the Copernican hypothesis could save the appearances of the comet. If we are therefore to reconstruct the arguments used by Mästlin in his lectures at Tübingen and their effect upon Kepler, then we must try to discover how it was that he reached this important conclusion. Can we be sure, for example, that Mästlin's decision to use heliocentric astronomy to find the comet's trajectory was the only viable alternative open to him? Could not the comet's motion have been saved by some combination of Ptolemaic epicycles? Was Mästlin's investigation of the comet in fact a necessary factor in converting him to Copernicanism? These are the main questions that we shall attempt to answer below.

Having determined that the comet showed no parallax, Mästlin's first task was to find its longitude and latitude with respect to the ecliptic[26]. This he did using only a piece of thread held up to the eye and without even the aid of an astrolabe. His method for locating the comet's position was to choose four reference stars whose coordinates were known, picking them in such a way that the comet lay on the great circle passing between pairs of them. The point of intersection of the two circles then set the true angular position of the comet. This point could be easily

---

[25] Observations of cometary appearances before 1577 had not been used to establish an orbit for any particular comet. However, Paolo dal Toscanelli (1397–1482), who observed Halley's Comet in 1456, left enough detailed observations for the comet of 1433 such that its orbit might have been calculated (Hellman, *op. cit.*, 74).

[26] In chapters 3 and 4, Mästlin proves to himself that the Comet cannot be located in the elementary region and then proceeds to argue against the many advocates of "Aristotelica Philosophia" who find nothing in the works of the Master in favor of this view. But, wrote Mästlin, using a familiar weapon against the Peripatetics, had Aristotle known of the method of parallax he would have changed his opinion (Mästlin, 16–17).

calculated employing a simple trigonometric theorem[27]. In Table I, a summary of Mästlin's actual observations is presented indicating the date and time of observation, the comparison stars used and the computed angular position of the comet. In all cases, four stars must have been employed although Mästlin does not always list that many. The longitudes and latitudes of the reference stars were taken directly from Copernicus' stellar tables[28].

TABLE 1[29]

MÄSTLIN'S OBSERVATIONS OF THE COMET OF 1577 WITH CALCULATIONS OF ITS LONGITUDE AND LATITUDE, NOVEMBER 12, 1577-JANUARY 8, 1578

Date	Time	Reference Stars		Comet	
		Long.	Lat.*	Long.	Lat.
Nov.					
12	6 a. m.	188°00′	44°30′	245°51′	7°05′
		251 00	1 30	(3 43 p)	
		270 40	7 30		
		319 20	4 30		
17	8 a. m.	263 00	15 30	262 58	15 26
		272 20	50 30	(20 50 þ)	
		270 40	7 30		
		300 20	23 00		
24	6 a. m.	—	—	278 35	21 18
				(5 47 ♒)	

[27] Ibid., chapter 5. In modern terminology: Given a spherical right triangle *ABC* with right angle *C*; then, $\sin b = \tan a \cot A$.

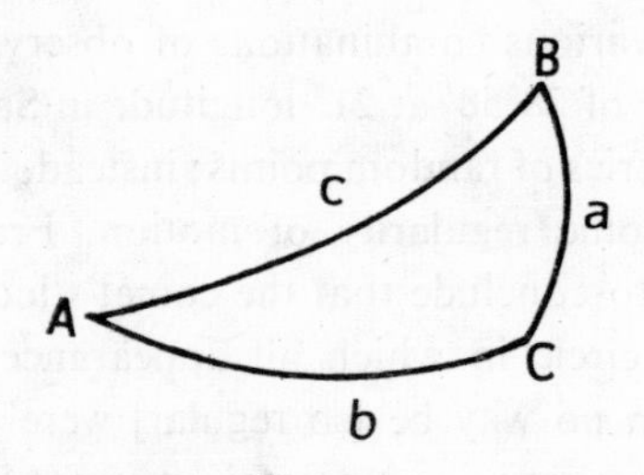

[28] Ibid., 31. See *De revolutionibus*, Book 2.

[29] Based upon Ibid., 28—34. Mästlin made a total of eight observations of the Comet *quantum mihi nubes coelum obtegentes concedebant* (30). At the end of his treatise, however, he presents a table of data which furnishes information on the comet's motion from November 5 (one week before Mästlin had even seen the object) to January 10. These figures were calculated from his hypothesis of the comet's orbit.

TABLE 1 (continued)

Date	Time	Reference Stars		Comet	
Dec.					
2	6 p. m.	267 10	29 10	289 25	24 46
		289 30	25 10	(17 17 ♒)	
		290 50	24 50		
	9 p. m.	—	—	289 33	24 47
				(17 25 ♒)	
7	6 p. m.	267 50	49 00	298 25	21 30
		298 40	21 30	(26 20 ♒)	
	9 p. m.	—	—	298 35	21 30
				(26 30 ♒)	
15	6 a. m.	319 30	29 00	301 48	27 20
		272 20	19 10	(12 40 ♒)	
		302 40	16 50		
		300 00	44 00		
31	6 p. m.	311 00	34 15	311 38	28 32
		312 10	18 00	(9 30)–()	
		313 50	16 00		
Jan.					
8	6 (?)	—	—	314 40	28 40
				(12 32)–()	

* Unless otherwise noted, all latitudes are North.

Now, any pairs of these observations would have allowed Mästlin to establish the great circle of the comet and the longitude of the intersection of that circle with the ecliptic together with its angle at that point. And clearly, if all points do, in fact, lie on the same great circle, then all pairs of data should give the same values for the angle of inclination of the comet's plane. This is essentially what Mästlin discovered when he tried various combinations of observations[30]. But the determination of a plane (an angle of 28°58′ at 21° longitude in Sagittarius)[31] suggested that he was not dealing with a series of random points; instead, it implied that the phenomenon possessed at least some regularity of motion. From this evidence Mästlin had an additional reason to conclude that the comet's location was not in the Elementary World: "... the circle in which all appearances occur regularly in time proves this: for it could in no way be [so regular] were it excited like those other exhalations which wander about in an uncertain manner illuminated in some higher region of the air[32]".

---

[30] The value for the inclination of the plane seems to have been found from the observations of Nov. 12 and Dec. 15.

[31] Mästlin, 35.

[32] Ibid., 34.

Having discovered the comet's plane, Mästlin could then turn to the next problem — the determination of its circle. Here he recognized at once that distortions would be created by the inclination of the plane: "... the obliquity brings forth a huge diversity in the phenomena as the Zodiac and Equinoctial testify [i. e. as one knows from relating angles between those two planes]; for the comet's motion will not be in the Ecliptic, but in its own obliquely-numbered circle"[33]. He resolved this question to his own satisfaction by dividing one quadrant of the comet's circle into 90°, measured from the node at 21° Sagittarius, and then computing the longitudes and latitudes of each of these points through spherical trigonometry. From these calculations, he constructed a table showing the correspondence between angles in the ecliptic and those in the cometary circle[34]. Using this table Mästlin could now convert each of his observations made between November 12, 1577 and January 8, 1578 into angles in the comet's circle. It should be noted that he excluded the observation for December 7 — no doubt because the computed latitude looked somewhat out of line[35]. The results are shown below.

TABLE 2[36]

RELATIONSHIP BETWEEN THE COMET'S ANGULAR POSITION IN THE ECLIPTIC AND ITS ANGULAR POSITION IN ITS OWN CIRCLE

Tabella motus Cometæ in suo Circulo.

	Locus Cometæ in Zodiaco.			Motus Cometæ in suo Circulo.	
	Sig.	g̃	scr	g̃	scr
Die 12. Nouembris.	♑	3	43	14	29
17	♑	20	50	33	15
24	♒	5	47	48	37
Die 2. Decembris.	♒	17	17	59	43
15	♒	29	40	71	10
31	♓	9	30	79	54
Die 8. Ianuarij.	♓	12	32	82	36

What struck Mästlin, upon examination of this scheme, was the great change in the velocity of the comet: it decreased from almost 4° /day to about 20′/ day in the space of nearly 90° of revolution[37]. How could this surprising, radical inequality

---

[33] Ibid.

[34] Ibid., 36.

[35] See Table 1.

[36] Reproduced from Mästlin, 37.

[37] *Hos motus si quis examini subijciat, animaduertet, etsi magna fuerit eorum inaequalitas, qua factum est, ut intra primos quinqs dies Cometa confecerit de suo circulo 18. gradus cum dodrante: in fine vero diebus octo non nisi 2. gradus cum besse: nihilominus tamen sub hac anomalia admirandum*

of motion be accomodated to the axiom of uniform, circular motion? Up to this point, Mästlin's procedure had been fairly straightforward; now, however, he found himself in the throes of a serious dilemma: "The difficulty, of course, is that I began to attack the problem from a very small arc so that I might consider what would be discovered in the entire revolution of the hitherto unknown body"[38]. But all of his efforts to find a geocentric orbit that would save these extraordinary appearances were in vain.

> Although I contemplated and played with the spheres with unwearying mediation, according to the customary hypotheses, I discovered nothing from all this work that might save the motion of the comet. Still, I could not give up in defeat: having put aside these [old hypotheses], I withdrew to the world harmony of Copernicus, after Ptolemy, truly the Prince of Astronomers ... Indeed, the usual hypotheses utterly reject the anomaly of this comet; but that other [hypothesis] does not; rather, it reconciles the greatest inequality of the appearances most remarkably well in order to yield the greatest equality...[39]

At this critical junction in his treatise, Mästlin keeps a virtual silence on the "usual hypotheses" that he evidently tried without success and the special problems which resisted solution. It is unfortunate that he does not tell us more here for we would then have a clear understanding of the causes for his abandonment of the Ptolemaic theory. That habit of complete candor, so characteristic of the student Kepler, was not a prime virtue of the teacher. Yet, we are not completely unenlightened. Mästlin has left to us a few threads of testimony from which we may hope to reconstruct the fabric of his unsuccessful efforts; and through our knowledge of the comet's actual orbit, we have some further basis for interpreting these fragmentary references.

Let us review the situation to this point. Mästlin had three parameters with which to work: (1) the plane of the comet's orbit; (2) the comet's longitudes and latitudes projected from the ecliptic into its own plane; (3) its angular velocity with radical variation over approximately 90° of motion in longitude. However, without the usual parameters of planetary computation – especially a value for mean motion, from which the eccentricity is derived – Mästlin would have lacked the basic material (and there is nothing which he could have withheld) necessary to save the motion of the comet. In the circumstances, there was only one remaining course of action: conjecture.

There exist two bare clues that suggest the possible direction of Mästlin's thought at this point. First, he inferred from the comet's variable motion that its appearances could best be saved with an epicycle. He wrote: "... it [the comet] had the most remarkable velocity in November which, little by little, grew slower until finally,

---

aequalitatem latere, quae non permiserit, ut velocitas illa subito quiesceret, sed sensim tantum, certa seruata proportione, remitteret (Ibid.).

[38] *Arduum sanè est, quod tentare coepi, ut ex paruulo arcu integram conuersionem corporis ante non cogniti, me inuenturum considerem...* (Ibid.)

[39] Ibid., 37–38.

just after this time, it came to rest entirely; this proves that its place was not in some concentric or eccentric orb travelling about the earth, but that is complied generally with the command of an epicycle"[40]. Secondly, Mästlin must have taken the important theoretical step of assuming that the comet's orbit was contained within one of the planetary orbs. This was certainly a natural assumption to proceed from insofar as it required no radical adjustment of the frame of the world (e. g. no additional orbs) and no discarding of the solid spheres. There was, however, another precious advantage to be gained from this supposition: if Mästlin could show that the parameters of one of the planets could be applied to his data from the comet, then he could borrow those values in determining the comet's path. This he ultimately succeeded in doing on the Copernican hypothesis by assigning the comet to the sphere of Venus. Why, then, had he failed to arrive at a similar interpretation utilizing the Ptolemaic hypothesis?

In the absence of more specific assistance from Mästlin, it seems most reasonable to conclude that he simply tried "all the spheres", as he said, attempting by trial and error to find a deferent and epicycle that would accord with the appearances. First, he must have placed the comet on the epicycles of the various planets. This hypothesis evidently did not work. For even if he had tried the eipcycle of Venus, he would have found that the comet was moving in the wrong direction and in the wrong part of the sky. Subsequently, he might have undertaken to add a separate epicycle for the comet to each of the planetary deferents. As we shall see later, such a scheme would have yielded satisfactory results in the case of Venus only with the use of a circle of libration, a device which he apparently did not use at this time. It seems likely, then, that Mästlin was truly forced to give up at some point upon realizing that his trials were leading nowhere. But since he had already developed some confidence in the Copernican system, while investigating the nova of 1572, it is perhaps not surprising that he should have turned to the heliocentric hypothesis in search of similar aid for the comet. Cautiously, he wrote: "I do not wish to approve it [the Copernican hypothesis] on the grounds that I have been deceived and fascinated by love of novelty, but rather, compelled by necessity, I came to it with reluctance..."[41] Mästlin's unwillingness to declare a public conviction in the physical reality of the Copernican system would not prevent him from using all of the computational benefits which it had to offer.

Our analysis has so far traced the main source of Mästlin's difficulties to his lack of sufficient information in order to construct an orbit for the "bearded star". Yet had his data been more complete he would still have expressed amazement at the great dissimilarity between the comet's motion and the normal trajectories of planets. Through modern cometary theory, hindsight reveals another source of the

---

[40] *Primo autem mira eius velocitas in Novembri, quae paulatim deinde remissior facta est, donec quàm proxime quiesceret totus, euincit, sedem eius non fuisse in aliquo orbe concentrico vel eccentrico terram ambiente, sed eum ommino ductu epicycli cuiusdam obseruare* (Ibid., 40).

[41] Ibid., 54.

obstacles which confronted Mästlin. It is now believed that comets, as members of the solar system, obey Newton's Law of Universal Gravitation and follow paths which are known, generally, as conics (i. e. the ellipse, parabola and hyperbola) with the sun located at one focus[42]. Unlike the planets, the orbits of comets are usually quite elongated and, upon approaching the sun, as Edmund Halley wrote, „by their Falls acquire such Velocity, as that they may again run off into the remotest

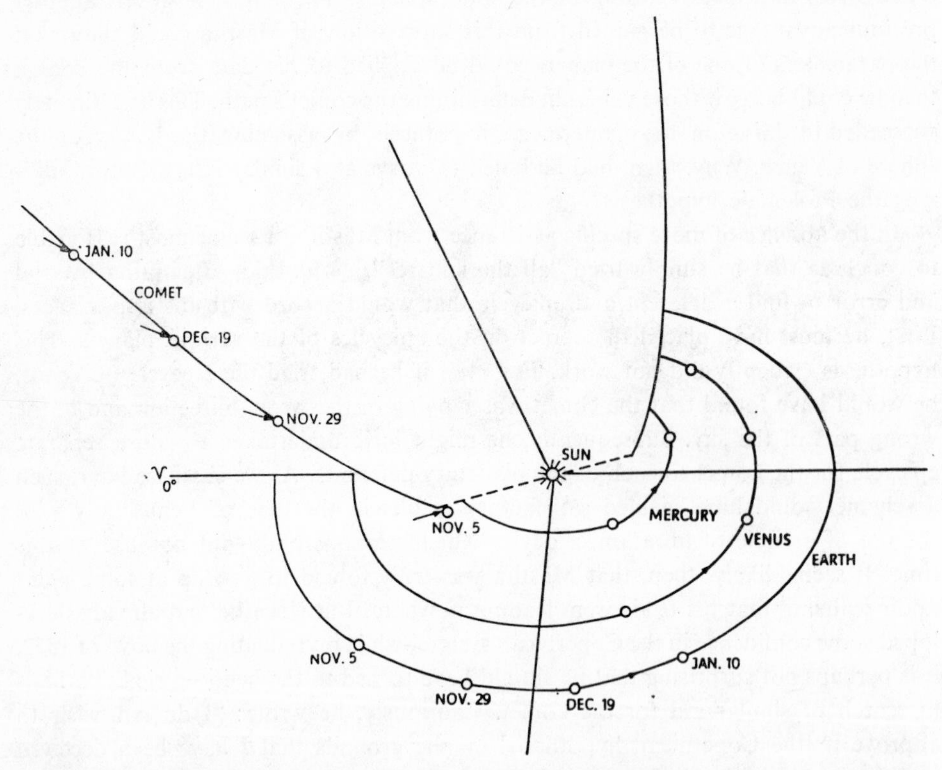

Figure 1. Three-Dimensional Reconstruction of the Orbit of the Comet of 1577. The axis of the orbit is inclined to the ecliptic at 104°. Several positions of the Earth are shown with corresponding positions of the Comet over the period of observation. Positions of the inferior planets are given for reference. All dates are on the Julian Calendar.

Parts of the Universe, moving upwards with such a perpetual Tendency, as never to return again to the Sun"[43]. And it was Halley who first claimed that the orbit of the comet of 1577 was *parabolic*. Using Tycho's figures, Halley concluded that the inclination of the orbit to the ecliptic was 74°32′ 45″[44] — a figure that was accep-

[42] Cf. M. Proctor and A. C. D. Crommelin, *Comets; their nature, origin, and place in the science of astronomy*, London, 1957.

[43] Edmund Halley, *A Synopsis of the Astronomy of Comets*, trans. from Original, printed at Oxford (London, 1705), 20.

[44] Ibid., 7.

ted by J. J. De Lalande in the eighteenth century[45] and evidently remained uncontested until F. Wolstedt's inaugural dissertation at Helsinki (1844)[46]. The most recent catalogue of cometary orbits, upon which our diagram is based (Figure 1), lists a value of 104° with perihelion on October 27, 1577[47]. Little did Mästlin know what monster he had by the tail.

A glance at Figure 1, showing a modern reconstruction of the orbit of the comet of 1577, should impress upon the reader the very formidable task facing Mästlin. This diagram demonstrates the quite steep angle of inclination of the comet to the ecliptic as well as the period of the comet's course during which Mästlin made his observations. The great inclination of the plane would explain the unusual changes in latitude found by Mästlin, even on the geocentric model. It is significant that he never bothered to work out the comet's angle of inclination on a heliocentric interpretation, for this would have ruined his final solution. Faced with these new apparent motions, plagued by the inadequacy of his data and limited by the Platonic axiom that all celestial motions must be circular and uniform, it is a wonder that Mästlin did not abandon his astronomical investigations and merely close with a dire astrological forecast. But he did not.

> Having examined everything in Copernicus' book, [wrote Mästlin] I discovered at last a certain orb, in Book 6, Chapter 2, where the latitudes of Venus are explained; and since I found out that its size and revolution corresponded to, and satisfied, the Comet's appearances exactly (as will be demonstrated in the following chapter): it was then established that the Comet chose no other place than the Sphere of Venus itself[48].

Now had Mästlin found in Copernicus' theory of Venus a solution to the problem which had eluded him using geocentric premises? Was his hypothesis entirely arbitrary? Naturally, we have more evidence for a reintegration of his successful approach than we do for his failures. Below, I shall suggest the most probable route which Mästlin followed. Of the possibilities open to him, there was at least one basic factor which pointed toward Venus. This initial clue was, in all likelihood, the comet's limited elongation from the sun — although one cannot absolutely ex-

---

[45] J. J. De Lalande, *Astronomie*, III (Paris, 1771), 365.

[46] F. Wolstedt, *De gradu praecisionis positionum Cometae anni 1577 a celeberrimo Tychone Brahe per distantias a stellis fixis mensuratas determinatarum et de fide elementorum orbitae, quae ex illis positionibus deduci possunt*, Helsingforsae, 1844.

[47] F. Baldet and G. de Obaldia report the following elements for the comet of 1577: Time of perihelion passage = Oct. 27, 4477 (Julian); argument of perihelion = 255°, 6400; longitude of ascending node = 25°, 3400; inclination of orbit = 104°, 8383; perihelion distance = 0, 17750 (*Catalogue général des orbites de Comètes de l'an 466 à 1952* (Paris, 1952), 12–13). These data represent corrections of information published in *Astronomische Nachrichten* (24, (1845), 7–8). The figures submitted to that journal were the results of Wolstedt's dissertation; hence, data from which the acutal orbit has been computed derive, ultimately, from Tycho Brahe.

[48] Mästlin, 38.

clude the possibility that he had found inspiration in a passage from the ninth-century Arab astrologer, Albumasar (786—886), in which the latter claimed to have observed a comet above the moon in the sphere of Venus[49]. With angular measurements alone, Mästlin found that, "the Comet submits itself to those same motions [passiones] to which Venus and Mercury are addicted, in digressing from the sun and returning in the same certain interval, and likewise borrowing the sun's mean motion in the limit of its greatest distance... Although a place was given to it in Mer cury's sphere, the narrowness of the sphere did not sustain it: for the Comet digressed perhaps 60° from the sun. Therefore, the only sphere that would suffice is that of Venus"[50].

From here Mästlin could postulate further parameters of the comet by identifying it fully with Venus — and this he did by adopting the magnitude of Venus' eccentricity (246/10000 parts) and its angular orientation (16°13′ II)[51]. Could a satisfactory heliostatic orbit now be found which would fit within the sphere of Venus? Consider

---

[49] In a very stimulating article, Professor Willy Hartner has suggested that Tycho Brahe's hypothesis on the comet was inspired by this passage from Albumasar, through an intermediary, Jerome Cardan (1501-1576). As Tycho is known to have placed considerable faith in astrological authorities, Hartner proposes that he turned to this account of Albumasar and used it as the initial basis for his own solution of the Comet's orbit. (Cf. W. Hartner, *Tycho Brahe et Albumasar*, in *La science au seizième siècle* (Paris, 1960, 137—150). Professor Hartner's essay caused me to wonder whether perhaps Mästlin, knew of Albumasar's report and whether it served him as a working hypothesis. A brief search of Mästlin's treatise on the Comet of 1577 turned up the following reference: "Albumasar, qui circa annum Christi 844 floruit, refert, suo aeuo quendam Cometam supra Veneris sphaeram conspectum esse. Quae huius fuisset causa Physica, si Cometarum alius nullus locus, quàm elementaris regio, nec alia materia, quam exhalationum fumi, credenda est?" (Mästlin, 19).

It is certainly reasonable to assume that Mästlin was encouraged by Albumasar's statement to view the comet as a planet — in any case, it would certainly have affirmed the *possibility* of a comet in one of the planetary orbs. But, in context, Mästlin cites Albumasar for the ostensible reason that he was an old and respectable authority (*ex historijs probari potest*) whose testimony lent support to the supra-lunar position of comets and the corruptibility of the heavens (*Ex his duobus exemplis* [i. e. Albumasar and the nova of 1572] *liquidò patet, coelum non planè omnis alterationis expers, nec tamen quicquam eius perfectioni, contagione peregrinae impressionis derogatum esse.* Ibid., 20). There was nothing unusual about this use of ancient authorities to support a position; it was a normal feature of Renaissance proofs. Elsewhere, when discussing his attempt to find the comet's orbit, Mästlin makes no mention of Albumasar; he merely says that he proceeded through trial and error (cf. pp. 00 and 00 above). Even after he had discovered a place for the Comet in the sphere of Venus he did not cite the authority of Albumasar in support of this conclusion. But this is perhaps understandable: presumably, Albumasar's comet, if it had an orbit, possessed a geocentric one; and Mästlin had already failed to discover such an orbit on the old scheme. Thus, had Mästlin interpreted Albumasar's claim as a rational, astronomical hypothesis on the Ptolemaic theory, he would have discoveıed it to be wrong! It was only when he turned to copernicus that the technical requirements (of which we shall speak) were at last satisfied.

[50] Mästlin, 40. Unlike Venus, however, the comet never returned — a conspicuouns fact which Mästlin never explained.

[51] Ibid., 43.

Figure 2. This diagram, based upon Mästlin's hypothesis, represents the various positions of the earth in its orbit together with the comet's angular positions observed with respect to the sun. This scheme actually shows the projection of the mean earth into the plane of the comet's circle — an incorrect handling of the latitudes. Fortu-

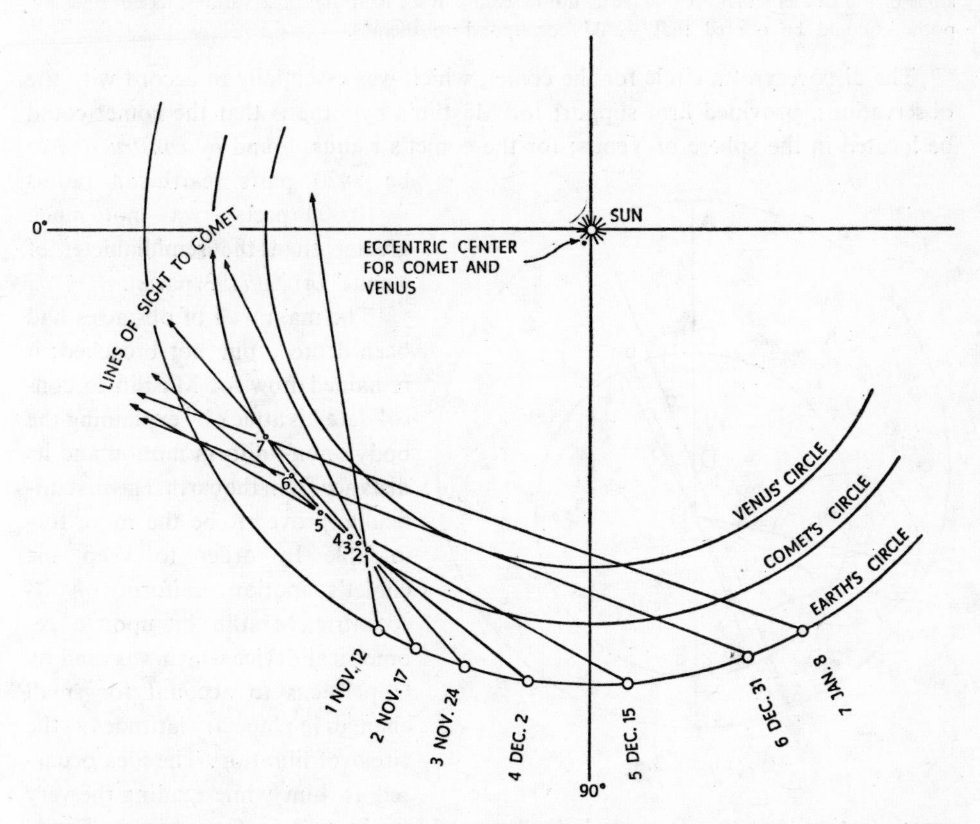

Figure 2. Mästlin's Discovery of the Comet's Circle, showing the position of Earth on dates when Mästlin made observations and lines of sight to the comet with respect to the sun. Since Mästlin had no way in which to determine the comet's actual distance from Earth, he first assigned to the comet the same eccentric center as Venus; then, by trial and error, he tried placing various points on the lines of sight until he obtained a smooth heliocentric curve.

nately, his ignorance of this error did not prevent him from continuing on his way to a personally-acceptable solution. His immediate problem, however was to find where on the lines of sight the comet should be situated. Mästlin probably tried several curves before he discovered that circle nearest in radius to Venus which would intersect the lines of sight in an orderly way. He describes the predicament of finding the comet's radius in the following manner:

Since, in other difficulties of this sort, I saw that the innermost secrets of nature might be approached through enigmas and, as it were, underground passages... I hoped that the same freedom might be granted to me [in solving this problem] if only I might set forth nothing against those most certain observations or the equality of motion. Therefore, weighing and balancing the calculations, I noticed that the phenomena could be saved in no other way than if HO and HF [the comet's radii][52] were assumed to be 8420 parts when DB, the semidiameter of the great orb, is 10,000; and likewise, when the semidiameter of Venus' eccentric is 7193... And from this, the true dimensions of the Comet's orb became clear, not deviating at all from the observations, as no other hypothesis could be offered that would correspond to them[53].

The discovery of a circle for the comet, which was essentially in accord with the observations, provided firm support for Mästlin's hypothesis that the comet could be located in the sphere of Venus; for the comet's radius, found *by construction* to be 8420 parts (earth-sun radius = 10,000 parts), was not much greater than the semidiameter of Venus' orbit (7193 parts).

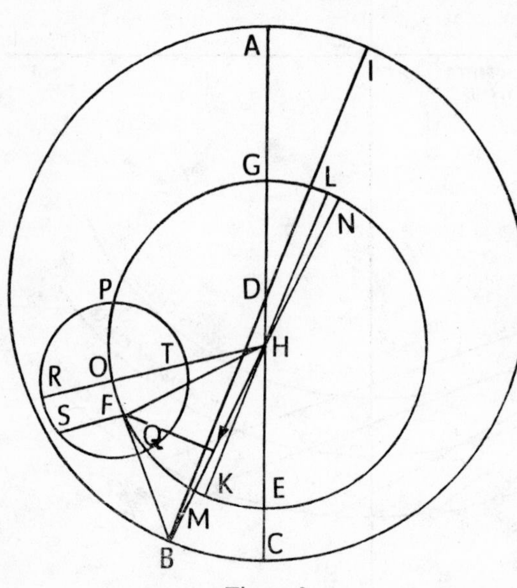

Figure 3.

The main wall of obstacles had been dented, but not breached; it remained now for Mästlin to consolidate his attack by explaining the body's non-uniform motion and its distance from the earth.The first difficulty proved to be the more formidable In order to keep the comet's motion uniform on its eccentric, Mästlin hit upon a geometrical device which was used by Copernicus to account for small changes in planetary latitudes — the circle of libration. The idea occurred to him while reading the very same section of the *De revolutionibus* (VI. 2) in which Copernicus discusses the latitudinal variations of Venus[54]. Mästlin's use of this geometrical technique is interesting because he applied it to an end different from the one for which Copernicus had initially intended it. In Figure 3, a reproduction of Mästlin's hypothesis, PRSQT is the circle in question. A point on this small circle revolves twice, with respect to the earth (*B*), for each revolution of the major circle (*GPEL*). The status of the small circle was purely computational since the comet did not move on it, but on the eccentric circle. The comet's position on the latter was determined by points projected onto it from the small circle, e. g. *R* to *O* and *S* to *F*. And the circle of libration was placed at right angles[55] to the plane

[52] See Figure 3 (ibid.)

[53] Ibid., 45.

[54] Ibid., 38–39.

[55] Ibid., 44.

of the orbit as required both by Copernicus' latitude usage and, apparently, in order to conserve space between the spheres of Venus and the Earth. The ratio of movements appears to have been a completely arbitrary decision since Mästlin was using it for a different purpose than Copernicus and lacked sufficient data to confirm its usage in detail. Indeed, if he had had further information his hypothesis would have been falsified. He was successful only because he had such a small piece of the comet's motion to fit; in retrospect, we have demonstrated that his construction did not reproduce the actual motion of the comet in space.

Now only a value for the mean motion of the comet in its own circle was necessary in order to plot the complete orbit. Once again, the insufficiency of the data forced Mästlin to make a desperate maneuver. Without a full revolution, as we have indicated, he had to take the comet's average motion over the short period of observation and to postulate that this value was constant for the entire orbit. To be sure, this step was not announced to the reader but it can easily be deduced from a table of the comet's motion in anomaly with respect to the mean sun[56]. How stubborn was Mästlin's determination to find a complete orbit for the ephemeral object!

With all of the necessary parameters finally at hand, Mästlin was able to work out computations for several dates which he found, to his delight, agreed quite well with his "certissimas observationes"[57]. And, from the relative distances of the comet, he could establish its actual distance from the earth by substituting Copernicus' figure of 1142 earth-radii as the mean distance of the earth to the sun[58]. At the end of his treatise, Mästlin presented a table summarizing both the comet's angular motions (in its own circle and with respect to the ecliptic) and its linear distance from the earth between November 5 and January 10. While the distances changed non-uniformly in equal times, his results showed that the body continually increased the space traversed from the earth, moving from 155 earth-radii to its last computed position of 1495.

Mästlin had cause to rejoice. "Could this calculation correspond in all its parts and with such perfection," he wrote, "if underlying it were false hypotheses?"[59] The answer was self-evident: if the hypothesis worked, then it must be correct. But his publicly proclaimed confidence in this conclusion leaves unanswered two important issues that question the credibility of his arguments. These problems are (1) the equivalence of the Ptolemaic and Copernican hypotheses and (2) the difficulty of accounting for the comet's retrograde motion using the sphere of

[56] Ibid., 42.

[57] Ibid., 46–48.

[58] Ibid., 48. For example, on Dec. 2, FB (Comet-Earth radius) = 4954 parts; DB (Earth-Sun radius) = 10,000 parts. Now, substituting Copernicus' figure of 1142 earth-radii for DB, Mästlin obtained the comet's semidiameter in identical units. Thus: 1142/x = 10,000/4594 = 565.7 earth radii. The calculation for Nov. 17 should give a result of 265. 5, but Mästlin entered 277 semidiameters-again, no doubt, in order to keep the figures smooth in the final table.

[59] Ibid. Cf. Copernicus (Pref., iij verso): "Nam si assumptae illorum hypotheses non essent fallaces, omnia quae ex illis sequuntur, verificarentur proculdubio".

Venus. It is widely recognized today that the Copernican and Ptolemaic hypotheses commute; thus, if one can save the phenomena using one model, then, of necessity, on has a geometrically-equivalent solution on the other[60]. And in the case in question, the comet of 1577, it can be shown that the appearances are saved on the geostatic theory by simply translating the Copernican solution into the former frame of reference. Figure 4 shows a schematic reconstruction of Mästlin's hypothesis for the comet's position on January 8, with the Copernican model for comparison. All of the angular parameters are constant; the circle of libration performs the same function in both reference frames; but, in the Ptolemaic version, the comet moves on an epicycle. Did Mästlin not realize that such an equivalent construction was possible? Was Kepler's teacher perhaps antecedently convinced, on philosophical ground, that the Copernican system was true and thus determined to show, at all costs, that it alone could save the appearances of the comet?

Upon reflection, this view seems quite unlikely. Sixteenth-century astronomers generally understood that either theory could readily *predict* planetary motions, but there is little evidence to suggest that they comprehended the *complete* interchangeability or formal equivalence of the two models[61]. Kepler was, in fact, the first to draw graphic comparisons between the major astronomies by showing that what the old theory could demonstrate with a plurality of initial assumptions, the new theory could establish with fewer. There is, however, a further argument against the possibility that Mästlin deliberately ignored a translation of his hypothesis into the Ptolemaic coordinate system. Above all, his treatise was a *demonstratio*[62], a dramatic description of the comet of 1577 and how its appearances in the aetherial region could be explained with geometry. Publicly, at least, it was not a defense of the Copernican system. Unlike Tycho, in his work on the comet, Mästlin put forth neither general cosmological arguments nor a schematic representation of the new cosmology. The Copernican theory was employed simply as a *calculational tool.* If, in fact, Mästlin's motive was to use the comet openly as a support for Copernicus, then he scarcely exercised his material to the greatest advantage. Since he was not a university professor at the time, he could have no fear of censure from those conservative quarters. In all likelihood, then, it was not preconceived judgment, but rather the existence of a convenient and viable alternative, that drove Mästlin to the Copernican hypothesis. And his success with it, in contrast to the earlier failures, must have confirmed his positive inclinations toward the new theory.

---

[60] Cf. D. J. de Solla Price, *Contra-Copernicus: A Critical Re-Estimation of the Mathematical Planetary Theory of Ptolemy, Copernicus, and Kepler,* [in] *Critical Problems in the History of Science,* ed. M. Clagett (Madison, Wisc., 1969), 202 ff.

[61] Dr. Price has only recently had to remind modern historians of science of this equivalence (Ibid., 197 ff).

[62] Cf. the definition of *Demonstratio* in Lee A. Sonnino, *A Handbook of Sixteenth Century Rhetoric* (London, 1968), 69—71.

Let us now turn to the final problem of the comet's motion — its motion retrograde with respect to the planets. Mathematically, at least, Mästlin's hypothesis had accounted for the comet's path; but did he also believe that the comet was transported on a solid sphere around the sun? The philological evidence points to an affirmative answer. In general, Mästlin speaks of the "orbis"[63] of the comet although he also refers to its "circulus"[64] and "cursus"[65]. He never speaks of a separate cometary "sphaera". That term is reserved for the planets. In an analysis of Copernicus' language, it has been shown that he sometimes regarded the planet as affixed to a three-dimensional sphere and sometimes simply as a two-dimensional great circle of the sphere[66]. The ambiguity arises from the term "orbis" which may possess both meanings. In spite of his varying terminology, Mästlin leaves no doubt concerning his position: "Cometa est in Veneris Sphaera"[67]. Here, it seems apparent, he wished to avoid the addition of an extra sphere to the universe thereby upsetting the other planetary dimensions. If, however, the comet were moving in an eastward direction within the westward-revolving sphere of Venus then Mästlin would have two bodies turning simultaneously in the same sphere — but in opposite directions. We find no comment from Mästlin upon this obvious contradiction. His diagrams of the comet, moreover, were purely mathematical construction which give no indication concerning the physical reality of the model. It was left for Tycho Brahe to take the next logical step: to abolish all crystalline spheres.

Tycho's solution of the comet's path was basically similar to Mästlin's. The comet circles the sun outside the orb of Venus, he wrote, "as if it were an adventitious and extraordinary planet"[68]. One wonders whether, indeed, Tycho had been led to his own hypothesis on the comet from reading Mästlin's work, for the quick praise he lavished on his German colleague may not be idle eulogy: "It is clear that this discovery of Mästlin shows great sagacity and industry and it smacks of the remarkable quality of genius"[69]. But, for the Dane, the only way in which to remove the discrepancy in Mästlin's hypothesis was to give up the physical basis upon which it rested. He wrote:

... in fact, there are really no Orbs in the heavens, which Mästlin openly thinks exist, and those which Authors have invented to save the Appearances exist only in the imagination, for the purpose of permitting the mind to conceive the motion which the heavenly bodies trace in their course... thus, it seems futile to undertake this labor of trying to discover a real orb, to which the comet may be attached, so that they would revolve together[70].

---

[63] Mästlin, 39 ff.

[64] Ibid., 37.

[65] Ibid., 40.

[66] Cf. Edward Rosen, *Three Copernican Treatises*, 2nd, ed. rev., (New York, 1959), 12 ff.

[67] Mästlin, 38. Tycho wrote of Mästlin: *Postea e COPERNICI Hypothesibus inuestigat Orbem quendam circa Sphaeram Veneris...* (Brahe, IV, 266).

[68] Brahe, Ibid., 190.

[69] Ibid., 266; see footnote 49. It would be useful to learn when he acquired Mästlin's work.

[70] Ibid. Also quoted in Boas, *op. cit.*, 114, and Rosen, *op. cit.*, 12 n. without the reference to Mästlin.

Using the benefits of our previous analysis, we are, at last, prepared to return to our brief reference from Kepler where he credits Mästlin with providing him a "special argument" in favor of the Copernican theory. In the first place, it is now more than evident why Kepler would have mentioned the comet in his defense of the Copernican system. Mästlin's own decision to adopt the new hypothesis had

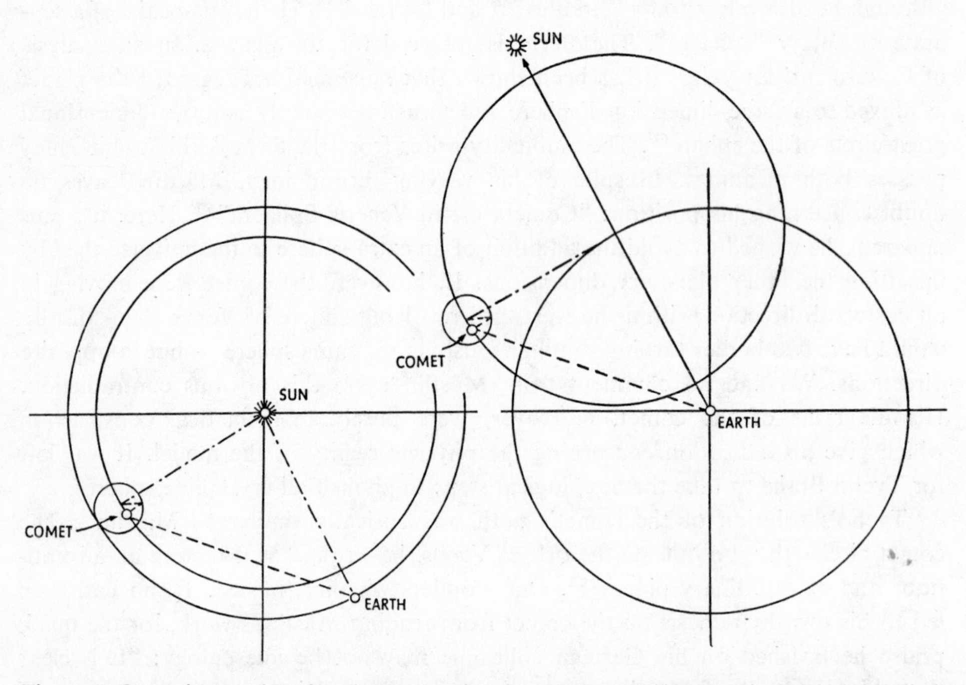

Figure 4. Copernican and Ptolemaic construction for the Comet (January 8) showing the Circle of Libration. The problem of the interchangeability of the two constructions is complicated by the fact that the three bodies (comet, sun and earth) do not move in the same plane, as is the case with the planets. Mästlin's Copernican solution did save the appearances *as he understood the problem*; and this is the significant point. But by failing to deal properly with the great heliocentric latitudes, the hypothesis was, in fact, wrong. The equivalent Ptolemaic solution, found in retrospect, is, ironically, closer to the mark.

been decisively confirmed by his investigation of the comet. In spite of the conjectural nature of his conclusions – founded often upon very questionable assumptions and inadequate evidence – Mästlin *thought* that the appearances could only be satisfied as he had done. Although he recognized it as a "conjectura", such a dramatic demonstration of the potency of the new hypothesis in accounting for the trajectory of this ephemeral phenomenon, where the old theory had failed, must have strongly affect the young and impressionable Kepler. In short, the system had proved its mettle. As Kepler wrote:

My first faith in Copernicus was sustained by the most admirable accord between his conceptions and all the phenomena which appear in the sky; an agreement which not only permits him to demonstrate previous motions, reported since antiquity, with absolute certainty, but in all cases, much more exactly than Ptolemy, Alfonso [the Wise] and the other astronomers... once this hypothesis [of Copernicus] is set forth, it will be able to demonstrate anything at all which truly appears in the sky, to go forward and backward, to deduce one [premise] fiom another, and to do anything you please which [agrees] with reality: ...the most intricate demonstrations always return to the initial hypotheses themselves[71].

If a true hypothesis must be true in *all* respects, as Kepler maintained, then the case of the Comet patently implied that the Ptolemaic hypothesis was founded upon false premises[72]. The Comet not only exposed the falseness of the ancient hypotheses, however, it also provided "the most important argument for the arrangement of the Copernican orbs". The meaning of this assertion can now be more deeply construed in light of our study of Mästlin and the comet. According to Kepler, Mästlin had established that the comet could only complete its orbit "in the same orb as the Copernican Venus". *To accept the place of Venus on this model, however, was to admit the new order of the inferior planets.* Mästlin had not only introduced Kepler to the basic tenets of the Copernican hypothesis, a theory whose harmonious treatment of the planetary distances had initially attracted for both of them; he had also furnished him with operational evidence that the Copernican order of the planets was the necessary and sufficient condition for saving the new appearance, and perhaps for saving *all* celestial phenomena. Yet where Mästlin had been outwardly timid in using the Comet as a specific argument in support of the Copernican theory, Kepler saw and proclaimed it as a profound confirmation of his belief.

What remained unstated by Kepler — and which the reader may already have wondered himself — was that the Comet could be viewed as an argument in favor of *either* the Tychonic or the Copernican hypothesis or, to be more precise, in favor of that part which the two held in common, namely, the order of the inferior planets. Mästlin, however, must surely have recognized this point. On May 15, 1588, Tycho had sent Mästlin a copy of his major work, *De Mundi Aetherei Recentioribus Phaenomenis* (See Plate No. 2), a gift which, as we learn in an irate letter from the former, was never acknowledged by the latter[73]. When, therefore, Kepler listened to Mästlin's lectures in the early 1590s, the Tübingen master would have been familiar not only with Tycho's system of the world, but also with Tycho's objections to his cometary hypothesis[74]. How much Mästlin spoke of the work of Brahe we do not know. What we can say from later evidence is that Mästlin completely opposed Tycho's arrangement of the universe on the grounds that the symmetry and proportionality of the world would be disrupted. In his important introduction to Rheticus' *Narratio Prima* (1596) he referred, critically, to Tycho's cosmology.

[71] G. W. I, 14: 29–35; 15: 15–19.
[72] See page 8.
[73] Tycho to Mästlin, April 21, 1598, G. W. XIII, 204.
[74] Brahe, *op. cit.*, 245 ff.

... in fact, these corrections to the ancient hypothesis do nothing but fix up the old, worn toga with a new patch, which shall probably be damaged in time. For I declare that by this arrangement [of Tycho] the centers of motion and motor virtues are destroyed and torn apart and that the remaining motions and orbs (or whatever else that has an orb) will be entangled by many very nonsensical intricacies; nor will magnitudes and motions be united by some proportion of order[75].

Thus, by rejecting Tycho's "new patch", it is probable that Mästlin also dismissed his scheme for saving the motion of the comet. In this same document, Mästlin gives us reason to believe that he had gone well beyond Tycho by espousing the Copernican theory of gravity[76].

Whether influenced by Tycho or whether, as seems more likely, he had arrived at this conclusion on his own, Kepler too had decided that the Adamant spheres were dispensable[77]. Hence, Tycho's objection concerning the existence of crystalline spheres would have formed no serious obstacle to Kepler's initial acceptance of Mästlin hypothesis. But neither would Brahe's purely geometrical hypothesis have bothered Kepler. Although the *Cosmographic Mystery* was

[75] G. W. I, 84: 41–47. Mästlin added further criticism of Tycho to the 1621 edittion (cf. 436-439).

[76] Ibid., 83: 12 ff.

[77] Kepler had abandoned solid spheres at least as early as October, 1595 (Kepler to Mästlin, XIII, 43: 389–99) and this view was reflected in his earliest work (I, 56: 15–20). Tycho was evidently unaware that Kepler had adopted this position when he wrote to the latter in April, 1598 (XIII, 198–99: 61–67) and Kepler's marginal note gives no indication that he had been influenced by Tycho: *Idem etiam per me licet, et per libellum meum*, he wrote in response to Tycho's admonition on the reality of solid orbs (Ibid., 201: 78–79). In a subsequent letter to Kepler, Tycho gives indications that he had read the *Cosmographic Mystery* more carefully, yet still criticized the younger man for ascribing "a certain reality" to the celestial orbs (Tycho to Kepler, Dec. 9, 1599, XIV, 94: 190 ff.). Our final evidence again supports the position that Kepler may have arrived at a rejection of the solid orbs independently of Tycho. In his unpublished *Apologia Tychonis Contra Ursum* (composed between 1600 and 1601), Kepler tells us: *Primum hoc illi* [to Ursus] *facile concessero, solidos orbes nullos esse. In quo et Tychonis rationibus facile subscribo et privatas etiam habeo* (*Joannis Kep leri astronomi opera omnia*, ed. Christian Frisch, I (Frankfurt/Main and Erlangen, 1858), 247).

Plate No. 1. Helisaeus Roeslin's diagram of the sphere and circle of comets, reproduced from his *Theoria Nova* Coelestium ΜΕΤΕΩΡΩΝ (1578). This is probably the most complex early representation of the comet of 1577 and illustrates an interesting, though crude, attempt to explain its unusual orbit. All the various circles of reference are drawn in the plane of the zodiac and are related to the line joining the two solstices, the solstitial colure. Roeslin's determination of the "Pole of Comets" seems to have been entirely arbitrary, for it is exactly opposite the "Pole of the World" and $23^1/_2$° from the "Pole of the Zodiac". The "Great Circle of Comets", which includes the comets of 1532, 1533, 1556, 1577, and the nova of 1572, passes through the aequinoctial points. There is also a "Sphere of Comets and Stars" which is concentric with the Circle of Comets and by means of which Roeslin hoped to offer a physical explanation for the comet of 1577 and the New Star of 1572. The sphere resembles the zodiac in that its band has the same width (16°), but Roeslin thought of it as a small tropical circle with a radius of 60°. Roeslin's hypothesis is interesting because it is basically unlike the final schemes of Mästlin and Tycho. Where the latter two had used planetary theory as a model for explaining the comet, Roeslin used only stellar coordinates.

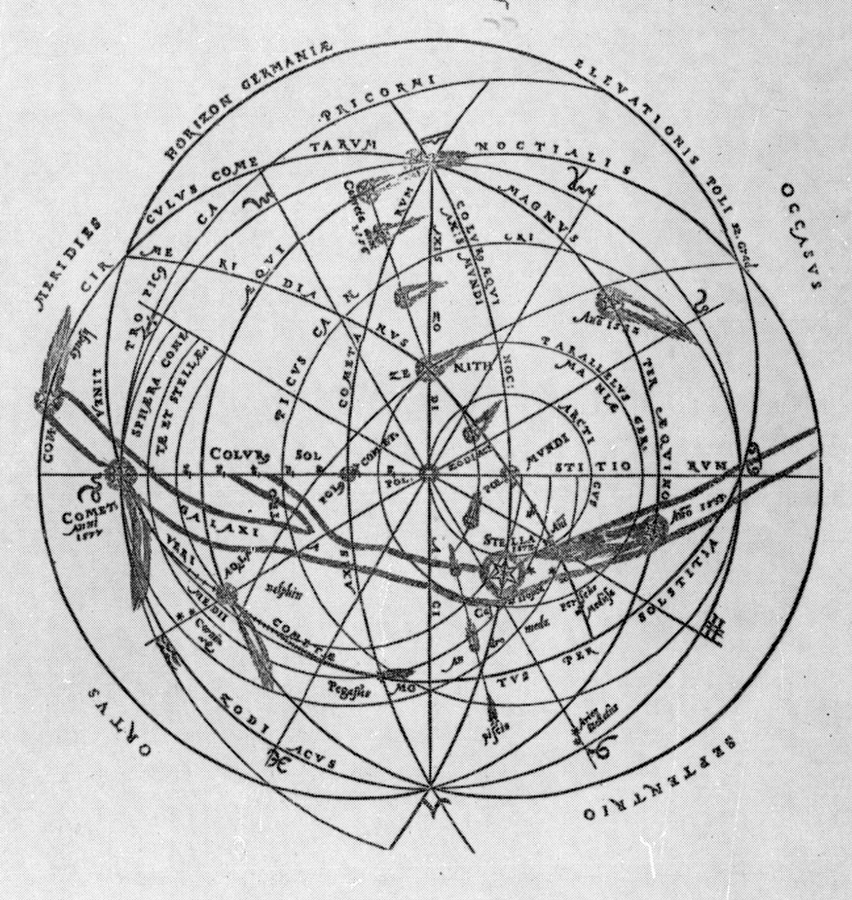
SPHÆRA NOVA COMETARVM ET MIRA
CVLORVM DEI, AVTHORE MEDICO HELISAEO ROESLIN.
HORIZON GERMANIÆ
ELEVATIONIS POLI
MERIDIES
OCCASVS
ORTVS
SEPTENTRIO
ZENITH
GALAXI

Plate No. 1.

Clarissimo Eruditissimoq Viro
M. Michaeli Moestlino Goeppingensi
Astronomo in primis excellenti
Artiumq Mathematicarum in
Inclyta universitate Heidelbergensi
professori ordinario
dono misit
Tycho Brahe Dan
Anno 1588. Maij 14.

TYCHONIS BRAHE DANI
DE
MVNDI AETHEREI
RECENTIORIBVS
PHAENOMENIS

LIBER SECVNDVS

QVI EST DE ILLVSTRI STELLA CAVDATA
ab elapso ferè triente Nouembris Anni 1577. vsq;
in finem Ianuarij sequentis
conspecta.

VRANIBVRGI
CVM PRIVILEGIO.

Plate No. 2.

primarily directed against Ptolemy, Kepler had already begun to attack the Tychonic system while still recognizing the element shared by both theories. He wrote:

> Although he [Tycho] was in disagreement with Copernicus regarding the place of the Earth, nevertheless, he retained the hitherto unknown cause of things: to wit, the fact that the Sun is the center [of motions] of the five planets[78].

Had he extended this argument to Tycho's hypothesis of the Comet, Kepler, I think, would have seen the Comet's trajectory as a confirmation of the *correct* part of Tycho's system of the world. Ironic though it may seem, it was the hypothesis of a circular, heliostatic orbit – erroneously conceived (as we now know) by both Brahe and Mästlin – which Kepler saw as a significant counter-instance to the Ptolemaic theory and which helped fundamentally to form his initial commitment to the Copernican system.

What became of this cometary argument in Kepler's subsequent writings? After a cursory mention in the *Cosmographic Mystery* Kepler seems to have abandoned it. What had impressed Kepler as a young initiate must have seemed less convincing to the mature man. Some eight years later, in a short digression within his *Ad Vitellionem Paralipomena* (1604) entitled "Appendix concerning the motion of Comets"[79], we encounter the first explicit indication of the Comet's fate.

> Those who demonstrated the comet's motion of the year 1577 undertook a most difficult task; nevertheless, it did not generally turn out to be successful because they did not suppose that it ought to be investigated more diligently. Much greater difficulties are encountered if they assume the same notions in demonstrating other comets.

After this poke at his predecessors, Kepler continues:

> In most [comets] for which I have obtained observations, the following path turned out to be easier: viz. if, because it suits the nature of things [quod natura rerum suadet], I attributed straight lines to them, then, frequently, they would travel equal [distances] in equal times – only a bit slower at the beginning and end and, when nearby, at rest, as is usual with other trajectories.

Thus, even before he had studied Halley's Comet (1607), Kepler had concluded that *all* comets have rectilinear paths. He now applied this theory to the comet of 1577.

---

[78] G. W. I, 16: 11–14.
[79] G. W. II (1938), 287–88.

Plate No. 2. Michael Mästlin's copy of Tycho Brahe's *De Mundi Aetherei Recentioribus Phaenomenis* (1588). This document proves that Mästlin was familiar with both Tycho's criticisms of his hypothesis on the comet of 1577 and with his world system several years before he lectured to Kepler at the University of Tübingen. Various underlinings in the volume indicate that Mästlin had actually read and studied it. The work, now in the British Museum, was signed by Tycho and inscribed to Mästlin. Underneath Tycho's inscription Mästlin wrote: "... oblatus est mihi liber hic die 15 Augusti 1588", with an additional reference to the messanger, Gellius Sascerides Hafnienses (d. 1612), who delivered the book. Sascerides, a medical student, was evidently an important link between the two men. The remaining notations are those of subsequent owners of the book.

> ... by introducing the Earth's motion, it [the Comet] easily reconciles circularity to itself. Thus, in the case of that Comet in 1577, if the straight line originating at the Tropic of Capricorn were perhaps to be inclined a bit towards the North Pole, then that straight line would have ascended at the same time that the Earth, while revolving about the Sun, induces a circular species of motion to it [the Sun] nevertheless, in the same way, it [the Earth] will unite the same circular species with the comet as though it were revolving about a motionless body...

One is impressed by the ease with which Kepler compounds distinct motions in the same body; and his hypothesis of rectilinear cometary motion comes much closer to a satisfactory explanation for the comet of 1577 than the earlier speculations of Tycho and Mästlin. A lengthy footnote in the second edition of the *Cosmographic Mystery* (1621), which refers to that now familiar passage on the comet, rejects the first efforts of his masters as "bare credulity and general conjecture"[80]. The comet had outlived its original usefulness as an argument in support of the new cosmology. By then, however, there was no need for it.

---

[80] G. W. VIII (1963), 41.

KRISTIAN PEDER MOESGAARD
University of Aarhus

COPERNICAN INFLUENCE ON TYCHO BRAHE

Preface

The reception of Copernicanism in Denmark and Norway was dominated by one man of genius, Tycho Brahe, and by two of great distinction, the astronomer Longomontanus and the philosopher Caspar Bartholin. The final acceptance ran parallel to the progress of Cartesianism at the time of the outstanding astronomer Ole Rømer, and was at last openly expressed by his pupil and successor Peder Horrebow.

So my story naturally falls in two distinct sections, viz. the present one treating how Tycho's astronomy developed through his acceptance and his rejection of Copernican ideas, and a later one (p. 117 ff.) dealing with the lesser figures, more representative of the educated community of their time. Among these, I give a fuller analysis of the works of the said authors, and describe positions taken by other Professors from the late sixteenth century to the early eighteenth century.

There still remain writings of students and others on astronomical and cosmological topics. For the sake of brevity I have omitted these from my contributions to this volume; but I retain them in the combined bibliography (p. 145 ff.) to the two essays.

My accomplishing these studies on the history of Copernicanism has been possible only due to a special grant offered by The Research Foundation of the University of Aarhus. I am greatly indebted to the History of Science Department for placing accomodations at my disposal, and I am particularly grateful to Mrs. Kate Larsen and to Prof. Olaf Pedersen for many valuable suggestions during the preparation of the manuscripts.

I. TYCHO THE COPERNICAN

Tycho's lecturing on 'Copernican' astronomy (1574–75)

From the 23rd of September 1574 until early in 1575 *Tycho Brahe* (1546–1601) gave, on the request of students, friends, and finally the Danish King, a series of lectures on astronomy at the University of Copenhagen. In his opening lecture *De disciplinis mathematicis oratio* [*19*: I, 143—173; cf. *119*: 73—78][1] he introduced Copernicus to the Danish world of learning in the following way:

In our time, however, Nicolaus Copernicus, who has justly been called a second Ptolemy, from his own observations found out that something was missing in Ptolemy. He judged that the hypotheses established by Ptolemy admitted something unsuitable and offending against mathematical axioms: he did not either find the Alphonsine computation meeting the heavenly motions. He therefore arranged, by the admirable skill of his genius, his own hypotheses in another manner and thus restored the science of the celestial motions in such a way that nobody before him has considered more accurately the course of the heavenly bodies. For although he devises certain features opposed to physical principles, eg. that the Sun rests at the center of the Universe, that the Earth, the elements associated with it, and the Moon move around the Sun in a triple motion, and that the eighth sphere remains unmoved, he does not for all that admit anything absurd as far as mathematical axioms are concerned. But if we examine the matter thoroughly it appears right to blame the current Ptolemaic hypotheses in this regard. For they dispose the motions of the heavenly bodies in their epicycles and eccentrics as irregular with respect to the centres of these very circles. This is absurd, and by means of an irregularity they save unsuitably the regular motion of the heavenly bodies.

Everything, then, which we now-a-days consider evident and well-known concerning the revolutions of the stars has been built up and transmitted by these two masters, Ptolemy and Copernicus" [*19*: I, 149].

In the subsequent lectures Tycho passed over the subject of the daily motion and chose to expound the doctrine of the *secunda mobilia* "according to Copernicus' mind and using his numerical values, but referring everything to a stable Earth". He carried through this programme for the fixed stars, for the Sun, and for the Moon explaining their models geometrically by means of triangles, as well as arithmetically according to the Prutenic Tables based upon a Copernican foundation — Tycho had, at his own expense, bought some copies of the Prutenic Tables for distribution among the poorest students [*19*: I, 172—173]. He had intended to deal with the models of the remaining planets in a similar manner, but his lecturing was interrupted by his extensive travelling in 1575, so he gave only a general exposition on these planets showing, however, how the phenomena might in this case, too, be brought into accordance with the idea of a stable Earth without deviating from the Copernican parameters. In addition he criticized Peucer and Dasipodius for their illegitimate mixing up Copernican calculation with Ptolemaic hypotheses.

This short account makes it evident that Copernicus and Copernican astronomy was introduced in Denmark in a thorough and competent way by an expert. It shows Tycho's genuine admiration of Copernicus as a mathematical astronomer, too, and reveals his general attitude towards Copernican cosmology as unacceptable for physical reasons. Although the Tychonian world system had not at that time come into existence it nevertheless appears that Tycho had already, within each separate planetary model, managed to get round the stumbling-block of the moving

[1] For Bibliography see pages 145—151.

Earth. Consequently astronomers were from now on free to make use of Copernican astronomy without troubling about his cosmology; on the other hand, natural philosophers could unscrupulously adhere to traditional views on these matters. And during the century following Tycho's lecturing a great many authors did so.

Copernican parameters in Tycho

The life and work of Tycho Brahe is excellently pictured by Dreyer [*119*]. I shall, accordingly, confine myself to bring only scattered details of the development of Tychonian astronomy to the extent I have thought them important in relation to my main theme.

You cannot give a satisfactory description of how Tycho, the founder of modern observational astronomy, was influenced by Copernicus without touching on problems of observation and numerical calculation. Everywhere in Tycho's books, letters, and diaries of observation one finds comparisons with Copernican or Prutenic numerical values abundantly spread in the form of inserted remarks, parallel computations or summarizing tables. In sum, it is clear that throughout his career Tycho considered the Copernican models and the Prutenic Tables to be, next to the heavens themselves, the most prominent touch-stone for his own work. He aimed at doing better himself, and so he did.

In his youth Tycho was surely convinced that the Prutenic Tables were far superior to the Alphonsine ones. Perhaps for a time he even hoped that astronomy might be restored by a redetermination alone of the parameters of the Copernican models. The origin of this position may be sought in the fact that already as a young student at Leipzig Tycho had observed the great conjunction of Saturn and Jupiter in 1563 and had found discrepancies of a whole month and a few days respectively on comparison with the Alphonsine and the Prutenic Tables [*19*: I. 40; V. 107; *119*: 18–19]. In his *De Nova Stella* (1573) he considered the Prutenic Tables to be superior to the Alphonsine "hodge-podge" because by numerous observations he had found that the motions set forth by Copernicus squared better with the heavens than did the Alphonsine or any other tables of the heavenly motions [*19*: I. 23]. He calculated all the data concerning the lunar eclipse in December 1573 according to the Prutenic Tables in order to invite astronomers to a more careful investigation of eclipses than could be obtained by ephemerides [*19*: I. 42]. It has already been mentioned that in 1574 Tycho lectured on Prutenic astronomy, and in his *Horoscope of Prince Christian* (1577) he preferred the Prutenic to the Alphonsine planetary positions when he had not himself made observations of the planets in question [*19*: I. 185–190].

However, as time went on and Tycho at Hveen observed and investigated one heavenly phenomenon after the other, the fact grew increasingly obvious that also the Copernican parameters were far from reliable. Moreover the Copernican long term partial motions, designed to save the supposed inequality of the precession of the equinoxes, the slow motions of the planetary apsidal lines, and the variation of the solar eccentricity, turned out faulty or fictitious. This meant that in all fields Tycho had to rebuild the very foundations of astronomy which, in turn, enormously increased the preliminary work of observation to be done. Consequently he did never achieve his main end, viz. a complete restoration of theoretical astronomy, but had to leave this task to his pupils, Kepler and Longomontanus. In 1598 Tycho in his *In solis et lunae motus restitutos prolegomena* admits that his own more accurate emendation had hardly been possible without the work of his predecessors from Hipparchus to Copernicus [*19*: V, 173]; but a little later he deplores that Copernicus did not arrive at more reliable measures of the year and the month since that would have spared Tycho a great toil and much money and time [*19*: V, 178–179]. In a letter to Holger Rosenkrantz from the same year Tycho, having noticed that Martin Everhart uses Prutenic stellar coordinates, puts the question: "How great troubles and how many thousand Joachimsdaler could I have spared, if these positions had been the true ones? [*19*: VIII. 62]. A similar line of thought is revealed already in 1588 in a letter to Caspar Peucer [*19*: VII. 139].

Morsing's journey to Frombork (Frauenburg) (1584)

Tycho's way from reliance on Copernican numbers in his youth to the attitude of annoyance and lenience revealed in his later writings has of course gradually taken him through a variety of intermediate stages. His emendation of the latitude of Frauenburg as determined by Copernicus deserves special attention because it is particularly illustrating of Tycho's way of working and of his position towards Copernicus.

In his *Progymnasmata* Tycho gives a detailed account on how careful investigations of the solar apogee and eccentricity during several years, and a comparison with the values set forth by Bernhard Walther, caused him to suspect that Copernicus had erred by about 0;3° in his determination of the polar altitude of Frauenburg. In 1584 one of Tycho's assistants, *Elias Olsen Morsing* (Cimber; died 1590, see *75*: 179, cf. however *119*: 123), was sent to Frauenburg and Königsberg in order to determine the latitudes of these localities. Morsing stayed at Frauenburg from the 13th May until the 16th June observing the solar height and the fixed stars. Much pains was taken to ensure that the sextant he used should not be damaged during the journey and, indeed, after his return to Hveen the latitude of Frauenburg was found to be greater than Copernicus' value by 0;2,45° [*19*: II. 29–32; cf. II. 156, VI. 58, 72, 103–104, VII. 298, and Dreyer's summary in *119*: 123–125; Morsing's observations are found in *19*: X. 345–348, and the calculations of the latitude of Frauenburg in *19*: V. 305–307].

This result obviously strengthened Tycho's selfconfidence and made him realize that he did better to trust his own results than those of his predecessors. Thus he frankly made similar emendations of the latitudes of Nürnberg [*19*: II. 39; cf. IV.

258] and Wittenberg [*19*: III. 146, 147], the places where Walther and Reinhold had made their observations; he considered his corrected values to be right within 0; 0,15°.

As another valuable result of his journey Morsing brought back with him a simple wooden triquetum offered as a present to Tycho from a canon at Frauenburg, Johannes Hannov. This parallactic ruler instrument had been used by Copernicus and was made by his own hands, and Tycho certainly found it surprisingly primitive. He guessed that Copernicus had also used other instruments [*19*: II. 32], but he could not help preparing immediately an heroic poem to the honour of his great predecessor who "by means of these puny cudgels surmounted the lofty Olympus" [*19*: VI. 265–267; cf. VI. 253]. Having referred to Copernicus' instrument Tycho describes in his *Astronomiae Instauratae Mechanica* a similar but improved copy made by himself [*19*: V. 44–47; cf.: V. 6, 144; VI. 103–104; VII. 133]; the *Epistolae Astronomicae* contains other Tychonian poems to Copernicus [*19*: VI. 270–271, 275].

So far Tycho had sufficiently explained why Copernicus went wrong in determining the solar apogee and eccentricity, but, moreover, in his *Progymnasmata* he was able to correct Copernicus' determination of the position of Spica, "since Copernicus' annotation rests upon careful observation of the heavens except that he does not use the right value of the polar height", while, on the other hand, a reasonable reconstruction of Werner's coordinates of the same star proved impossible, because "the Wernerian annotation is composed conjecturally, not to say in order to fit his prescriptions" [*19*: II. 222–226; cf. II. 156, 212, 253].

Because he was able to explain why Copernicus' data of observation were unreliable, Tycho retained or even strengthened his earlier deference to Copernicus. It was a pity that Copernicus had used so imperfect instruments. He had not considered the effect of refraction, and it was questionable whether he had taken into account the effect of parallax or not [*19*: II. 18, 31, 89; cf. VI. 69; VII. 279–280]. Tycho's principal charge, however, against Copernicus regarded his far too uncritical reliance upon the records of the ancient astronomers, e. g. Ptolemy's Hipparchian star catalogue, in which case Copernicus ought to have known better since he could not have been completely ignorant of observational errors and faulty transcriptions [*19*: II. 229, 281; III. 337–338; IV. 20]. Tycho had himself created new standards of observational precision and thus, for the first time, freed astronomers from their dependence on earlier records. To Copernicus, on the other hand, it formed an integral part of his scientific method and was the very sign of honesty in working on astronomy "to follow in their (i. e. the ancients') footsteps and hold fast to their observations, bequeathed to us like an inheritance" [*93*: 99].

However much Tycho might have been disappointed in Copernicus he could, nevertheless, continue to compare with Prutenic values to show to what extent he had himself surpassed even the most prominent results of earlier astronomy. And in 1602 his heirs could at last, quite in Tycho's spirit, bring the following appreciation in the dedicatory preface of the *Progymnasmata* addressed to the Emperor Rudolph the 2nd:

> Several distinguished men, equally ingenious, indeed, in studying this discipline, but of unequal means, ventured to supply the said deficiency (of the Alphonsine Tables) particularly, however, that great and never sufficiently appreciated Nicolaus Copernicus from Thorn. But the same scarcity of observations and instruments which we said had obstructed the Alphonsines did not yet permit him to possess the universal mastery he aimed at, although his restoration of the motions does not deviate from the heavenly precept to the same extent as does that of his predecessors [*19*: II. 6].

Copernican geometry of planetary models in Tycho

As to the mathematical formulation of the astronomical models Copernicus replaced the Ptolemaic equant by two additional partial motions going on uniformly in two small circles; these may conveniently be named epicyclets to distinguish them from the more place-demanding – and hence offensive – epicycles and deferents of the outer and inner planets respectively. Further, Copernicus identified the partial motions, which the last-mentioned circles were designed to save, with each other and with the annual motion of the Earth, thus making them superfluous. In this way he had established a unified mathematical description of the planetary system as distinct from a set of separate models of the planets which in turn enabled astronomers to calculate the relative planetary distances without resorting to speculations of the type: *what* must be situated *where* in the Universe on the assumption that *everything* demands a place and *every place* is to be occupied by something? Moreover, Copernicus showed that two partial motions of the uniform circular type may produce a socalled libration, being in fact a linear harmonic oscillation. He utilized this libration in his theory of the precession of the equinoxes, in his model of Mercury, and in his theory of the planetary latitudes.

As early as in his lectures in 1574 [cf. p. 32] Tycho made it clear that he fully adopted the Copernican geometry of astronomy. The present section is going to review how this adoption in several situations of Tycho's career overshadowed his reasons for rejecting the Copernican system. These reasons shall in turn be the subject of the following two chapters.

Having been asked by his friend, the royal physician Peder Sørensen, to explain Copernicus' hypothesis of the triple motion of the Earth Tycho, in a letter dated 1576 September 3rd, does so without any criticism of Copernican ideas, and he goes on: "In the said third motion, however, he (Copernicus) makes the terrestrial equator perform an unequal motion 'in antecedentia' by means of an admirable arrangement of two small circles described around the axis of the terrestrial equator; thus he saves the unequal motion of the fixed stars, too, as this motion has hitherto appeared to us. By this very mechanism he also explains another motion of the equator by which it approaches and recedes from the ecliptic". Tycho thereafter admits that all these matters are hard to grasp unless they are visibly exposed in the form of a mechanical model, and he promises some day to prepare such a machine [*19*: VII. 39–40].

This letter shows that the young Tycho did really accept Copernicus' mathematical reformulation of the astronomical models down to the last detail. As to librations, however, later on he realized that the Copernican long term libration-corrections did not agree with his own observational results, and so came to doubt the necessity of using librations. Maestlin had introduced a librational motion into his theory of the comet of 1577. Tycho takes this as an occasion to discuss librations in his *De Recentioribus Phaenomenis.* Although Copernicus has ingeniously invented this libration Tycho nevertheless doubts that it really occurs in the heavens. The Copernican model for the precession of the equinoxes and the variation of the obliquity of the ecliptic does not agree with the phenomena. He admits that in the theory of Mercury the libration meets certain short-comings of the Ptolemaic model and brings about a better agreement with phenomena. Still, the theory has grown so complex that Tycho considers it less fit to explain heavenly motions, believing himself to be able to dispense with both the Ptolemaic equant and the Copernican libration. Since librations are unfit for the heavens, and for the planets, they will be far less suitable for explaining the motions of comets [*19*: IV. 224].

Further discussions on libration are found in Tycho's *Progymnasmata* [*19*: III. 78] and in his correspondence with Kepler [*19*: VIII. 45], Rothmann [*19*: VI, 160, 197–198, 217, 220–221], and Scaliger [*19*: VIII. 328–329]. It is to be noted that Tycho never completely abandoned the invention, and in one place he used a libration mechanism in order to explain an anomaly in the motion of the Moon [*19*: II, 101]. Thus it might well be Tycho or one of his pupils who praised the libration in the copy of the *De Revolutionibus* belonging to The Royal Library at Copenhagen by the marginal note: "Novam inventum Copernici ... Demonstratio agregiae, utilis et inauditae rei ..." [cf. *133*: 451–452].

However, apart from the criticism of librations Tycho accepted Copernicus' mathematical ideas and made them the indispensable basis of his own formulation of new hypotheses. This implies that the Tychonian models must necessarily be in one way or another "inverted Copernican models". Tycho is generally quite conscious of this fact and in his MS *De Marte* (1585) some calculations occur relating to the Copernican model whereafter the very same calculations are performed in relation to an "inverted Copernican model" which shows, then, how a Tychonian model came to be [*19*: V. 276–288, see Figures 1 and 2]. On one occasion, in a letter to Rothmann, Tycho denies that his models should be inversions of the Copernican ones [*19*: VI, 178]; but this alleged independence concerns the proof of the superiority of Tycho's world system rather than the origin of his models.

In order to expose further to what extent Tycho attached importance to mathematical reasoning as compared with physical arguments it may be profitable to quote from another letter, dated 1587 January 20th, to the abovementioned Christopher Rothmann, astronomer, to the Landgrave Wilhelm the 4th of Hessia:

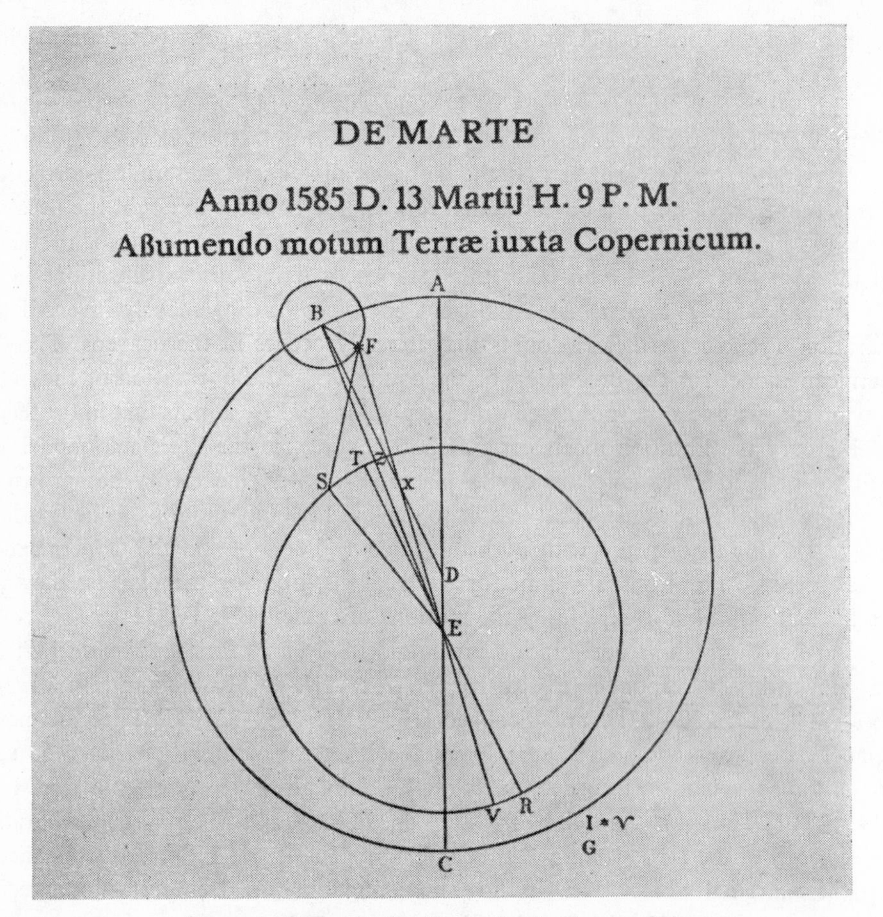

Figure 1. Tycho Brahe, *Opera Omnia* V, page 281.

If Nicolaus Copernicus, the distinguished and incomparable master, in this work (the restoration of astronomy) had not been deprived of exquisite and faultless instruments, he would have left us this science far more well-established. For he, if anybody, was outstanding and had the most perfect understanding of the geometrical and arithmetical requisites for building up this discipline. Nor was he in this respect inferior to Ptolemy; on the contrary, he surpassed him greatly in certain fields, particularly as far as the device of fitness and compendious harmony in hypotheses is concerned. And his apparently absurd opinion that the Earth revolves does not obstruct this estimate, because a circular motion designed to go on uniformly about another point than the very center of the circle, as actually found in the Ptolemaic hypotheses of all the planets except that of the Sun, offends against the very basic principles of our discipline in a far more absurd and intolerable way than does the attributing to the Earth one motion or another which, being a natural motion, turns out to be imperceptible. There does not at all arise from this assumption so many unsuitable consequences as most people think.

Tycho thereafter notices that Buchanan's criticism against the motion of the Earth was unsuccessful since he failed to see that the sea and the lower layers of the atmosphere would, of course, perform a natural motion together with the Earth.

Inuerſio Hypotheſis Copernianæ in motu ♂, vt centrum vniuerſi TERRA quieſcens iuxta veterum ſententiam occupet, et nihilominus motus æquales circulorum propria ſua reſpiciant centra, repudiata Ptolemaica diſcohærentia.

Hæc ratio inuertendi Copernianam Hypotheſin ſufficit tribus ſuperioribus planetis ♄ ♃ ♂.

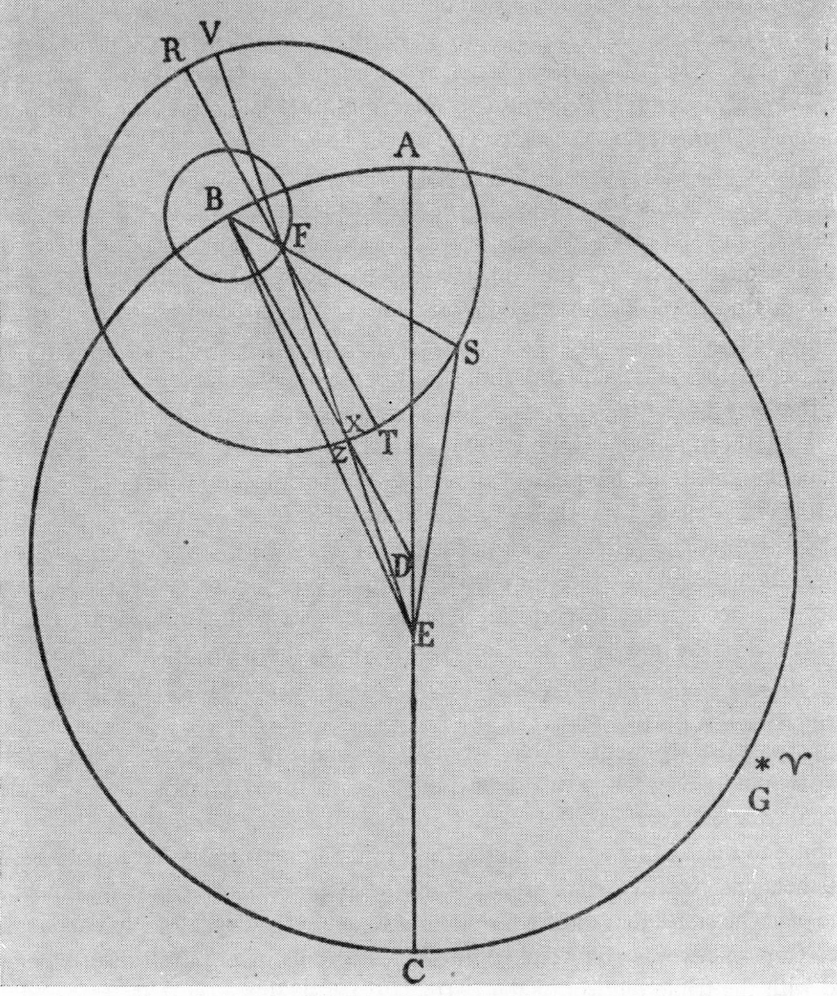

Figure 2. Tycho Brahe, *Opera Omnia* V, page 284.

Tycho promises, as he does repeatedly elsewhere [see e. g. *19*: II. 14; IV. 156] to discuss the matter in greater detail in his *Opus Astronomicum*. Moreover he announces that he himself hopes to introduce new hypotheses acceptable from a mathematical as well as from a physical point of view [*19*: VI. 102–103].

II. TYCHO'S ALTERNATIVE SOLUTION

Tycho between the Ptolemaic Scylla and the Copernican Charybdis

Tycho's plan of composing an all-embracing *Theatrum Astronomicum* [*19*: VII. 132; cf. VII. 292; VIII. 199] was only partially carried out through his works on the new star, *Astronomiae Instauratae Progymnasmata* (1602), on the comet from 1577, *De Mundi Aetherei Recentioribus Phaenomenis* (1588), and on his instruments, *Astronomiae Instauratae Mechanica* (1598). So he never kept his promise of giving a systematic and detailed account on cosmology and on the question of the motion of the Earth. But his own arrangement of the planetary system had been worked out already in 1583 [*119*: 167], and it was published in 1588 as a prerequisite of his discussion on the place of the comet in the Universe [see Figure 3].

On this occasion Tycho criticizes once more the Ptolemaic system because of the equant and the numerous and great epicycles. Having noticed that the Copernican system meets these mathematical difficulties he points out that it conflicts with physical principles by letting the dense and sluggish Earth move. It is also at variance with the authority of the Holy Scripture which in several places maintains the stability of the Earth, not to speak of the immense space, totally void of stars, between the orbit of Saturn and the eighth sphere, or of other inconveniences connected with this hypothesis. Thereafter Tycho accounts for his own system which has proved right, mathematically – since all partial motions were just as circular and uniform as they had been in the Copernican description – as well as physically, the Earth being at rest in the middle of the Universe. Nor was there any reason to fear theological judgements, and at the same time the system was, of course, in full accordance with all celestial phenomena [*19*: IV. 156–158].

This passage represents Tycho's typical account on the world systems, and it is easy to find in his works a wealth of similar formulations doing away with Ptolemy, for mathematical reasons, and with Copernicus, referring briefly to the physical absurdity, to the conflict with Scripture, and to the unreasonably huge space between the planets and the fixed stars arising from the assumption of a moving Earth. And in one place he states that during his attempt to avoid the Scylla of Ptolemaic mathematics Copernicus was shipwrecked on the Charybdis of physical absurdities associated with the triple motion of the Earth [*19*: IV. 473].

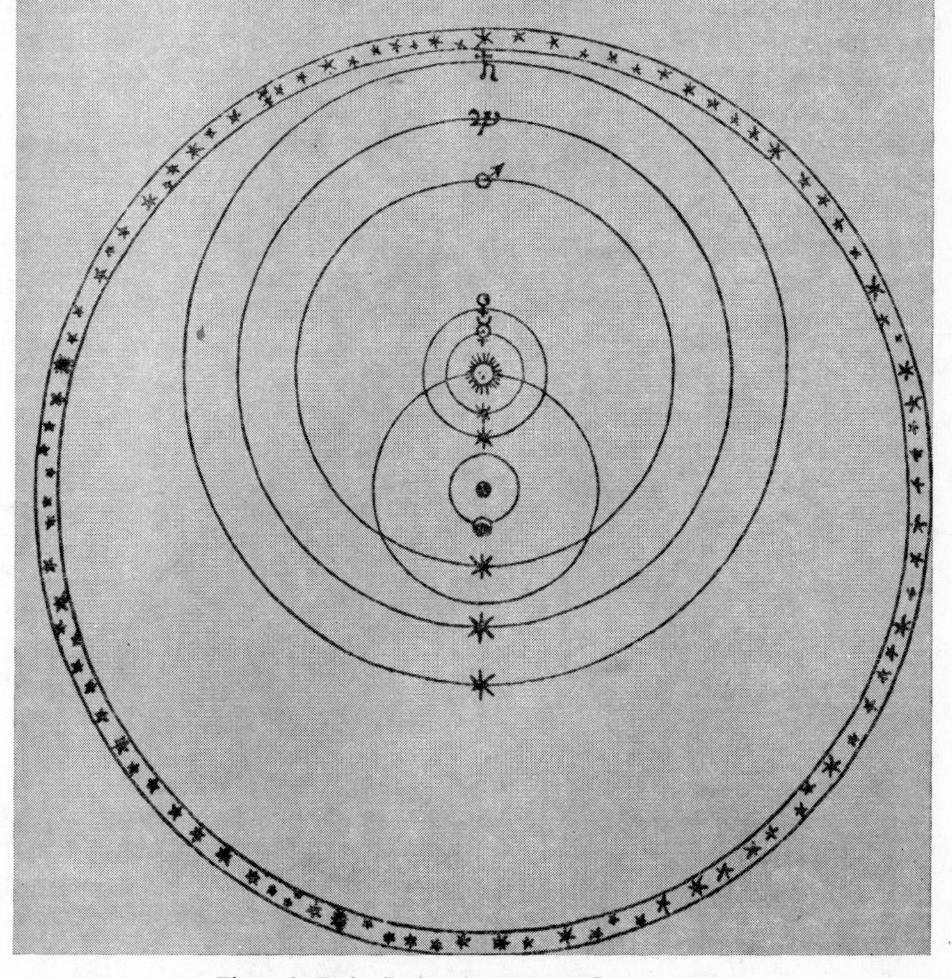
158 DE MUNDI ÆTHEREI

NOVA MVNDANI SYSTEMATIS HYPOTYPOSIS AB AUTHORE NUPER ADINUENTA, QUA TUM VETUS ILLA PTOLEMAICA REDUNDANTIA & INCONCINNITAS, TUM ETIAM RECENS COPERNIANA IN MOTU TERRÆ PHYSICA ABSURDITAS, EXCLUDUNTUR, OMNIAQUE APPARENTIIS CŒLESTIBUS APTISSIME CORRESPONDENT.

Figure 3. Tycho Brahe, *Opera Omnia* IV, page 158.

Tycho against Aristotle

Tycho's repeated, but vague references to physical absurdities connected with the Copernican system suggest that he was simply an Aristotelian. But that conclusion would be premature. Generally, he accepted traditional views on physical and cosmological matters. But he had himself destroyed the unalterability of the heavens by his proof that the New Star of 1572 was actually situated among and was similar to the fixed stars; and he had, through his work On the Comet of 1577 once and for all destroyed the solid spheres of planetary models. He could therefore certainly not be an orthodox Aristotelian, but had once more to direct his course between the Scylla and the Charybdis. Hence his general caution of expression and scarcity of detailed arguments. A special challenge was needed to make him show his hand more clearly. It happened when Rothmann after having studied the Tychonian world system became a convinced Copernican which, indeed, made Tycho use Aristotelian arguments.

But another challenge was John Craig's attack on Tycho's work on the Comet of 1577. The *De recentioribus Phaenomenis* brings Tycho's account and criticism of Maestlin's Copernican theory of this comet [*19*: IV. 221—237]. Maestlin's theory as well as Tycho's own opinion on the comet had been attacked by Craig, to whose criticism Tycho replied in his *Apologetica Responsio ad Craigum Scotum de Cometis* (1589) [cf. *119*: 208]. And this "apologetic answer" reveals Tycho's critical attitude towards the Aristotelians.

According to Tycho Maestlin is not to be blamed because he employed the Copernican assumption of the moving Earth, to devise his model for saving the phenomena of the comet. His diligence in this respect is highly recommendable, although the Copernican hypothesis does not sufficiently explain all the special motions of the comet from first to last. The very fact, however, that Maestlin, following Copernicus, assigns to the Earth an annual revolution and makes the Sun rest, reflects a legitimate freedom of expression. For the phenomena of the planets may conveniently by saved on this assumption, too; the great Copernicus himself has shown that sufficiently by his excellent work on the heavenly revolutions. No matter whether you assume a moving Sun or a moving Earth, the phenomenological consequences will be the same, so it is only to physicists, but not to mathematicians, that something absurd arises from the latter opinion.

Tycho himself had supposed the Earth to be at rest in order to allow a physical as well as a mathematical consideration. Maestlin, on the other hand, wrote as a mathematician to mathematicians; he studied the heavens themselves, the open Book of Nature, and not the cloudy nonsense offered for sale as the truth at the universities. So his work is greatly preferable to the fantasies of those who merely parrot Aristotle [*19*: IV. 446—447; cf. VII. 209].

Later in the same treatise Tycho discusses at some length his own opinion on comets and on the world system [*19*: IV. 472—476]. In his apology he asserts that no doubt Aristotle would himself have changed his mind not only about comets but about the whole nature and essence of the haevens, too, if he had acquired some knowledge of parallax from reliable observations [*19*: IV. 472]. Tycho has done away with the reality of the spheres, so he affirms that the planets move freely by themselves through the clean and clear ether, and that is not at variance with the more reasonable physics. For the fact that Aristotle imagined real spheres in the heavens does not make them actually be there since reason and experience prove that in reality there are none [*19*: IV. 474]. Elsewhere Tycho records as a generally accepted opinion that Copernicus had supposed real solid spheres to exist [*19*: II. 397–398; cf. III. 173]. In conclusion Tycho states that Craig's doubt concerning the new Tychonian hypotheses is altogether unfounded, and without any mathematical basis; regarding Aristotle one had better correct *his* physics by means of observations and geometry than the other way round [*19*: IV. 476].

The vain search for proofs

If Tycho were to accept a proof of the superiority of one world system to another it had to be, indeed, a mathematical proof including the use of observational results.

Thus the detection of an annual parallax of the fixed stars would settle the matter in favour of the Copernican system. Thomas Digges had promised to determine stellar parallax by means of some new instrument, and Rothmann had entered upon speculations concerning an annual variation of the polar height and of the distances between certain stars. Tycho, however, was unable to establish variations of the kind, and he guessed that the alleged results were due to observational errors or to some effect of refraction. Hence Digges and others might construct whatever instruments they wanted; the better and more exactly they observed the less would they find any parallax or any annual motion of the center of the Earth [*19*: III. 197–198; cf. VIII. 209].

To Tycho it was, then, impossible to prove the truth of the Copernican system from any determination of stellar parallax; but he realized, too, that the mere absence of any observable parallax did not disprove the Copernican system either, since Copernicus had, so to speak, a priori pushed the stars so far away as to make any such effect disappear.

Everything that can be said of the fixed stars in this respect is true concerning the new star, too, in so far as it behaves like the fixed stars and is placed among them. Tycho is painstaking to show that this is the case no matter whether the Ptolemaic, the Copernican, or the Tychonian system is applied [*19*: II. 378–379, 397–398].

Consequently, the new star could not either provide arguments to decide the question of world systems, and Tycho obviously felt annoyed at the attempts made by others. Thus Maestlin had maintained the very great distance of the new star by assuming the Earth to be moving; to this Tycho remarked that there was no reason at all to demonstrate something true from dubious, not to say plainly false and absurd premisses when, after all, far better and even correct arguments might easily be found [*19*: III. 63]. Digges had proposed to settle the question of the motion of the Earth from the varying luminosity of the new star which caused Tycho to prepare a critical discussion on this problem [*19*: III. 172–175, 203].

Tycho's 1588 proof that his own world system is true

Tycho himself imagined the fixed stars to be placed rather near to the outermost limit of the sphere of Saturn, so he did not believe them to be as distant as Copernicus had supposed in order to make the annual orbit of the Earth insensibly small as compared to the distance of the stars [*19*: II. 430]. In many places he offered arguments of due symmetry and orderliness within the Universe to support his view [cf. p. 53-54]; but in order to find compelling proofs he had to concentrate upon nearer objects. In 1588 at latest Tycho believed to have found a real proof doing away with the Ptolemaic as well as with the Copernican system and hence demonstrating the truth of his own arrangement of the heavenly motions.

This demonstration he communicated in letters to some of his colleagues, to witt one letter of 1588 September 13th to Caspar Peucer [*19*: VII. 127–130], another of 1589 February 21th to Christopher Rothmann [*19*: VI. 178–179], still another of 1589 November 1st to Thaddaeus Hagecius [*19*: VII. 199–200], and finally a letter of 1590 December 1st to Johs. Ant. Magini. In the last-mentioned letter Tycho promises explicitly to disprove the triple motion of the Earth by irrefutable arguments, not only theologically and physically but also mathematically, although Copernicus had hoped that his system had been adequate and irreprehensible to mathematicians. [*19*: VII. 294–295].

The most complete account of Tycho's proof is found in the said *letter to Peucer*. To some German astronomers Henric Rantzov had distributed delineations of the new Tychonian world system, which had caused Peucer to put some questions. Tycho then wrote to answer Peucer's questions and make his doubt disappear.

Already as a student at Leipzig Tycho had been aware of the discrepancies between the actual phenomena and the Copernican as well as the Alphonsine computations; he therefore began to consider whether this calamity was caused by improper observations or by erroneous hypotheses. For several years he explored the matter thoroughly and found the Ptolemaic hypotheses to offend against the first principles of astronomy because they admitted a circular motion to go on uniformly with respect to some other point than the center of the circular motion. Furthermore the numerous and great epicycles demanded much place and were superfluous. And no explanation was offered to account for the mutual connection which evidently had to exist between the Sun and the planets because they had an annual partial motion in common.

But the absurd triple motion of the Earth, on the other hand, at once deterred

Tycho from adhering to the great Copernicus' new arrangement. For although his solution met the above-mentioned difficulties of the Ptolemaic system, the mere attribution of a regular, perfect and not very complex motion to the sluggish and rather obscure Earth made this view rather suspicious, in particular because it was also in open contradiction of several places of the *Holy Scripture.*

Hence Tycho came to doubt if any of the two views was right, but since no acceptable possibility besides the two had ever been proposed he decided to find a crucial test as to which of these came nearest to the truth. And this problem could most conveniently be settled if the parallax of Mars was determined when this planet was in opposition to the Sun. For according to the Copernican description Mars should then be nearer to the Earth than the Sun while the Ptolemaic system assumed Mars to be always more remote than the Sun.

Mars was therefore, especially in the year 1582, observed very carefully and at great expense by means of different astronomical instruments which made it possible to establish stellar positions accurate to 0;0,15°. Numerous observations of this kind were performed when Mars was rising or setting or when it passed the meridian, and they showed the parallax of Mars to be greater than that of the Sun. Hence Mars really had to be situated nearer to the Earth than the Sun. Moreover Tycho found the daily rate of the retrograde motion of Mars to be greater than it ought to be according to Ptolemy and the Alphonsines, but in better agreement with the Copernican parameters. Similar results had been brought about in the case of Venus and were finally, on the 24th of February 1587, affirmed by some very rare observations of Venus performed in the evening and the following morning.

Since all these results did not at all agree with the Ptolemaic hypotheses I was urged afterwards to put more and more confidence in the Copernican invention. The exceedingly absurd opinion that the Earth revolves uniformly and perpetually nevertheless made up a very great obstacle, and in addition the irrefutable authority of the *Holy Scripture* maintained the opposite view.

During his further considerations on the problem Tycho therefore made persistent attempts of establishing new hypotheses. At first his plan appeared to be impracticable, mainly because the only possible solution caused the orbits of the Sun and Mars to intersect each other, and at that time Tycho was still imbued with the belief that real orbits existed in the heavens and carried with them the heavenly bodies. But having finally by diligent research on the motion and the parallax of several comets been led to the conclusion that these were situated far beyond the Moon Tycho completely abandoned the assumption of any hard and impervious substance for the celestial orbits.

Add to this that two comets moving in opposition to the Sun quite evidently showed that the Earth does not actually perform any annual revolution, in so far as the supposed 'commutation motion' of the Earth did not detract from their predetermined and proportionate rate of motion any single jot, as it happens in the case of the planets which, according to Copernicus, retrograde just because of the annual motion of the Earth.

This amounts to saying that from the absence of a yearly partial motion within his well-established models of the motions of some comets Tycho argued that any supposed annual revolution of the Earth was fictitious. His argument is formally sound since a motion of the Earth ought to be reflected in the motion of every celestial body above the Moon; but it depends upon a gross extrapolation from observational results obtained during the very brief visibility of the comet. Anyhow it was the decisive excuse Tycho offered at last for being a Tychonian and not a Copernican.

The problem of Tycho's 1584 version of the same proof

So far the fully developed Tychonian proof of his own world system has been sketched, and further references are found above. It is, however, worth noticing that one part of his proof, viz. the argument concerning the parallax of Mars, is mentioned separately by Tycho on several occasions, thus in his *Apologetica Responsio* [*19*: IV. 474—475], in his *Progymnasmata* [*19*: II. 383], in a letter of 1587 January 18th to the Landgrave Wilhelm the IVth of Hessia [*19*: VI. 70], and in a letter from 1584 to Henric Bruceus [*19*: VII. 78—82].

The *letter to Bruceus* is especially interesting because here Tycho, concerning the parallax of Mars, arrives at a conclusion unique in his authorship:

Tycho asserts to have found towards the end of 1582 and in the beginning of 1583, from numerous, most exquisite and mutually concordant observations of Mars in opposition to the Sun that the parallax of Mars was far smaller than that of the Sun. Therefore, the whole sphere of Mars must necessarily be remoter from the Earth than that of the Sun. Accordingly he altogether abandoned the Aristarchian doctrine of the annual motion brought to light anew by Copernicus. He moreover rejects the concept of a "motion in declination together with its librations", while he admits that a daily rotation might perhaps be ascribed to the Earth, although he considers it more likely that such a rotation went on in the heavens.

On the whole Tycho takes up in this letter an unusually severe attitude towards Copernican ideas. Thus Bruceus had asked him to send "some sort of instrument showing the Copernican hypotheses", which Tycho would have done with pleasure if he had ever prepared such a device. He had, however, never dreamt of constructing instruments of that kind to explain the inventions of others when these might easily be understood by means of a plane delineation alone and when, moreover, it was still uncertain whether they corresponded to reality or not. A little later Tycho points out how difficult in would be to reproduce all the motions within the Copernican theory by means of one single mechanical machine, especially because of the incomparably great distance between the Earth and the eighth sphere. And he concluded this section of his letter saying: "In case you want, after all, to imbue the minds of your students with his (Copernicus') inventions you may easily prepare a plane delineation, even if that is quite a job, too".

Note that in 1576 Tycho did not scruple to prepare a mechanical representation of the Copernican system; on the contrary he intended to do so [cf. p. 37].

Concerning the problem of the parallax of Mars several questions arise. The modern values of the parallaxes of Mars and the Sun being 0;0,23° and 0;0,9° respectively, it seems mysterious how in 1584 Tycho was able to find a parallax of Mars much smaller than that of the Sun. But Tycho believed the solar parallax to be 0;3°, so it appears even more suspect that later on he could find a greater parallax for Mars. Kepler has suggested that one of Tycho's pupils had by a mistake calculated the parallax of Mars from Copernican parameters instead of from observational results [cf. *119*: 179]. It seems to me incredible that Tycho should ever have entrusted a pupil with the task of performing so crucial calculations without himself controlling carefully the results. In fact, all values of parallax except that of the Moon are far too small to be investigated to any reasonable degree of certainty by Tycho. Probably he did never fully realize this fact, but it would be too hard to blame him for that.

However, at different times Tycho did arrive at opposite conclusions from the very same observations. A few glimpses from Tycho's development in the 1580'es may perhaps throw some light on this offensive fact.

At first it is worth noticing that in 1584 Tycho had worked out a delineation of his world system which showed the orbit of Mars surrounding the solar orbit instead of intersecting it, and the fact that in 1588 he described the said sketch as a mistake [*19*: VI. 179] does not prove that it was not meant seriously in 1584. On the whole it begins to look as if in 1584 Tycho did not at all realize how the restoration of astronomy he had planned was to be completed. After having spent eight years at Hveen with observations he had been brought to doubt the results of his predecessors, including Copernicus, to an unexpected extent. Evidently this made him impatient to arrive at results which could guide his further work. Hence Morsing was sent to Frauenburg, and from this expedition Tycho learned why Copernicus' results were uncertain. He also realized that Copernicus had been an honest scientist as well as how to correct his results [cf. p. 35]. Accordingly in 1585 he ventured to construct his own models as inverted Copernican models [cf. p. 37]. And in 1587 he is on the point of accepting the motion of the Earth as a natural motion [cf. p. 38].

But his preconceived scruples carried too much weight and after having found the argument from the absence of any annual partial motion within his theories of the comets he dared to reject completely the motion of the Earth. Accordingly, in 1588 he published his own world system. And thenceforth he adhered unscrupulously to traditional views on the site and the state of rest of the Earth. That is particularly clear from his correspondence with Rothmann on the world systems from the autumn 1588 until in August 1590 Rothmann visited Hveen where the discussions were continued orally.

III. TYCHO AGAINST COPERNICUS AFTER 1588

The discussion with Rothmann, 1588–1590

Among Tycho's comprehensive correspondence the letters exchanged between Hveen and Cassel during the period 1585–92 deserve a special attention because of their size as well as of their contents. Many of the letters are real astronomical treatises, and in 1596 Tycho published them as his *Epistolarum Astronomicarum, Liber I* [cf. *119*: 133–136, 228–230]. For the present treatise it is particularly important that in their letters between 1588 and 1590 Tycho and Rothmann entered into details concerning the Copernican and the Tychonian world systems and thus enabled us to imagine the eventual contents of the systematic and detailed account regarding the motion of the Earth, which Tycho repeatedly promised to give but never gave. To show how the discussion developed the letters in question will be reviewed below in chronological order.

In a letter dated 1587 September 21st Rothmann writes that he has himself established "inverted" Copernican hypotheses. But since by his observations he has found the Prutenic Tables to deviate too much from the truth he would be glad to substitute a Tychonian computation for the Prutenic one in order to make the former familiar to his students [*19*: VI. 118]. In 1588 August 17th Tycho answers that a Prutenic calculation might well be published for educational purposes. He invites Rothmann to do so since the attempts of others within this field had illegitimately connected Ptolemaic hypotheses and Copernican numbers [cf. p. 32]. Besides Tycho calls attention to his own new hypotheses as described in his *De Recentioribus Phaenomenis*. A copy of this book had recently been sent to Rothmann, and Tycho expresses his belief that Rothmann would be the more pleased with this new arrangement the more diligently he studied it [*19*: VI. 146–147].

Rothmann's 1588 letter

However, the very opposite was to happen. In his next letter dated *1588 September 19th* Rothmann discusses at some length the problem whether Tycho's new models were identical with Rothmann's own inverted Copernican models and with the hypotheses underlying a certain small automaton which the Landgrave had got constructed the foregoing year. Having admitted his unability to answer definitely this question Rothmann enters upon a criticism of the general disposition of the spheres within the Tychonian system for which he has a dislike the more marked the longer he considers it.

His *first* charge against Tycho regards the retrogradations and stationary points within the planetary theories. The Tychonian arrangement correctly describes *how* these phenomena go on, but not *why* they do so, which the Copernican system had done in a very beautiful manner. *Secondly* he doubts that anybody will ever believe the center of the major solar epicycle to be so powerful as to draw with it all the planets, there being no material connection at all between the heavenly bodies. Thus the Landgrave had jokingly said: "Good Heavens! That solar circle has to be far stronger than copper to be able to carry with it so many planets". *Thirdly* this inverted Copernican ordering of the spheres re-introduces among those spheres a confusion which had just been abolished by Copernicus, and which moreover conflicted with the divine orderliness of the heavens. *Fourthly* the Tychonian description is not economic to the same extent as the Copernican one. And *fifthly* a variation of the solar eccentricity will within this system affect the models of the remaining planets, too.

"Therefore, over and over considering the said arguments and many others, too, I cannot help infering that no hypothesis can be true except the Copernican one".

Thence Rothmann argues that the Copernican system does not conflict with physical principles when it adduces the similarity between the Earth and the heavenly bodies following from the universal conception of gravitation as supposed by Copernicus, and from the spherical shape common to the Earth and the planets. And from the stability of the artificial rotation of a globe he argues that a natural rotation of the Earth must be even more stable. The system is not in opposition to the Holy Scripture either, since the Bible has not been written for Tycho and Rothmann alone, but for all people, and accordingly must speak a language intelligible to everybody. This point is further exposed by examples from the Scripture and ends up with the assertion that sure knowledge concerning astronomical subjects must necessarily be obtained from mathematical demonstrations.

Gellius Sascerides (1562–1612) had, indeed, told Rothmann that Tycho possessed geometrical proofs of his system, but Rothmann hardly dared believe that. Finally he judges the Copernican hypothesis unsuitable for students since even professionals, e. g. Maestlin, did not quite understand it. For this very reason Rothmann had prepared an inverted Copernican system of his own for educational purposes [*19*: VI. 156–160].

Tycho's two 1589 letters

In his next letter dated *1589 February 21st* Tycho is evidently faltering at this unexpected attitude of Rothmann. At that time he is unwilling to bring forward many arguments to defend and approve his own hypotheses. That would be too wearisome and demand a complicated demonstration from certain observational results. He prefers to apply his efforts to a proper restoration of the heavenly motions to the effect that his hypotheses may in due course show themselves to square exactly with the heaventy phenomena, hence being far superior to the Ptolemaic as well as to the Copernican ones, and thus coming much nearer to truth. If, however, Rothmann dislikes these hypotheses so much that he prefers, with Copernicus to move the Earth, the oceans, and the Moon together — and that with a triple motion —

and if he wants to raise the Earth beyond the rank of the stars in spite of its obviously inferior nature and lesser mobility, that is all the same to Tycho and must be Rothmann's private affair.

Nevertheless Tycho wonders if that does not mean a confusion of inferior and superior matters, and a total inversion of the order of natural things? In Tycho's own words: "You must not be mistaken; your conviction that the physical absurdities of the Copernican hypothesis have been sufficiently rejected by Copernicus himself is plainly false. And your own arguments regarding this case are also unable to remove every doubt on the subject which I shall prove elsewhere discussing these and others of the kind". Further, the arguments from the Holy Scripture are far less capable of settling the matter. For the reverence to the Divine records and their authority are and ought to be too great to be interpreted like a lofty tragedy. After going into details concerning some scriptural passages Tycho concludes that the authority of the Holy Scripture has until this very day remained unshaken. So Rothmann will not at all be succesful in defending the Copernican point of view from that quarter, either.

Tycho thereafter answers some of the objections to his own system by stressing that the planetary motions are not due to any involuntary carrying off, but are caused by a natural and divinely ingrafted observance. And if magnetite and iron, terrestrial and lifeless as they are, may, naturally and without violence, attract each other even through an interposed body, why should such things not be likely to happen in the heavenly bodies which by the Platonists and the more judicious philosophers were believed to be animate? Rothmann is urged to study diligently Plinius' account on the subject. And after doing away with the real solid spheres Tycho imagines a single uniform heavenly matter stretching from the Moon to the eighth sphere, in which matter the planets are free to move up and down while performing their most beautiful and harmonious revolutions around the revolving central Sun. Hence no planetary confusion is found within the Tychonian system.

Tycho then denies that his system should have come into existence by an inversion of the Copernican hypotheses, and he briefly accounts for the arguments on the parallax of Mars and on the motions of comets [cf. p. 45] claiming explicitly that these two arguments alone suffice to prove the truth of his system [*19*: VI. 176–179].

Tycho finally reveals how he had at first wondered about the origin of the automaton mentioned by Rothmann, but that now he is quite sure that it is due to Nicolaus Reymers who must have copied one of Tycho's drawings during his stay at Hveen in 1584 [*19*: VI. 179–180; cf. *119*: 183–185].

This letter did not reach Rothmann until August 17th, so his answer dated *1589 August 22nd* mostly contains short remarks and no significant new material regarding the wordl systems [*19*: VI. 183].

Tycho's next letter dated *1589 November 24th* opens with a thorough discussion on physics and theology [*19*: VI. 185–187] and later it ensures Rothmann his freedom to accept Ptolemaic or Copernican hypotheses as he likes. In return Tycho must also be permitted to announce freely what he has learned on the subject from the heavens themselves. He repeats his plan of composing a connected account on the planetary motions and will until then be glad to receive objections of his system.

However, since Rothmann has grown so fond of the Copernican idea of the triple motion of the Earth Tycho will readily adduce a single argument against each of its three partial motions.

Re the *daily* motion: If the Earth rotates, how is it to be explained that a leaden globe falling from a high tower travels vertically and strikes the surface of the Earth perpendicularly? For the falling body does not accompany the atmosphere but moves violently through the air.

Re the *annual* motion: The eighth sphere should be placed vastly far away to make the orbit of the Earth vanish in comparison with the distance of the fixed stars. On the assumption that 0; 1° is a possible value of stellar parallax, the distance between Saturn and the fixed stars must necessarily be 700 times the distance between the Sun and Saturn. Moreover, a star of third magnitude having an apparent diameter of 0;1° in that case had to be as great as the whole annual orbit of the Earth, i. e. its diameter should equal 2284 terrestrial radii. And what to say about the stars of first magnitude with an apparent diameter of 0;2—0;3°? And what to say if the eighth sphere is to be imagined still farther away to make the annual orbit of the Earth vanish? [*19*: VI. 196—197].

Here Tycho has amplified his often repeated [*19*: II. 376, 430; III. 198; IV. 155—156] argument on the immense and superfluous void between Saturn and the eighth sphere. He thus formulated what was later going to be the most well-known Tychonian argument against the motion of the Earth. Another elaborate version of this argument forms part of Tycho's criticism of Maestlin in the *Progymnasmata* [*19*: III. 63—64]. I call attention to the fact that both versions are found with polemic contexts.

The *third* motion collapses when the annual motion has been explained away. But even if the annual motion of the center of the Earth be retained, by what means should its axis turn round in the opposite direction exactly to the effect that it appears nevertheless to be at rest? And how can the center and the axis of a single simple body perform two mutually contrary motions, not to speak of the third, i. e. the daily motion? In addition the perplexing librations do not agree with the motion of the fixed stars and are highly inexpedient.

"Consider carefully these and similar arguments... you will get others on another occasion" [*19*: VI. 197—198].

Rothmann's visit to Hveen in August 1590

In his letter dated *1590 April 18th* Rothmann answers re the *daily* motion: The falling globe partakes in the daily rotation of the Earth, too. And this circular partial motion of falling bodies is as natural a motion as the rotation of the Earth as a whole, just as small portions separated from a nugget retain the nature of gold of the whole nugget. Otherwise the tower from which the globe was thrown must necessarily be left behind, too, the ground moving away beneath it. In addition it is far more natural and economic to let the small Earth rotate than to force all the planets and stars to perform such a multitude of incomprehensibly fast revolutions.

Re the *annual* motion Rothmann puts the question why on earth the Universe and the stars must not be great? Although it might appear unreasonable this is nevertheless a far too flimsy excuse for restricting the Creator and His creation.

Re the *third* motion Rothmann answers the Tychonian *how* by asking *why not*? After all the Earth moves freely and is not attached to any material sphere. Yet he admits that Copernicus' own account is somewhat obscure on this point, and moreover he maintains that it is quite possible to dispense with the third motion. Librations are accepted by Rothmann as principally natural motions, but Copernicus' perplexing exposition regarding this subject may easily be criticized, too [*19*: VI. 215–217].

Tycho did not receive this letter until Rothmann in *1590 August 1st* arrived at Hveen where the two astronomers carried on their discussions orally. This continuation is extant in Tycho's reproduction: Having discussed for some weeks Rothmann at first hesitated, then he put forward his arguments still more timidly, and finally he appeared quite opposed to his own former opinions. He even declared that he had not till then published anything in that direction and would never do so in the future. He had taken the Copernican standpoint rather for the sake of argument [*19*: VI. 218].

Perhaps Rothmann was so impressed by his being defeated that afterwards he dared not face the Landgrave. Anyhow, on September 1st he left Hveen to go to Cassel, but after changing his mind he went to his native Place Bernburg, and he never returned to Cassel [*119*: 208–209].

The discussion on the *daily* motion *firstly* includes some further fencing with the question of the falling globe by arguments having, of course, no demonstrative force, but merely revealing that Tycho did accept the Aristotelian doctrine of motion, while Rothmann did not. *Secondly* Tycho offers a lengthy account on artillery shots maintaining that, other things equal, their ranges are the same in every direction at all geographical latitudes, which he deems incompatible with a rotation of the Earth. Tycho's *third* argument regards an arrow ejected vertically upwards and depends on its returning exactly to the starting point. By the way he denies the assertion of those who maintain that an arrow thrown in the same way from a moving ship should return to the starting point. *Fourthly* Tycho brings a brief and obscure allusion to an argument concerning the natural motion of the oceans.

The daily rotation having thus been rejected the annual motion and the third motion together with its librations sink to the ground by themselves and turn out to be useless. However, to provide a sufficient reply to Rothmann's objections Tycho at this point inserts *general considerations* beginning with arguments of economy in the explanation of nature which apparently favoured the Copernican system as compared to the Tychonian. But, according to Tycho, God the Creator manifested His inscrutable wisdom and omnipotence just by His Will and Power of attributing to the vaste heavenly bodies their swift and uniform motion, while He retained the body of the Earth, sluggish and dense, and hence unfit for continuous motion, in an everlasting and stable position. From this resting center he thus made it possible to contemplate the stupendous and indefatigable courses of the heavenly bodies, or in other words, to regard and admire in a mirror, as it were, the Grandeur of the Divine Creator.

And since the Earth rests patiently at the center of the Universe it may most conveniently receive the powers and influences tending towards the center from the actively revolving heavens. Having further elaborated the last point Tycho ends this section by consenting to the Platonic doctrine of living heavenly bodies within an animate heaven.

Thus prepared Tycho returns to the problem of the *annual* motion, and now he is enabled to maintain that, in a Copernican description the huge and superfluous void between Saturn and the eighth sphere would bear witness of an otherwise quite unknown and alien irregularity and lack of orderliness and of symmetry within the work of the Creator. Even the Sun would vanish and be next to nothing as compared to the fixed stars, although it is evidently the most prominent body of the Universe. It is true that the finite world bears no proportion to the infinity of its Creator, but all the same there must be some definite scale which is applicable to all the different parts of the created world. This is obviously the case of terrestrial creatures, e. g. within the microcosmos of the human body whose due harmony is excellently painted by Dürer [*19*: VI. 218–222].

Tycho the Pythagorean

It has been made clear above [cf. p. 42] that Tycho was by no means an orthodox adherent of the Aristotelian cosmology. But the foregoing debate with Rothmann reveals that when facing the dogged resistance against his own fully developed world system, he felt the need for resorting to Aristotelian bastions. However, it also appears that Pythagorean and Platonic arguments of symmetry and harmony, intimately connected to religious considerations, and even thoughtst ouching astrological matters had far more weight with Tycho than had the Aristotelian doctrine of motion.

Further evidence on this point is found in several places within Tycho's authorship. Already in his *lectures in 1574* he claimed the Earth to be situated at the center of the Universe, in agreement with God's revealed truth, and so as to be fit for receiving the influence of the stars [*19*: I. 152]. According to a letter to Thaddaeus Hagecius, dated *1589, November 1st*, a similar motive makes Tycho advance new hypotheses enabling the Earth to produce, so to speak, terrestrial stars nicely analogous with the heavenly ones [*19*: VII. 202—203]. Further arguments of symmetry and harmony directed against the Copernican system are found in *Progymnasmata* [*19*: II. 435], and they play a central role in a couple of letters dated *1598 April 1st* and *1599 November 29th/December 9th* (old and new style) which are worth mentioning because of their addressee, Johannes Kepler.

In the former letter Tycho admits that the daily rotation of the Earth appears quite reasonable [*19*: VIII. 45—46]. He had perhaps finally realized that against this special motion he had no real arguments other than those drawing on Aristotle. Thus he prepares the way for Longomontanus' Semitychonian system [cf. p. 131]. In the other letter Tycho stresses the necessity of retaining uniform circular partial motions. For otherwise the courses of the heavenly bodies could not be perpetual nor return to their starting points; and besides that they would destroy eternity itself. In addition their motions would be less simple, more irregular and not very fit for theoretical or practical use [*19*: VIII. 208—209].

Tycho therefore throughout his career adhered faithfully to the principle of uniform circular motions as integral parts of the said complex of Pythagorean-Platonic, religious and astrological convictions deeply rooted in the mind of this great scientist. It seems a paradox that substantially the same complex of convictions, including until a certain point of time the axiom of uniform circular motions, guided Kepler's construction of his "elliptical" astronomy, and that the same belief, apart perhaps [cf. *128*: 136] from the astrological elements, had earlier governed Copernicus' restoration of astronomy thus providing the basis of a moving Earth. Of course every argument drawn from that kind of source is very general and proportionately devoid of concrete scientific contents. Accordingly we need not give a thought to the possibility of arriving from such arguments at opposite conclusions concerning specific astronomical questions. But to understand why the said arguments were considered so important we have better search for common basic conditions for the working on theoretical astronomy in the period between Copernicus and Kepler, otherwise so rich with innovations.

Thus it must be remembered that all astronomers of the said period not satisfied by "saving phenomena" only, had to accept the Aristotelian cosmology and doctrine of motion, or else to make the best they could of what may be called a "vacuum of dynamics". Descartes and Newton still belonged to the unknown future. Galileo, towards the end of the period, did establish new and sound kinematical concepts which were, however, not immediately fit for explaining the causes of heavenly motions. And Gilbert's *De magnete* which proposed the heavenly globes to be loadstones was not published till the year 1600, i. e. one year before Tycho's death.

Gilbert's ideas were utilized by Kepler to explain the planetary motions, so Kepler, according to Dreyer, made "the first attempt to interpret the mechanism of the solar system" [*126*: 394 ff.]. A common feature of most other explanations was their fundamentally mathematical nature, be it hypotheses regarding the five regular solids [*126*: 374 ff.], musical consonant intervals [*126*: 405 ff.], or plain number mysticism [cf. p. 127]. Until the invention of Kepler's elliptical, planetary orbits, the cornerstone of most theories was the principle of uniform circular partial motions. This principle was raised above its context within natural philosophy, and given a new status as a sort of mathematical axiom; i. e. the most divine of all sciences, namely mathematics, was to replace the Aristotelian physics which had shown itself unable to provide reliable explanations of astronomical phenomena. Thus astronomers began to hope that their science, *mathemata superiorum*, might one day attain a perfection of formulation and accuracy similar to that of Euclidian geometry [cf. Tycho's discussion with Petrus Ramus; *19*: VI. 88–89; *119*: 33–34].

Even these brief considerations suggest profound reasons why Tycho regarded Ptolemy's equant to be so fatal, why he felt himself congenial with Copernicus in several respects, and why he — and Copernicus as well — believed to have proved the truth of their models, although to 20th century minds they have only demonstrated them to be mathematically and phenomenologically possible. Whatever Tycho's attitude towards Copernicus at various stages of his career, it is true that attention should be paid to every single occurrence of Copernicus' name in Tycho's work. The science of astronomy made progress through Tycho's accepting, and through his rejecting Copernican ideas.

HANS BLUMENBERG
Westfälische Wilhelms-Universität,
Münster

DIE KOPERNIKANISCHE KONSEQUENZ FÜR DEN ZEITBEGRIFF

I.

Als fast schon unbestreitbarer Typos der Wissenschaftsgeschichte gilt die Behauptung, dass Kopernikus für die Veränderung des Raumbegriffes der Neuzeit eine entscheidende Rolle gespielt hat. Der absolute unendliche Raum Newtons erscheint als die deutliche Konsequenz der Ausweitung des kosmischen Raumes, die infolge der kopernikanischen Voraussetzungen vorgenommen werden musste. Ebenso deutlich ist die kritische Rückbildung dieser Konsequenz auf dem Umweg über die Idealisierung des Raumes seit Leibniz.

Meine These ist nun, dass auch die für die Neuzeit charakteristische Verschärfung der Problematik der Zeit ohne die Veränderung nicht begriffen werden kann, die Kopernikus am Weltmodell vorgenommen hatte. Die Problematik der Zeit ist freilich weniger spektakulär, weniger affektiv wirksam als die des Raumes; das mag erklären, dass die kopernikanische Konsequenz für den Zeitbegriff bisher nicht dargestellt worden ist.

Man muss bei dieser Thematik freilich methodisch vorsichtig genug sein hinsichtlich dessen, was wir als historische „Wirkung" behaupten können. Denn schon die blosse Möglichkeit der kopernikanischen Reform beruht auf bestimmten Veränderungen im scholastischen System hinsichtlich des Zeitbegriffs, die für den Raumbegriff nicht gefordert zu werden brauchen, weil es sich hier zunächst nur um notwendige quantitative Veränderungen handelte. Gerade weil ich im Folgenden zeigen muss, dass die antiken Implikationen des Zeitbegriffs für die kopernikanische Reform keinen Spielraum enthalten hätten, wird es entscheidend, schon in der Bewegung des mittelalterlichen Denkens bestimmte Erweiterungen, Variabilitäten des Zeitbegriffes aufzuweisen, ohne die der kopernikanische Grundgedanke nicht durchführbar oder auch nicht zu verteidigen gewesen wäre. Denn genau dies ist die Frage, in deren Zusammenhang überhaupt das Zeitproblem aufgeworfen werden kann: nicht wie Kopernikus auf den Gedanken der strukturellen Reform des Planetensystems kommen konnte, sondern wie der astronomisch zwingende Gedanke dieser Reform unter den Voraussetzungen der zeitgenössischen Vorstellungen vom Ganzen der Welt zu behaupten und durchzusetzen war. Unter diesem Gesichtspunkt erscheint das, was man als „Wirkung" der kopernikanischen Reform behaupten möchte, nur als der letzte Schritt der Konsequenz ihrer Voraussetzungen. Für Kopernikus war es nicht nötig, den Gedanken des absoluten Raumes und der absoluten Zeit seinem System zu integrieren; aber ohne ein gewisses Mass an Veränderungen der antiken und hochscholastischen Positionen hinsichtlich dieses Problems in Richtung auf Newtons absolute Begriffe wäre die kopernikanische Reform nicht durchsetzbar gewesen. Der theoretische Erfolg des neuen Systems treibt den in ihm schon vorausgesetzten Prozess voran, bestätigt ihn, bringt ihn zur vollen Explizität. Darin besteht das, was „Wirkung" genannt werden darf.

Hier nun is est wichtig zu beobachten, wie spät Kopernikus ausdrücklich der Vorwurf gemacht worden ist, er habe bei seiner astronomischen Reform auf die Frage nach der kosmologisch-naturphilosophischen Möglichkeit keine Rücksicht genommen. Zweifellos schwelt diese Frage in den meisten Reaktionen auf das Werk des Kopernikus, ohne dass sie zureichend hätte artikuliert werden können. Das mag daran liegen, dass die Leser des Kopernikus im 16. und teilweise noch im 17. Jahrhundert keinen ganz bestimmten Eindruck von dem Wahrheitsanspruch des Autors und seines Werkes hatten, weil dieser durch das illegitime Vorwort des Osiander, das die Leser dazu aufforderte, das neue Modell als rein technische Hilfskonstruktion des astronomischen Kalküls zu betrachten, eingeschränkt oder aufgehoben erschien. Kopernikus hatte diese Vorrede des Osiander nicht gebilligt, weil er den Wahrheitsanspruch seines Werkes nicht nur für astronomisch ausreichend begründet, sondern auch für integrierbar in das naturphilosophische System hielt, das ihm vertraut war. Aber gerade diese Integrierbarkeit war durch sehr sparsame Mittel hergestellt und für den nicht sehr gründlichen Leser nur schwer erkennbar angelegt. So konnte noch zu Beginn des folgenden Jahrhunderts Francis Bacon nicht ohne Berechtigung den Vorwurf gegen Kopernikus erheben, er habe sich um die naturphilosophische Möglichkeit seiner Reform nicht gekümmert. Eine flüchtige Lektüre der *Revolutiones* scheint Bacon Recht zu geben. Der Nachweis, dass Kopernikus zwar sparsam, aber doch sehr sorgfältig auf naturphilosophische Systemelemente Rücksicht genommen hat, soll hier für das spezielle Problem des Zeitbegriffs geführt werden. Dazu ist allerdings vor allem darzustellen, dass das Zeitproblem für die kopernikanische Veränderung des Weltsystems überhaupt Bedeutung hatte, denn eben dies liegt nicht ohne weiteres auf der Hand.

Nun kann man einwenden, für Kopernikus und die ihm vertraute Tradition habe sich der Begriff der Astronomie gerade dadurch definiert, dass sie naturphilosophische Fragestellungen nicht nur aufwerfen oder berücksichtigen *konnte*, sondern dies nicht einmal *durfte*. Denn die Astronomie der Tradition seit der Antike war eben keine physikalische Disziplin, die Fragen nach der Kausalität, der Realität des Raumes und der Zeit, der tatsächlichen Beschaffenheit der Gestirne und ihrer Sphären, den tatsächlichen Entfernungen und Grössen zu klären hatte, sondern die ausschliesslich der phoronomischen Darstellung und Berechnung der am Himmel erscheinenden Phänomene diente. Physik im Sinne dieser Tradition galt also nur im Bereich unterhalb der Sphäre des Mondes. Die Erde war kein Stern, so lässt sich dieser Sachverhalt auch formulieren.

Aber die derart beschriebene Sonderstellung der Astronomie und ihrer Methodik – oder auch der Erde und der in ihrer Region geltenden Physik – war in dem Augenblick nicht mehr durchzuhalten, in dem Kopernikus die Erde selbst zu einem Himmelskörper und damit zum Thema der Astronomie – und umgekehrt die anderen Himmelskörper potentiell zum Gegenstand der bisherigen Physik – machte. Dies war ja einer der Gründe dafür, dass die Zeitgenossen im Werk des Kopernikus die Erfüllung des Ideals der Pythagoreer sahen, die bereits die Erde als Stern unter Sternen bezeichnet hatten. Das war nicht nur eine Formel für eine neue Würde, die schon Nikolaus von Cues als Nobilitierung verstanden hatte, sondern auch eine Veränderung der theoretischen Situation. Wenn die Erde ein Stern unter Sternen war, mussten sich generelle und austauschbare Aussagen in diesem neuen homogenen Gegenstandsuniversum machen lassen. Die Astronomie konnte fortan nicht so tun, als liesse sich von dem kosmischen Reich der Gestirne nichts mehr als das wissen, was sich dem Auge des Beobachters in Gestalt von bewegten Lichtpunkten, wie eine Kulisse vor einer undurchdringlichen Wirklichkeit, darbietet, und als brauche sie für die Stringenz ihrer Erkenntnisse von den möglichen realen Verhältnissen hinter dieser Kulisse nicht Notiz zu nehmen. Die Erhebung der Erde zum Stern in der kopernikanischen Reform stellte prinzipiell das methodische Werkzeug der Extrapolation zur Verfügung. Es war freilich in Vergessenheit geraten, dass die Unüberschreitbarkeit der Grenzen der irdischen Physik ursprünglich als ein Mangel empfunden worden war; wie so oft hatte sich der Mangel der methodischen Möglichkeit im Laufe der Zeit in einen Kanon der sachlichen Unzulässigkeit transformiert. Nun hat sich Kopernikus erkennbar daran gehalten, den Kanon der traditionellen astronomischen Disziplin möglichst nicht zu verletzen. Mit diesem Bestreben hängt zusammen, dass er sein eigenes Hauptwerk mit dem klassischen Standardwerk des Ptolemäus streng zu parallelisieren suchte. Gerade deshalb diskutiert Kopernikus nicht ausdrücklich die naturphilosophischen Möglichkeiten seiner astronomischen Konstruktion, soweit dies ausserhalb der Argumentation liegt, die Ptolemäus ihm vorgegeben hatte. Aber selbstverständlich musste er bei seinen Lesern – vielleicht weniger bei den Fachgenossen als bei den allgemeiner Interessierten – mit einer Prüfung seiner Thesen unter Gesichtspunkten des geltenden scholastischen Systems der Naturphilosophie rechnen. Diese beiden Voraussetzungen bestimmen die Eigenart seiner Interpretation. Innerhalb des kanonischen Rahmens argumentiert er zugunsten der drei von ihm behaupteten Bewegungen der Erde so, dass diese wenigstens den möglichen Einwänden entzogen werden, wenn nicht gar zusätzliche Begünstigung und Plausibilität erhalten.

Zu diesem Argumentationskonzept gehört vor allem, dass die der Erde zugeschriebenen Bewegungen eben die Funktionen übernehmen können, die in dem traditionellen System der Naturphilosophie durch die jetzt als scheinbar erklärten Bewegungsphänomene des Himmels realisiert worden waren. Nun steht aber gerade die Tagesumdrehung des Fixsternhimmels, die jetzt zum phänomenalen Äquivalent einer realen Achsendrehung der Erde erklärt worden war, für die naturphilosophische Tradition seit der Antike im engsten Zusammenhang mit dem Problem der Fundierung von Realität und Messbarkeit der Zeit.

II.

Wenn die Achsendrehung der Erde die scheinbare Tagesbewegung des Fixsternhimmels zu verantworten hatte, dann bestand die am schwersten wiegende Konsequenz dieses Funktionstausches der extremen kosmischen Instanzen des aristotelisch-scholastischen Systems darin, dass die Erde zu übernehmen hatte, wofür sie in der Tradition als am wenigsten disponiert erschienen war, nämlich die schnellste und gleichförmigste kosmische Bewegung, die den Erfordernissen der Manifestation der Zeit und ihrer Messung zu genügen hatte.

Unter keinem anderen Aspekt wird die Radikalität der kopernikanischen Reform so handgreiflich wie unter dem des Problems der Zeit. Unter keinem anderen Gesichtspunkt wird so deutlich, weshalb die Erde Stern unter Sternen nicht nur werden *durfte*, sondern notwendig werden *musste*. Und diese Notwendigkeit ist es, die uns bestimmte argumentative Anstrengungen und Formulierungen des Kopernikus, die sonst eher ornamental wirken, sachlich erst verstehen lassen. Dazu gehören vor allem die Aussagen über die Äquivalenz der Gestalt der Welt im ganzen (*forma mundi*) und des Erdkörpers (*figura terrae*), mit denen sich zwei eigene Kapitel im ersten Buch der *Revolutiones* beschäftigen. Damit die Erde die Funktion der Realisierung der Zeit übernehmen kann, die sie doch notwendigerweise unter der Voraussetzung der aristotelisch-scholastischen Definition übernehmen muss, hat sie den von Aristoteles für den ersten Himmel aufgestellten Bedingungen der unmöglichen Abweichung von der Gleichförmigkeit in anderer Weise zu genügen. Eben darauf besteht Kopernikus durch die Behauptung der Äquivalenz der beiden Gestalten als der Voraussetzung für die Austauschbarkeit ihrer Funktionen. Die Erde musste ein Stern im traditionellen Sinne werden, damit sie im neuen System das leisten kann, was in dem altem nur den Himmelskörpern, oder sogar nur einem von diesen, zugetraut werden konnte — in der „Metaphysik" des Aristoteles noch nicht einmal dem Himmelskörper ohne die metaphysisch erweisbare Antriebsleistung des unbewegten Bewegers.

Es besagt also wenig, wenn wir von dem, was man den „Zeitbegriff" des Kopernikus nennen könnte, fast nichts wissen. Entscheidend ist, dass Kopernikus hinsichtlich dieses Begriffs unter den neuen Voraussetzungen auf die alten Erfordernisse erkennbar Rücksicht nimmt.

Im Prooemium zum zweiten Buch der *Revolutiones* gibt es wenigstens eine Andeutung zum Zeitbegriff des Kopernikus. Er resümiert hier die im ersten Buch vorgelegte Argumentation zugunsten einer dreifachen Bewegung der Erde und kündigt an, nun im einzelnen zu zeigen, wie diese Bewegungen die Erscheinungen am Himmel erklären. Dabei sei mit dem zu beginnen, was am deutlichsten in die Sinne fällt und den höchsten Grad von Öffentlichkeit hat, mit der Tag-Nacht-Umdrehung (*notissima omnium diurni nocturnique temporis revolutione*). Dieses sei die Bewengung, die dem Erdkörper hauptsächlich und unmittelbar zukomme (*maxime ac sine medio appropriatam*). Diese Formulierung ist von besonderer Bedeutung, denn sie besagt einmal, dass die Tagesrotation der Erde wesentlich und naturgemäss zukomme, und sie besagt als Folgerung daraus zum anderen, dass sie ihr unmittelbar und ohne sekundäre Faktoren zukomme. Das wird sich als Bestätigung der Überlegungen erweisen, die im ersten Buch im Zusammenhang mit der Kugelgestalt der Erde angestellt worden waren. Die Formulierung kann nur bedeuten, dass – im Gegensatz zur Jahresbewegung um die Sonne und zur Stabilisierungsdrehung der Achse in der Ekliptik – die Tagesrotation der Erde mit dem Übertragungssystem der Bewegungskräfte im traditionellen Universum nichts zu tun hat. Für die Tagesdrehung der Erde gilt nicht, dass sie, wie die Umläufe der Planeten, der Sonne und des Mondes um das Weltzentrum im aristotelischen Modell letzlich durch die Umdrehung der ersten Sphäre in absteigender Kausalität bewirkt wird. Die Tagesbewegung der Erde wird also gesehen als ein in dem traditionellen Gesamtsystem nicht vorkommendes und daher auch nicht begründetes Spezifikum der Erde. Diese Bewegung ist nicht gewaltsam, sondern natürlich; das ist die wichtigste Voraussetzung dafür, dass sie gleichförmig und unveränderlich und damit ein Garant für die Produktion des zeitbegründenden Phänomens der Fixsternbewegung sein kann.

Entsprechend wird der Begriff der Zeit ausschliesslich von dieser Tagesbewegung abgeleitet, und zwar nach Analogie des Verhältnisses von Einheit und Zahl. Die Zusammenfassung grösserer Einheiten zu Monaten, Jahren und anderen Komplexen wird ausdrücklich als eine nominelle Operation angesehen, während die reelle Einheit der Zeitbestimmung weder im Sonnentag noch im Mondmonat oder im Sonnenjahr steckt, sondern ausschliesslich im Sternentag: *quoniam ab ipsa menses, anni et alia tempora multis nominibus exurgunt, tanquam ab unitate numerus*. Es ist für Kopernikus an dieser Stelle entscheidend, dass er unter den von ihm veränderten konstruktiven Voraussetzungen des Weltmodells die Versicherung geben kann, für das Problem der Zeit bedeute dies keine radikale Neuerung. Bei völliger Äquivalenz der Bedingungen habe es nichts auf sich, ob jemand mit der ruhenden Erde und der kreisenden Sphäre arbeite oder, wie er selbst, mit der umgekehrten Konzeption: *Nihilque refert, si quod illi per quietam terram, et mundi vertiginem demonstrant, hoc nos ex opposito suscipientes ad eandem concurramus metam: quoniam in his, quae ad invicem sunt, ita contingit, ut vicissim sibi ipsis consentiant*. Man sieht wie Kopernikus eine der wesentlichen naturphilosophischen Fragen behandelt, die er selbst nicht ausdrücklich stellt, die aber von ihm bei der Verteidigung seines Systems gewärtigt werden musste. Und genau bei der hier eingeführten Äquivalenzthese muss man ansetzen, um die Schwierigkeiten zu verstehen, die im Vorwurf der mangelnden naturphilosophischen Rücksicht stecken und die unausbleiblich gerade beim Begriff der Zeit zu schwerwiegenden Konsequenzen innerhalb der Neuzeit geführt haben.

Vom Prooemium des zweiten Buches her wird der argumentative Zusammenhang durchsichtig, in welchem sich Kopernikus veranlasst gesehen hatte, das erste Buch seiner *Revolutiones* mit der Erörterung der Frage zu beginnen, welche Form der Kosmos im ganzen einerseits und die Erde im besonderen andererseits haben, und für beide die Vollkommenheit der Kugelgestalt zu behaupten.

Man darf nicht vergessen, dass bei Aristoteles der Erdkörper sich gerade deshalb in der absoluten Ruhelage befindet, weil die Kugelgestalt der Erde das Resultat ihrer Lage in der Weltmitte ist[1]. Die Kugelgestalt ergab sich bei Aristoteles „beiläufig" aus der Tendenz des Elementes Erde zur Weltmitte hin. Da die punktuelle Weltmitte nicht von allen Teilen des Elements Erde auch erreicht werden kann, ergibt sich eine stabile Indifferenz aller von der Mitte gleichermassen entfernten Partikel zu ihrem natürlichen Ort. Die Gleichmässigkeit der Umlagerung des einen Mittelpunktes verbürgt die Ruhelage des Ganzen. Für Aristoteles war die Kugelgestal der Erde das Resultat eines Prozesses, in welchem dieser zwar faktisch, aber nicht ideal abgeschlossen war, denn ideal wäre nur die Lokalisierung aller Erdmaterie im Weltmittelpunkt, also die Absurdität des Aufgehens der Kugel in einem Punkt. Die Kugelform war das Stillhalten des Prozesses dort, wo er physisch deshalb nich weitergehen kann, weil es mehr Erdmaterie als „natürlichen Ort" für sie gibt. In der Pariser Occamisten-Schule des 14. Jahrhunderts führt dieselbe Prämisse zu der These einer Unruhe des Erdkörpers, als einem ständig vibrierenden Ausgleich kleiner Ungleichgewichte. Diese Art von „Bewegung" der Erde hat mit einer Vorbereitung des Kopernikanismus nicht das Geringste zu tun.

Im Vergleich zu der aristotelischen Vorstellung von der Kugelgestalt der Erde muss der Ausgangspunkt der Erörterung des Kopernikus eher platonisch genannt werden, obwohl die Tradition von der essentiellen Selbstbewegung der Kugel aus der Atomistik und deren Lehre von der Kugelförmigkeit der Feueratome — und damit der Bestandteile der Seele — bei Demokrit stammt. In der Atomistik mussten zuerst alle physischen Eigenschaften der Elementarkörper aus ihren geometrischen Gestalten abgeleitet werden: die Disposition der Kugel zur leichtesten Beweglichkeit geht in ihrem Grenzwert über in die Selbstbewegung. Daran erinnert noch in der Sprache des Kopernikus die ständige Wendung von der *mobilitas terrae*, in der schon sprachlich die Bedeutungen von Beweglichkeit und Bewegtheit erkennbar ineinander übergehen sollen.

Kopernikus demonstriert mit der Behandlung der Erdgestalt in einem astronomischen Traktat, dass er nicht mehr Astronomie im traditionellen Sinne betreiben kann, denn für die astronomische Tradition war die Erde gerade nicht Gegenstand ihrer Theorie; sie war nichts als der für sich genommen irrelevante Nullpunkt eines

---

[1] *De caelo* II 14; 297a 30—297b 14.

Bezugssystems. Die These von ihrem Stillstand im Mittelpunkt der Welt bedeutete nur ihre Neutralisierung für die phoronomische Theorie. Indem Kopernikus die Erde mit der blossen Thematisierung ihrer Gestalt zum Himmelskörper qualifiziert, hat er das engere Argumentationsziel am Anfang seines Werkes noch nicht erreicht. Dieses besteht darin, den Vorrang der Himmelssphäre im Hinblick auf das Prädikat der Bewegung aufzuheben und die Entscheidung, ob die Tagesbewegung jener letzten Sphäre oder der Erde real zugeschrieben werden muss, für eine konstruktive Beweisführung im weiteren Verlauf des Werkes offen zu halten, oder besser: allererst zu öffnen.

Dass Kopernikus mit seiner Argumentation aus der Äquivalenz von Kugelform der Welt und der Erde über die Indifferenz der ausstehenden Entscheidbarkeit hinsichtlich des Prädikats der Bewegung nicht hinaus kommt, liegt gerade daran, dass er den nächsten Schritt nicht getan hat oder tun konnte, der sich aus der Phänomenalisierung der Himmelsbewegung ergeben hätte. Dieser Schritt konnte nur darin bestehen, nach dem Schein der Bewegung auch den Schein der Kugelgestalt der äussersten Sphäre zu entdecken. Aber mit der Auflösung der festen Sphäre des Fixsternhimmels sollte in der kopernikanischen Konsequenz erst Thomas Digges beginnen. Dieser weitere perspektivische Schritt hätte in der Argumentation über die Äquivalenz schon hinausgeführt, aber zugleich zu weit vom Kanon des Ptolemäus entfernt. Solange freilich auch der Fixsternhimmel seine reale Kugelgestalt behielt, wurde das Argument aus der Kugelgestallt der Erde neutralisiert. So bleibt die *mobilitas terrae* eine blosse naturphilosophische Schutzbehauptung in einem theoretischen Selbstverständnis, das immer nur mit dem kleinsten Schritt der Entfernung von der Tradition auszukommen sucht. Noch der frühe Galilei arbeitet mit diesem Prinzip der kleinsten möglichen Distanzierung von der Tradition, indem er die kosmologische Problematik auf die Tagesrotation der Erde beschränkt sieht. Dann aber wäre die weiterhin im Mittelpunkt der Welt befindliche, nun aber als rotierend gedachte Erdkugel der ideale Ausgangspunkt für den ersten Schritt in Richtung auf den Kopernikanismus. Diese Kugel brauchte wegen ihrer vollkommenen Indifferenz gegen jede mögliche Bewegung nur als einmal in Bewegung versetzt gedacht zu werden, um die Dauer dieser Bewegung annehmen zu können. Schaltet man die gesamte aristotelische Elementenlehre aus, so wäre die ideale Kugel im Weltmittelpunkt resistent gegen alle Änderungen ihrer Zustände, auch wenn ihr faktischer Zustand der der Bewegung wäre. Die Voraussetzung, dass die Erdkugel auch in einem Bewegungszustand verharren kann, dann aber doch einmal in Bewegung versetzt worden sein muss, lässt sich im Gegensatz zum Aristotelismus im christlichen Sprachraum unwidersprochen einführen; für den Aristoteliker wäre hier geradezu eine Sperre eingebaut gewesen. Galileis Abhandlung über die Kreisbewegung aus der Pisaner Zeit enthält vom Kopernikanismus also nur das Minimum unter Auslassung seiner heliozentrischen Hauptthese, mit der zur Achsenrotation die Bahnbewegung um die Sonne hinzukommen muss. Man sieht, dass diese beiden Schritte so lange einander ausschliessen als das Beharrungsprinzip nich über die Ableitung aus der Indifferenz der Mittelpunktlage hinaus entwickelt ist[2].

Nun sagt sich dies leicht, im christlichen Sprachraum lasse sich die Bewegung der Erdkugel im Mittelpunkt des Weltalls unwidersprochen einführen, weil ein Anfang hier ohnehin angenommen werden muss. Im Zusammenhang mit dieser frühen Form des Trägheitsprinzips, angewendet auf die Beharrung eines an seinem „natürlichen Ort" kreisenden Kugelkörpers, lassen sich die Schwierigkeiten freilich umgehen, die sich aus der Fortgeltung der aristotelischen Physik und Metaphysik ergeben mussten. Denn diese Kreisbewegung bedarf nach ihrer einmaligen Auslösung am Anfang der Welt keiner weiteren Kausalität, also nicht des Systems der Übertragung von Bewegung innerhalb des Kosmos durch Berührung der Sphären oder durch Beseelung der Sphärenkörper, das der Aristotelismus geschaffen hatte und innerhalb dessen Fernwirkungen ausgeschlossen waren.

Das kopernikanische System hat seine Schwierigkeit eben darin, dass es weit von der äussersten Sphäre entfernt in der Nähe des Weltzentrums das Höchstmass an Bewegung erfordert, also in der weitesten Entfernung von der im scholastischen System angenommenen höchsten Kausalität der Bewegung. Wir können nicht davon ausgehen, dass Kopernikus denselben Weg zu seiner Theorie gefunden hatte, den Galilei zum Kopernikanismus einschlagen sollte, nämlich den von einer Achsendrehung der Erde im Mittelpunkt des Weltalls. Alles deutet vielmehr darauf hin, dass Kopernikus zuerst die Jahresbewegung um die Sonne gefunden hat, indem er das zunächst angenommene System einer Kreisbewegung des Merkur and der Venus um die Sonne derart erweiterte, dass er auch die Erde und die anderen Planeten um die Sonne kreisen liess[3]. Die Tagesbewegung könnte in diesem theoretischen Prozess die sekundäre Annahme sein, die sich freilich nach der einmal durchbrochenen Blockade gegen eine Bewegung der Erde überhaupt nahelegte. Die Darstellungsfolge in den *Revolutiones* muss also keineswegs die Genesis der Theorie widerspiegeln. Lichtenberg hat sogar versucht, noch einen anderen Weg für Kopernikus nachzuweisen, auf dem der erste Schritt die Entdeckung der unermesslichen Grösse des Universums gewesen wäre und die Annahme der Tagesrotation erst die Folgerung aus der angenommenen Unermesslichkeit, weil für den immensen Sphärenkörper des Fixsternhimmels die Rotation unwahrscheinlich orde gar widerspruchsvoll

---

[2] E. Wohlwill, *Die Entdeckung des Beharrungsgesetzes*, [in:] „Zeitschrift für Völkerpsychologie und Sprachwissenschaft" XV, 1884, 81–83. Diesen Stand der kosmologischen Reformation, die Tagesrotation der Erde im Zentralpunkt, hatte Celio Calcagnini in Ferrara schon um 1520 erreicht. Aber der Übergang von der geozentrischen zur heliozentrischen Konstruktion wird durch diesen „Fortschritt" geradezu behindert, solange die Annahme der Beharrung in der Rotation an die der Mittelpunktlage gekoppelt bleibt, also Rotation Revolution auschliesst. Es ist daher äusserst wichtig zu sehen, dass Kopernikus seinen Weg gerade nicht primär über die Lösung der Frage der Tagesbewegung genommen hat.

[3] Diese These habe ich zuerst ausgesprochen und begründet in *Kopernikus im Selbstverständnis der Neuzeit* (Abh. d. Akademie der Wissenschaften und der Literatur Mainz, Geistes- und sozialwiss. Kl., Jg. 1964, Nr. 5, 355 Anm.).

erschien[4]. Tatsächlich aber muss umgekehrt gesagt werden, dass die Annahme der Unermesslichkeit des Universums für Kopernikus erst eine Folgerung aus der Erprobung des primären Schrittes zur Jahresbewegung um die Sonne gewesen ist, weil dieser Schritt ihn in die Verlegenheit brachte, keine parallaktischen Erscheinungen am Fixsternhimmel vorweisen zu können, die sich doch aus der Grösse der Jahresbahn um die Sonne hätten ergeben müssen — es sei denn, man musste der Fixsternsphäre eine bis dahin unvermutete, ungeahnte und erschreckende Grösse zusprechen.

Man kann sagen, dass der Weg, der Galilei zum Kopernikanismus führte, in seinem ersten Schritt am ehesten die aristotelischen Prämissen festzuhalten sucht. Für Kopernikus kann das nicht in gleicher Weise gelten. Die Kugelgestalt der Erde ist für ihn nicht Ausdruck der Indifferenz ihrer Teile im Verhältnis zum Weltmittelpunkt und Weltschwerpunkt. Der Nachdruck, den Kopernikus auf die *Vollkommenheit* der Kugelgestalt der Erde in Äquivalenz zu der des äussersten Himmels legt, scheint auf einen platonischen und teleologischen Hintergrund hinzudeuten. Die Kugelgestalt ist ein Sachverhalt, der sich so teleologisch interpretieren lässt, dass dieser Körper die Bewegung ausführt, für die ihn seine Form disponiert. Bewegung und Ruhe sind hier nicht, wie sie es bei Galilei sein werden, äquivalent. Schon Plato hatte im *Timaios* der Erde eine Bewegung um die Weltachse zugeschrieben, obwohl er mit dieser Bewegung kosmologisch gar nichts anzufangen weiss,

---

[4] Nach der Aufzeichnung von Gottlieb Gamauf (*Erinnerungen aus Lichtenbergs Vorlesungen über Astronomie*, Wien und Triest 1814, 94–97) hat Lichtenberg die Folge der Überlegungen des Kopernikus folgendermassen dargestellt: „Copernicus ging in Ansehung der Bewegung der Erde, sowohl um ihre Axe als um die Sonne, von einem der grössten und kühnsten Gedanken aus, den der Mensch je gewagt hat, von dem Gedanken: nicht bloss der Halbmesser unserer Erdkugel, sondern auch die Distanz derselben von dem Mittelpunkt der Welt – in welchen er nachher die Sonne legte – sey in Vergleichung mit der Distanz der Fixsterne ein unmerklicher Punkt, ein blosses Nichts". Kopernikus sei zu dieser Annahme durch die elementare Tatsache gekommen, dass der Horizont für jeden Beobachter die Sphäre genau halbiert, was nur daraus resultieren könne, dass die Grösse der Erde die eines blossen Punktes sei. „Nun dachte er weiter: Sollten wohl alle die grossen Himmelskörper sich um den einzigen unbedeutenden Punkt unserer Erde herumdrehen; sollte dies vollends mit der ungeheuren Himmelskugel jede 24 Stunden einmahl geschehen? Dass die tägliche Bewegung blos scheinbar sey und durch die Umdrehung der Erde um ihre Axe bewirkt werde, drang sich ihm so gleich unwiderstehlich auf; und bald geschah es auch so mit der jährlichen Bewegung". Hier erst lässt Lichtenberg die Erwägung Platz greifen, dass die Bahnen von Merkur und Venus monströse Schwierigkeiten bereiten, wenn man sie über oder unter der Sonnenbahn ansetzt, statt sie um die Sonne kreisen zu lassen. „Das that er denn. Eben so liess er ferner Mars, Jupiter und Saturn gleichfalls um die Sonne als den Mittelpunkt ihrer Bahnen laufen". Bis hierher hat Lichtenberg das regressive System des Tycho Brahe als Vorstufe des kopernikanischen Endausbaus rekonstruiert, wohl indem er Kopernikus die Hauptsorge des dänischen Astronomen unterstellt, nämlich an der geozentrischen Struktur um jeden Preis festzuhalten. Nur das Ärgernis des leeren Raumes, der bei dieser Konstruktion zwischen den Bahnen von Venus und Mars bleibt, soll zum letzten Schritt geführt haben, in diesen Raum die Bahn der Erde um die Sonne zu legen. „Und so stand dem nun sein herrliches Gebäude symmetrisch und ordnungsvoll da und wird da stehen bis an das Ende der Zeiten".

wie sich daraus ergibt, dass er sie durch eine Ausgleichsbewegung des Fixsternhimmels phänomenal wieder aufzuheben gezwungen ist. Die Erde muss sich drehen, weil sie Kugelgestalt besitzt. So weit ist Kopernikus nicht gegangen, aber eine Disposition zur Drehung um sich selbst ist für ihn mit der Kugelgestalt gegeben. Weiter konnte er eben deshalb nicht gehen, weil er sonst den Stillstand der ebenso kugelförmigen Fixsternsphäre nicht hätte verteidigen können.

III.

Für die Wissenschaftsgeschichte mag die Herkunft eines bestimmten gedanklichen Elements — noch dazu, wenn es nicht zur Begründung einer Theorie, sondern zu deren Abschirmung dient — von keinem oder geringem Interesse sein. Hier sollte man aber bedenken, dass das, was die Leistung der Abschirmung zu erbringen hatte, im Horizont seiner Herkunft zugleich auch die Antwort auf die Frage nach seiner Wirksamkeit enthält. Denn nur wenn der abschirmende Gedanke den Zeitgenossen nicht zu abwegig und zu abgelegen erscheint, kann er leisten, was er leisten soll und was allemal ein gewisses Mass an Plausibilität für ein Publikum voraussetzt. Hier ist es nicht wichtig, was und woher der Autor gefunden und aufgenommen hat, sondern was er hinsichtlich seiner Adressaten an reellen und vermeintlichen Voraussetzungen und Evidenzen erwarten darf. Dies ist zwar noch keine Wirkungsgeschichte, aber deren unmittelbare Vorfrage: die Besetzung des Wirkungsfeldes mit potentiellen Plausibilitäten. Betrachtet man die Absichten eines Autors und die Funktion der von ihm aufgenommenen und eingesetzten Materialien so, dann wird man auch nicht leicht geneigt sein, die Provenienz dieser Materialien als Indiz für Einflussverhältnisse zu nehmen und auf breiter gelagerte Substrate zu schliessen. Wie platonisch zum Beispiel das Element der vollkommenen Kugel bei Kopernikus auch immer erscheinen mag, auf einen Platonismus in den Voraussetzungen seiner Theorie lässt es keinen Schluss zu. Was Kopernikus zur Abschirmung seiner Theorie verwendet hat, lässt sich nicht gleichsetzen mit dem, was ihn zu dieser Theorie motiviert, in ihrer Behauptung bestärkt und zu ihrer Integration in einen übergeordneten Zusammenhang von Einstellungen und Wertungen disponiert hat.

Es sagt also noch nichts über Einflüsse und dogmatische Dominanz, wenn im Zusammenhang mit der vollkommenen Kugel darauf hinzuweisen ist, dass Kopernikus die Lehre von der höchsten Mobilität der genauen Kugelgestalt bei Nikolaus von Cues gefunden haben kann. D. Mahnke, L. A. Birkenmajer und R. Klibansky haben faktische, wahrscheinliche und mögliche Kontakte des Kopernikus mit dem Werk des Cusaners aufgedeckt. Als Vermittlung kommt auch das Werk des Carolus Bovillus in Frage, das in einem mit authentischen Marginalien versehenen Exemplar aus der Bibliothek des Kopernikus in Uppsala erhalten ist.

Einschlägig für die These von der Beweglichkeit der vollkommenen Kugel ist vor allem der erste der beiden Dialoge *De ludo globi*. Hier ist die Rede von einer Art Kugelspiel, dessen Kugel infolge einer konkaven Asymmetrie auf einer gekrümmten Bahn geworfen werden kann. Dieses Spielinstrument führt die Diskussion auf das Problem der absolut genauen Kugelgestalt. Im platonischen Sinne wird dabei behauptet, dass die geometrisch genaue Kugel überhaupt nicht materielle Erscheinung annehmen kann. Die Begründung für diese Behauptung liegt nicht auf der Linie der platonischen Tradition. Wenn auch die Rundung jedes realen Körpers genauer als gegeben vorgestellt und gemacht werden kann, so wäre der unerreichbare Extremwert dieser Steigerung doch gerade nicht der vollkommene, aber nicht mehr reale Körper, sondern der Widerspruch eines zwar realen, aber nicht mehr erfahrbaren Körpers. Die Aufhebung der Sichtbarkeit ergibt sich daraus, dass die vollkommene Kugel von Geraden oder Ebenen nur noch in einem Punkte berührt werden kann, dass ihre *extremitas rotundi* mit dem Punkt auf der Kugeloberfläche identisch wird, eine Fläche aber nicht aus Punkten konstituiert gedacht werden kann: *Nam rotunditas, quae rotundior esse non posset, nequaquam est visibilis… non enim rotunditas ex punctis potest esse composita.* Das Rezept des Cusaners ist hier wie sonst deutlich erkennbar: die Steigerung einer Eigenschaft zu absoluter Reinheit transzendiert die physische Gegenständlichkeit nicht in der Weise des platonischen Chorismos, sondern durch Sprengung des Gegenstandes, an dem sie imaginativ durchgeführt wird. Was ganz und gar nicht dem platonischen Konzept entspricht und das Eigentümliche der Methode des Cusaners ausmacht, ist die Anwendung dieser Grundfigur auf die wirkliche Welt. Er greift dabei die Vorstellung der traditionellen Kosmologie auf, die äusserste Himmelssphäre müsse von vollkommenster körperlicher Rundung sein. Dann aber muss sie nach dem vorausgesetzten Gedankengang ihre Wahrnehmbarkeit von aussen verlieren: … *si possibile foret quem extra mundum constitui mundus foret illi invisibilis ad instar indivisibilis puncti.* Die platonische Dualität von Idee und Erscheinung gilt nur für die einzelnen, die Welt aufbauenden Gegenstände; die Totalität dieser Erscheinungen ist ihrerseits etwas, was nicht mehr als Erscheinung erfasst werden kann. Diese Sonderstellung der vollkommenen Kugel wird ausdrücklich als Ausnahme vom platonischen Dualismus charakterisiert. Der im Dialog auftretende Kardinal sagt zu seinem Gesprächspartner, dem Herzog von Bayern: *Obwohl du im platonischen Sinne das Richtige sagst, besteht dennoch ein Unterschied zwischen der Rundheit und einer anderen Form. Wenn es auch möglich wäre, die Rundheit im Stoff zu realisieren, wäre sie dennoch unsichtbar.* Und genau dies gilt für das Weltall im ganzen. Die stoffliche Realität der vollkommenen Kugel hat hier eine Eigenschaft zur Folge, die innerhalb der physischen Welt nicht vorkommen kann und dennoch dieser Welt zukommt. Für die Funktion, die die genaue Kugelgestalt in den Gedankenexperimenten der frühen Mechanik haben sollte, muss die Eigenschaft der Welt als mögliche Eigenschaft in der Welt gedacht werden können.

Die Bedeutung dieses Gedankenganges liegt nun darin, dass die Eigenschaften der vollkommen runden Kugel spekulativ aus der Rücksicht auf die Bedingungen der tradierten Physik und Kosmologie herausgenommen werden können. Die Transzendenz ist etwas, was der Welt selbst zugute kommen kann. So ist der Zusammenhang von Kugelgestalt und Bewegung im Kontext des Dialogs zu verstehen. Der Cusaner lässt seinen Kardinal die Behauptung aufstellen, dass die vollkommen runde Kugel eine ihr einmal verliehene Bewegung niemals verlieren würde. Was sich in Bewegung befindet, kommt niemals zur Ruhe, wenn es sich nicht zu einem Zeitpunkt anders verhalten kann als zu einem anderen. Die vollkommene Kugel auf der vollkommenen Ebene müsste sich wegen der Gleichheit der Bedingungen ihres Verhaltens zu jedem Zeitpunkt immerfort bewegen, wenn sie einmal in Bewegung versetzt worden wäre: *Perfecte igitur rotundus... postquam incepit moveri, quantum in se est numquam cessabit, cum varie se habere nequeat. Non enim id, quod movetur aliquando, cessaret nisi varie se haberet uno tempore et alio.* Der Gedanke von der Beharrung der rotierenden Bewegung einer vollkommenen Kugel ist schon hier über den Stand hinaus vorangetrieben, den der frühe Galilei ihm geben wird mit Bezug auf die um die Weltmitte gelagerte Kugel. Für den Cusaner ist die Form der Rundheit die zur Beharrung in der Bewegung schlechthin geeignete, wobei er freilich die aristotelische Bedingung hinzufügt, dass die Bewegung dem Körper „natürlich" sein müsse: *Forma igitur rotunditatis ad perpetuitatem motus est aptissima. Cui si motus advenit naturaliter numquam cessabit.* Auch im Zusammenhang aristotelischer Vorstellungen ist es keineswegs selbstverständlich, dass eine natürliche Bewegung beharrt, denn im Erfahrungsbereich der Elemente sind alle natürlichen Bewegungen zum eigentümlichen Ort zielgebunden und endlich.

Aber welche Bewegung sollte dem vollkommen runden Globus sonst natürlich sein können als die um seine eigene Achse? Die Natürlichkeit der Bewegung der Himmelskörper hatte sich schon für die astronomische Tradition nicht daraus ergeben, dass sie sich auf Kreisbahnen bewegen, sondern dass diese Bahnen nur die Erscheinungsform der Rotation von Kugelsphären um ihre Achsen waren. Für den Cusaner entstehen alle Bewegungen, sofern sie „natürlich" sind, durch Teilhabe an der Bewegung der äussersten Sphäre: *... quem motum omnia naturalem motum habentia participant.* Gerade deshalb könnten hier die Prädikate der äussersten Sphäre nicht auf die Erde übertragen werden. Aber eben diese Übertragung ist der Gebrauch, den Kopernikus von den spekulativen Prädikaten der vollkommenen Kugel machen sollte.

Die entscheidende Folge der Theorie von der möglichen immerwährenden Bewegung einer vollkommenen Kugel ist die Indifferenz gegenüber der aristotelisch-scholastischen Metaphysik des unbewegten Bewegers, die auf der Voraussetzung beruhte, dass für die Prozesse der Welt eine ständig von aussen einwirkende bewegende Kraft notwendig sei. Der Cusaner kann sich damit begnügen, dass die Kugelsphäre und ihre Bewegung in einem Schöpfungsakt ihren Anfang genommen haben. In ihrem Fortbestand wird die Welt von Gott so wenig bewegt wie die Kugel des Spielers, nachdem sie die werfende Hand verlassen hat. Der *impetus*, den der Wurf ihr verleiht, erlahmt nur dann, wenn die Kugel nicht vollkommen rund ist. In diesem Zusammenhang nennt der Cusaner die Platoniker, weil der Gedanke von der immerwährend sich bewegenden Kugel die platonische Vorstellung von der Seele als einer geschaffenen Selbstbewegung veranschauliche: *Sed creatus est in te motus se ipsum movens secundum Platonicos, qui est anima rationalis movens se et cuncta sua.* Dieser Vergleich ist auch für die Kosmologie der frühen Neuzeit nicht ohne Ertrag, denn die organische Auffassung des Weltalls als eines beseelten Ganzen nach stoischer wie platonischer Tradition wird entbehrlich durch die Annahme der beharrenden Rotation der äussersten Sphäre — oder des Erdkörpers.

Die Bedeutung, die Kopernikus so offenkundig dem Nachweis beimisst, dass die Erde nicht nur Kugelgestalt, sondern diese in Vollkommenheit habe, wird jetzt verständlicher. Er gewinnt diesen Nachweis insbesondere aus der Gestalt des Erdschattens bei Mondfinsternissen: *absoluti enim circuli amfractibus Lunam deficientem efficit.* Daraus ergibt sich für ihn die Ablehnung aller anderen Theorien über die Erdgestalt, die aufgezählt werden, und als Schlussfolgerung: ...*sed rotunditate absoluta, ut philosophi sentiunt*. Sogleich zu Anfang des vierten Kapitels leitet Kopernikus ganz allgemein die *mobilitas sphaerae* aus der Kugelgestalt ab. Für die Himmelssphären und die ihnen zugeschriebene heterogene Substanz war das eine vertraute Behauptung. Aber deren Bewegung wird für die Erfahrung des irdischen Betrachters gar nicht als Drehung einer Kugel phänomenal, sondern durch die linearen Bahnen der an die Sphären gebundenen Gestirne. Die Kugel ist nur die hypothetische Hilfskonstruktion für die Darstellung der Himmelsbewegungen in Kreisen und für die Übertragung der Bewegungskausalität von aussen nach innen. Solange von Sternen und Planeten gesprochen wird, bleibt die Frage nach der Realität jener als Träger der stellaren Erscheinungen angenommenen Hohlkörper irrelevant. Kopernikus aber hält in diesem vierten Kapitel des ersten Buches der *Revolutiones* seine Formulierungen ganz so, dass der Ausdruck *motus circularis* jederzeit auch auf die Achsenrotation eines soliden Kugelkörpers angewendet werden kann, wie es die Erde ist. Dabei versteht er im Sinne der von Demokrit begründeten und von Plato sanktionierten Tradition die rotierende Bewegung als die Verwirklichung der sphärischen Form. Sie wird ihm so etwas wie der „angemessene Ausdruck" der eidetischen Einfachheit der Kugel: *Mobilitas enim sphaerae est in circulum volvi, ipso actu formam suam exprimentis in simplicissimo corpore, ubi non est reperire principium et finem nec unum ab altero secernere, dum per eadem in se ipsam movetur.* Bevor Newton die Kreisbahn als das dynamische Produkt von Kräften zu betrachten lehrte, galten Kugel und Kreis als die einfachsten und schlechthin rationalen Formen der Körper und ihrer Bewegung. Der in sich selbst kreisende Körper vereinigt die Antithesen von Bewegung und Ruhe, von Verschiedenheit und Identität, so wie die Kreisbahn in jedem ihrer Punkte Anfang und Ende zur Identität bringt. Indem Kopernikus dieser Erörterung mehr Raum gibt als Ptolemäus, der sie nur in einem einzigen Satz im dritten Kapitel des dritten Buches seines *Almagest* erwähnt hatte, nähert er sich seinem Argumentationsziel, der Erde diejenigen Eigenschaften kraft ihrer Kugelgestalt zuzueignen, die bis dahin unbestritten nur den Himmelssphären zugeschrieben worden waren.

Das Argumentationsziel ist die „Natürlichkeit" der Bewegung der Erde im

Weltraum. Natürlichkeit der Bewegung ist ein Begriff, der nur im System des Aristotelismus verständlich ist. Kopernikus aber versucht, die nur in einem aristotelischen Kontext definierbare Qualität der Natürlichkeit mit platonisierenden Mitteln für die Erde zu vindizieren. Das ist nur in bezug auf die Argumente verständlich, die Ptolemäus gegen die Behauptung der Erdbewegung geltend gemacht hatte und die Kopernikus genau parallelisierend im siebenten Kapitel des ersten Buches seines Werkes behandelt. Wenn er die Natürlichkeit der Erdbewegung beansprucht, hält er sich streng an die Argumentationsweise des Ptolemäus, der die Erdbewegung gerade wegen der Konsequenzen ihrer Gewaltsamkeit ausschliesst. Gegen Aristarch von Samos und dessen heliozentrisches Modell hatte Ptolemäus nämlich einzuwenden gehabt: *Wenn der Erde eine tatsächliche Bewegung zukäme, so müsste sie wegen ihrer alles übertreffenden Ausmasse allen anderen Körpern in der Bewegung weit voraus sein, so dass die Lebewesen auf ihr und die nicht an ihr befestigten Gegenstände weit hinter ihr her in der Luft schweben würden; schliesslich müsste die Erde selbst durch ihre grosse Geschwindigkeit aus dem Weltall herausfallen.* Die Gegenwehr des Kopernikus gegen dieses Argument ist sparsam und von scholastischem Typus. Man dürfe selbstverständlich nicht von der Voraussetzung ausgehen, dass die vorgeschlagene Bewegung der Erde den Charakter einer „gewaltsamen" Bewegung habe. Wenn man einmal zugestehen wolle, die Bewegungen der Erde seien „natürliche" Bewegungen, so würden auch jene misslichen Konsequenzen entfallen, denn was von Natur geschieht, verhält sich angemessen und bewahrt seinen vollkommenen Zusammenhang: *Quae vero a natura fiunt recte se habent et conservantur in optima sua compositione*[5].

Man ist enttäuscht und doch zugleich entzückt von der Zweckmässigkeit dieser Argumentation. Man bewundert die Ökonomie, mit der Kopernikus nach der einfachsten Schutzbehauptung für seine These sucht. Nicht die Beweise also für das neue System sind philosophisch, sondern nur die Abschirmungen gegen die Prämissen des alten und die Erfüllung einiger seiner Zusatzforderungen. Dazu gehört, dass die Möglichkeit der Bewegung der Erde dieselbe Sicherung des Zeitbegriffs ergibt, die Aristoteles und die Scholastik in ihrem Weltmodell gegeben hatten.

Resultat der Erörterungen im vierten Kapitel des ersten Buches ist folgerichtig nur die *mobilitas* der Erde, nicht die Behauptung ihrer tatsächlichen Bewegung. Am Schluss des Kapitels ist die weitere Aufgabe ernsthaft durchführbar geworden, sorgfältig zu untersuchen, welche Stellung die Erde zum Himmel einnimmt. Die Gefährdung durch das konstitutive Vorurteil der Astronomie seit dem Sturz des Thales von Milet ist behoben, dass wir bei der Erforschung der entfernten erhabenen Gegenstände des Himmels das Nächstliegende unbeachtet lassen und dem Himmel zuschreiben, was doch der Erde zukommt: *quae telluris sint attribuamus caelestibus.* Das nächste Kapitel beginnt Kopernikus nun konsequent mit der Erinnerung daran, dass auch die Erde die Gestalt einer Kugel habe: *terram quoque globi formam habere.*

[5] *Revolutiones* I 8.

Weil sie diese Gestalt habe, müsse untersucht werden, ob ihre Beweglichkeit nicht in faktischer Bewegung realisiert sei. Aus der Äquivalenz der Kugelgestalt von Himmel und Erde folgt eine bestimmte Beweislast, die demjenigen zufällt, der von vornherein für die Erde die Realität der Bewegung negiert. Diese beweistaktische Wendung ist für den Gang des Werkes überaus wichtig. Sie charakterisiert mit einem weiteren Zug den noch ganz mittelalterlichen Typus der kopernikanischen Reform. Der gut scholastische Fundierungszusammenhang von *forma* und *actus* manifestiert sich als noch tragfähiges Denkschema. Die spekulative kosmologische Erörterung dieser ersten Kapitel des ersten Buches führt zu dem Punkt, an dem die Erde als Weltkörper ihre primäre Thematisierung für die Astronomie erfahren hat, nicht *obwohl*, sondern gerade *weil* sie der Standort des Beobachters und damit die wichtigste Bedingung für das ist, was sich ihm zeigt: *Terra autem est, unde caelestis ille circuitus aspicitur et visui reproducitur nostro.*

Auch die Wissenschaftsgeschichte hat ihre Ironien. Eben die vollkommene Kugelgestalt der Erde, die Kopernikus die naturphilosophische Abschirmung seiner Reform ermöglichte, wurde kaum zwei Jahrhunderte später der Preis, um den der endgültige physikalische Beweis für die erste kopernikanische Erdbewegung erbracht werden konnte. Es ist bezeichnend für den kopernikanischen Ansatz bei der Erde selbst und ihrer Stellarisierung, dass dieser astronomisch für die zweite kopernikanische Bewegung so lange gesuchte und erst im 19. Jahrhundert durch Bessels Entdeckung der Fixsternparallaxe entdeckte Beweis bereits in der Mitte des 18. Jahrhunderts auf und an der Erde selbst durch die empirische Erforschung ihrer Form — genauer ihrer Deformation — geführt wurde. Als Maupertius in einem der ersten weltweit organisierten Forschungsunternehmen die Abplattung der Erde an den Polen bestätigte, verhalf er nicht nur Newtons Himmelsdynamik zum Durchbruch auf dem Kontinent, sondern demonstrierte fast anschaulich die Bewegung der Erde im Verhältnis zum absoluten Raum. Die unangefochtene Geltung des Kopernikanismus gründete sich gerade darauf, dass eine seiner wesentlichen Voraussetzungen zerstört wurde.

IV.

Das theoretische Programm Newtons, reduziert auf seine einfachste Formel, ist immer noch das des Kopernikus, nämlich durch die scheinbaren Bewegungen hindurch zu den wahren Bewegungen zu gelangen. Wie bei Kopernikus, so ist auch bei Newton dieses Programm ohne Tangierung des Zeitproblems nicht durchführbar. Die Unterscheidung von Erscheinung und Wirklichkeit hinsichtlich der Bewegung bei Kopernikus ebenso wie bei Galilei und Newton ordnet die naturwissenschaftliche Leistung der frühen Neuzeit dem übergreifenden und schliesslich homogenen Programm der europäischen Philosophie seit Parmenides und Heraklit — man möchte sagen: aufs Wort — ein.

In Newtons Vorrede zu den *Principia* von 1686 wird deutlich, dass die Frage nach den scheinbaren und den wahren Bewegungen in letzter Tendenz auf die

Kräfte der Natur zielt, dass aber diese Frage nur als beantwortbar erscheint, wenn man den kräftefreien Zustand materieller Systeme definieren kann. Man sieht leicht, dass Newton hier in die Nähe derselben Frage kommt, die in der Vorstellung des Kopernikus von der Bewegung der vollkommenen Kugel impliziert war. Wenn das so ist, ergibt sich für die Frage nach der Zeit, dass die Möglichkeit ihres Begriffs auf der Möglichkeit einer von Kräften unbeeinflussten Bewegung, als der garantierten Gleichförmigkeit, beruht. Die Rationalität dieser Bewegung kann für Newton nicht mehr in der Rotation der Kugel oder in der Revolution der Himmelskörper um ein Massezentrum liegen, sondern nur in der geradlinigen Trägheitsbewegung. Die kopernikanische vollkommene Kugel war das Resultat der Anwendung des Prinzips des zureichenden Grundes: die absolute Homogenität der Kugel bedeutet, dass es in ihrer Gestalt keinen Grund für die Veränderung der von ihr einmal eingenommenen Zustände gibt, dass sie also Ruhe wie Bewegung absolut konserviert, von Natur jedoch zur Bewegung disponiert ist. Newtons Kraftbegriff erfordert dieselbe Anwendung des Prinzips des zureichenden Grundes, indem ein von Kräften nicht beeinflusster Körper seine einmal eingeschlagene Bewegungsrichtung geradlinig und gleichförmig absolut einhält. Die Schwierigkeiten, die wir sehen können, liegen sowohl in der Definition der Geradlinigkeit als auch in der der Gleichförmigkeit. Die Schwierigkeit, die in der Bedingung der Geradlinigkeit liegt, konnte von Newton nicht wahrgenommen werden; die Schwierigkeit der Gleichförmigkeit liegt darin, dass sie nur mit Hilfe des Zeitbegriffs bestimmt werden kann, indem „Gleichförmigkeit" bedeutet, dass gleiche Strecken in gleichen Zeiteinheiten zurückgelegt werden. Newton hätte sagen können, dieses Erfordernis sei rational gerade dadurch erfüllt, dass die Voraussetzung gemacht wird, keine Kräfte wirkten auf den bewegten Körper ein. Ein solcher Körper könnte Zeitmesser sein.

Nun war aber diese Definition die für Newton uninteressante, weil für die ihm bekannte physische Wirklichkeit ziemlich sicher feststand, dass es eine solche Trägheitsbewegung im absoluten Sinne nicht gab. Newton kam es daher nicht auf die Definition der Trägheitsbewegung durch Kräftefreiheit an, sondern umgekehrt auf die Definition der Kraft als Veränderung von Trägheitszuständen. Damit verlor er jedoch die Definition der Zeit, die im Begriff der Gleichförmigkeit für die Bestimmung des Kraftbegriffs schon enthalten sein musste. Die Ausflucht aus diesem Dilemma ist die Absolutsetzung der Zeit nach Analogie der Absolutsetzung des Raumes. Die Einsicht, die Newton zu diesem Schritt hinsichtlich der Zeit veranlasste, ist in dem Satz seiner „Principia" ausgedrückt: *Es ist möglich, dass es keine gleichförmige Bewegung gibt, mit der die Zeit genau gemessen werden könnte. Es kann sein, dass alle Bewegungen beschleunigt oder verzögert verlaufen* (*das heisst, dass immer und überall Kräfte auf Körper einwirken*), *aber der Strom der absoluten Zeit kann nicht verändert werden*[6].

---

[6] *Opera omnia*, ed. S. Horsley, II 8: *Possibile est, ut nullus sit motus aequabilis, quo tempus accurate mensuretur...*

Der „Anblick" der Himmelsphänomene hat sich von dieser Einsicht her völlig verändert. Rotationen und Revolutionen als die Grundphänomene des gestirnten Himmels sind nicht mehr die idealen Grundzustände und Normvorgänge der gleichsam verkörperten Rationalität. Sie sind immer schon Resultate des Ineinandergreifens von Kräften, der Verwischung der elementaren Normalität, komplexe Verstellungen einer nur noch schwer zugänglichen Ausgangssituation. Als Newton 1666 an das Problem der Mondbahn heranging, nahm er die traditionelle Vorstellung einer Sphäre auf und betrachtete den Druck, den der Mondkörper auf die Innenfläche dieser Sphäre ausübt, während er in ihr um die Erde herumgeschleudert wird. Diesem zentrifugalen Druck musste eine kompensatorische Kraft entsprechen, die den Mond genauso wie einen zur Erde niederfallenden Stein erreichen und beeinflussen konnte. Diese erste und elementare Überlegung ergab das gegen Kopernikus und Kepler entscheidende Neue: die Bahnen der Himmelskörper um ihre Zentralkörper, seien es Kreise oder Ellipsen, waren aus mindestens zwei verschiedenen Krafteinwirkungen zusammengesetzte Figuren, die nichts mit der Realisation einer geometrischen Idealität zu tun hatten. Die Geometrie war hier nur ein Teil der allgemeinen Mechanik, und nicht umgekehrt diese die Theorie der Erscheinungen einer geometrischen Rationalität.

An die Stelle der Idealität des Kreises, den es nun auch in der stellaren Wirklichkeit nicht mehr gibt, ist die Idealität des absoluten Raumes und der absoluten Zeit selbst getreten, jener beiden Bastarde der philosophischen Tradition, denen einerseits der absolute Charakter der Reinheit der traditionellen Idealität zugesprochen wird, andererseits eine nur erschliessbare, aber nicht empirisch in Messungen vollziehbare physische Realität. Man kann die Hypothek, die Newton für die Geschichte der neuzeitlichen Physik aufnahm, nur dann als eine nicht bloss dogmatische Vorentscheidung und willkürliche Belastung begreifen, wenn man klar im Blick behält, dass es Newton von dem Augenblick seiner Überlegungen zur Bewegung des Mondes an und entgegen vielen seiner Schutzbehauptungen zur Abwehr der Unterstellung scholastischer Entitäten ausschliesslich darauf ankam, Kräfte nachzuweisen und in ihrer Grösse bestimmbar zu machen, wie die Kraft, die dem zentrifugalen Druck des Mondes auf seine Sphäre entgegen wirken konnte und in einem Kontinuum der Bestimmbarkeit zu derjenigen stehen musste, mit der ein Körper in der Nähe der Erdoberfläche beschleunigt wird. Der Vorwurf der empirischen Unbestimmbarkeit des absoluten Raumes und der absoluten Zeit trifft Newton deshalb nicht, weil diese Unbestimmbarkeit für ihn die unüberschreitbare Voraussetzung dafür war, die Bestimmbarkeit von Kräften im Hinblick auf den kräftefreien Trägheitszustand zu erreichen und zu definieren.

Zugleich ist aber klar, dass für das Problem des Raumes und der Zeit die Richtung auf die Idealisierung, wie sie schon Leibniz alsbald in der Auseinandersetzung mit Newtons Sprecher Samuel Clarke einschlung, vorgegeben war. Die philosophische Tradition wurde gerade in dem Augenblick übermächtig, in dem bestritten werden musste, dass in der Realität diejenige Bedingung anzutreffen war, die nach den

Forderungen der Tradition für die Realisierung der im Zeitbegriff gegebenen Bedingungen anzutreffen sein musste und die Aristoteles in der Tagesbewegung des Fixsternhimmels, Kopernikus in der Rotation der vollkommenen Erdkugel als physiche Realität angeben zu können geglaubt hatten. Aber man sieht von hier aus, dass Kopernikus den entscheidenden Schritt zur Zerstörung dieser Tradition getan hatte, als er den Garanten für die Realität der Zeit, für das *fundamentum in re* des Zeitbegriffs, in der Erde selbst aufgewiesen und damit der empirischen Observation in tödlicher Weise zugänglich gemacht hatte. Der Funktionstausch von Himmel und Erde erwies sich schliesslich nicht als theoretische Äquivalenz, sondern als radikale Veränderung der Bedingungen theoretischer Objektivierung. Das ist es, was vor allem bei Newton der historischen Explikation unterworfen wird.

Diese Veränderung der Voraussetzungen zugunsten der theoretischen Dignität der Erde und ihrer nächsten Umgebung gestattete Newton ein gegenüber der traditionellen Kosmologie so neuratiges Verfahren, in der terrestrischen Provinz mit der Lösung der grossen Probleme des Universums zu beginnen und schliesslich das Sonnensystem als blosse Erweiterung dieser Provinz anzusehen. Aber das Sonnensystem war nicht mehr, wie für Kopernikus, der wesentliche und zentrale Teil dieses Universums. Gerade in der Konsequenz der kopernikanischen Reform war die Erweiterung des kosmischen Raumes, bestätigt durch den Umgang mit dem Fernrohr Galileis und durch Olaf Römers Entdeckung der endlichen Lichtgeschwindigkeit erstmals in den Griff der Messbarkeit gerückt, zur Insularisierung des Sonnensystems innerhalb der kosmischen Totalität geworden. Newtons absoluter Raum ist auch dadurch motiviert und charakterisiert, dass in ihm das Auftreten von materiellen Systemen ein sporadisches und rares Ereignis geworden ist. Welten sind in der nachkopernikanischen Welt trotz der ungeheuren Steigerung ihrer Zahl relativ zur Erweiterung des Raumes immer seltener geworden, der Raum immer leerer, auch wenn man in die Betrachtung einbezieht, dass Welten jetzt nicht mehr nur aus Körpern, sondern auch aus der Realität ihrer Massenkräfte bestehen. Die Welt von Welten entsteht erst durch die im 18. Jahrhundert immer präziser formulierte Annahme, dass Sonnensysteme nicht die letzten Systemgrössen dieses Universums sind, sondern dass sie Teile übergeordneter Systeme und Übersysteme immer höherer Ordnung sein müssten. Der Raum wird nicht nur durch die Körper und ihre Massenkräfte, sondern durch Systeme und Übersysteme und deren Bewegungen strukturiert.

Dass der Weltraum leer ist, bedeutete jetzt nicht nur die Seltenheit von Körpersystemen im Verhältnis zur Weite des Raumes, sondern auch die Geringfügigkeit der Massen und damit die regionale Begrenztheit der von diesen Massen ausgeübten systembildenden Wirkungen. Der Gedanke, dass die diffuse Gravitation sehr ferner Massen das Führungsfeld der Trägheitsbewegung erzeugen könnte, also der Gedanke einer Einheit des Weltraums als Wirkung der in ihm bestehenden Massenkräfte, lag völlig fern. Wie man den absoluten Raum als Konsequenz der relativen Seltenheit der in ihm auftretenden Körper und Massenkräfte ansehen kann, so muss man umgekehrt die von Ernst Mach eingeleitete Destruktion des absoluten Raumes als

Konsequenz der Erkenntnis von der Bedeutung der Massenkräfte im Universum betrachten.

So wie das Sonnensystem als ein Komplex von Körpern und Kräften in grenzenlosen absoluten Raum ein *insulares* Vorkommnis ist, so ist die Gesamtheit der Welt in der absoluten Zeit ein *episodisches*. Schöpfung und Untergang der Welt sind für Newton systemgemässe – fast möchte man sagen: vertraute und nahe – Daten. Durch die Schöpfung ist die Welt in den Raum und in die Zeit gesetzt, und durch die apokalyptischen Ereignisse wird sie aus Raum und Zeit wieder getilgt. Raum und Zeit gehören nicht zur Kontingenz der Welt, und sie werden damit zwangsläufig auf die Seite der Gottheit, wenn nicht ihrer Attribute, so doch ihrer Organe, gebracht. Für den Raum hat Newton das mit der umstrittenen Formel vom *sensorium divinum* ausdrücklich vollzogen.

Für Leibniz wurde durch die Vorgegebenheit des absoluten Raumes und der absoluten Zeit mit der notwendigen Indifferenz aller Raum- und Zeitstellen die Schöpfung zu einer schlechthin unbergründbaren Handlung eines Willkürherrschers, zu einem *decret absolument absolu*. Die Seinsgrundfrage ist dazu die entschiedenste Gegenposition. Leibniz hat durch seinen Widerspruch, vor allem im Briefwechsel mit Samuel Clarke, den stringneten Zusammenhang von theologischem Absolutismus und Raum-Zeit-Absolutismus als das bohrende Ärgenis der neuzeitlichen Rationalität herauspräpariert und verschärft. Die Heterogenität der systematischen Elemente springt heraus, wenn man den Sachverhalt so formuliert: der heidnische unendliche Raum und die ebenso heidnische absolute Zeit enthalten das christlich interpretierte Ereignis Welt nur als ein in jeder Hinsicht provinzielles Faktum. Hier erst endet Augustins entscheidende Koppelung, durch die die Zeit in die Schöpfung der Welt hineingezogen und an ihre Vergänglichkeit gebunden worden war, genauso wie die Materie im Gegensatz zur griechischen Tradition mit der Welt geschaffen war und damit nicht mehr ihr blosses bestimmungsloses, indifferentes Material darstellen konnte. Da aber die von Augustin gesetzte Zäsur doch nur eine christliche Konsequenz aus der antiken Prämisse der Abhängigkeit der Zeit von der Bewegung des Kosmos (unter der freilich widerspruchsvollen Zusatzannahme von Anfang und Ende) war, kann Newton keineswegs zur antiken Ausgangsposition zurückkehren. Im Grunde demonstriert Newton mit seinen Widersprüchen und der langwierigen Geschichte ihrer Auflösung nur, wie konsequent Spinoza die Rationalität der neuen Epoche zu Ende gedacht hatte.

Wenn Augustin gesagt hatte, die Zeit würde auch bei Stillstand des Himmels fortdauern, sofern nur noch eine Töpferscheibe sich drehte, muss Newton folgerichtig den weiteren Schritt tun zuzulassen, dass die Zeit ohne Rücksicht auf irgendeine reale Bewegung fortdauert: *Dieselbe Dauer und dasselbe Verharren findet für die Existenz aller Dinge statt, mögen die Bewegungen schnell oder langsam oder Null sein. Daher sollte diese Dauer von ihren durch die Sinne wahrnehmbaren Massen unterschieden werden.*

Bemerkenswert am Zusammenhang von absoluter Zeit und absolutem Raum bei Newton ist, dass er für die Bestimmung des absoluten Charakters des Raumes den Zeitbegriff benötigt, dass es aber keinen Ansatz für das umgekehrte Fundierungsverhältnis gibt. Newton bestimmt den Begriff der absoluten Bewegung durch ihre Relativität zu „unbewegten Orten", die den unbeweglichen Raum konstituieren. Diese unbewegten Orte werden in bezug auf die absolute Zeit definiert: *Unbewegte Orte sind aber nur solche, die sämtlich von Unendlichkeit zu Unendlichkeit dieselben gegenseitigen Lagen bewahren, also immer unbewegt bleiben und den Raum konstituieren, den ich unbeweglich nenne*[7]. Identität der Bestimmbarkeit im absoluten Raum besteht nur in bezug auf die Unendlichkeit der absoluten Zeit.

Hier überall tauchen für Raum und Zeit die klassischen Attribute des Göttlichen auf, durch die ihre Realität jedem Verdacht entzogen wird, sie seien nur relative, psychologische oder gar konventionelle Grössen. In der Sprache der Scholastik hätte man diesen Sachverhalt einer erkenntnistheoretisch notwendigen, selbst aber nicht objektivierbaren Voraussetzung der Erkenntnis so bestimmen können, dass sie auf die Seite der *veritas ontologica* gehört.

Dass die absolute Zeit im Hinblick auf ihre metaphysische Dignität noch eine Stufe hinter der Funktion des absoluten Raumes steht, wird weiter dadurch greifbar, dass es indirekte Erfahrung vom absoluten Raum gibt, weil es eine ausgezeichnete Art der Bewegung im Verhältnis zum absoluten Raum, nämlich die Achsendrehung der Weltkörper, gibt und die bei dieser auftretenden Fliehkräfte eben jenes Verhältnis der Bewegung zum absoluten Raum objektivieren. Absolute und relative Bewegung sind nur unterschieden durch die Wirkungen, die als Fliehkräfte von der Achse der Rotationsbewegung auftreten. In diesen Fliehkräften wird das Moment der verhinderten Gradlinigkeit der Trägheitsbewegung im Verhältnis zum absoluten Raum erfassbar, und in eben dieser Verhinderung eine zweite Kraftkomponente, die in die Trägheitsbewegung eingreift und die kreisförmige Bewegung zu einer beschleunigten Bewegung macht.

Die Realität der absoluten Zeit ist der Ausdruck dafür, dass die kopernikanische erste Bewegung der Erde um die eigene Achse kein absolut homogener Naturvorgang mehr sein kann, genauso wenig wie die Jahresbewegung der Erde um die Sonne, deren Abweichung von der Kreisform und deren unterschiedliche Bahngeschwindigkeiten bereits durch Kepler entdeckt worden waren. Es gibt in der Natur die Gleichförmigkeit offenbar nicht mehr, die notwendig wäre, um den Zeitbegriff als einen empirischen Begriff zu retten. Die Kreisbewegung, Rotation wie Revolution, die in der ganzen aristotelischen Tradition die Auszeichnung der im strengen Sinne „natürlichen" Bewegung erfahren hatte und mit dieser Qualität bei Kopernikus dem Erdkörper seine astronomische Integration und fundierende Funktion für den Zeitbegriff verleiht, ist auch bei Newton noch von einzigartiger Bedeutung, weil sie die einzige empirisch nachweisbare Bewegung im Verhältnis zum absoluten Raum

[7] *A. a. O.* 9: *Loca autem immota non sunt, nisi quae omnia ab infinito in infinitum datas servant positiones ad invicem; atque adeo semper manent immota, spatiumque constituunt quod immobilem appello.*

bleibt. Aber sie bleibt es um den Preis eben jener homogenen „Natürlichkeit", die die kopernikanische Reform abzusichern vermochte. Der Verlust der Qualität der „Natürlichkeit" für die Kreisbahnbewegung der Planeten war bereits mit den Gesetzen Keplers, vor allem mit dem zweiten Gesetz über die ungleichmässigen Bahngeschwindigkeiten, entschieden gewesen.

Damit ist aber auch implicite entschieden, dass kein astronomisch fassbares Phänomen fortan die Zeit in dem strengen Sinne manifestieren und messbar machen kann, wie es unter den aristotelischen Voraussetzungen gefordert gewesen war. Man könnte dies so ausdrücken, dass in der vermeintlich natürlichen Kreisbewegung durch Kepler und Newton der Wurm der Gewaltsamkeit – also der Beeinflussung durch Kräfte – entdeckt werden ist. Selbst wenn wir im Gedankenexperiment einmal annehmen, eine Veränderlichkeit und Unregelmässigkeit etwa der Tagesrotation des Erdkörpers hätte bis zum heutigen Tage empirisch nicht verifiziert werden können, so könnte dies an der grundsätzlichen theoretischen Lage nichts ändern, dass eine solche Unregelmässigkeit erwartet, vermutet und weiter gesucht werden müsste und dass damit die Erfordernisse, die Aristoteles im Zeitbegriff an die Regelmässigkeit des manifestierenden Phänomens der Himmelsbewegung gestellt gesehen hatte, nicht mehr erfüllt sein können.

Der absolute Raum und die absolute Zeit haben ihre Funktion bei Newton in der Herstellung der Definierbarkeit des Trägheitszustandes. Das Trägheitsprinzip ist eine kopernikanische Konsequenz, denn es behebt fast alle Einwände physikalischer Art, die gegen die Bewegung der Erde erhoben werden konnten. Dabei geht es immer nur um die *Möglichkeit* des kopernikanischen Systems, nicht um seine *Wahrheit* im Sinne der empirischen Beweisbarkeit. Aber gerade um die durch die Vorrede Osianders verschleierte Strenge des kopernikanischen Wahrheitsanspruches ging es auch Newton. Die geradlinige Trägheitsbewegung im Verhältnis zum absoluten Raum, deren Definition den Begriff der absoluten Zeit erfordert, gibt es in der physischen Realität nicht, und dem entspricht der Sachverhalt, dass sie auch schlechthin nicht erweisbar wäre. Aber im System der Mechanik Newtons ist die einzig erweisbare Bewegung im Verhältnis zum absoluten Raum die Kreisbewegung, und zwar sowohl die Rotation um die eigene Achse als auch die Bahnbewegung in der Ellipse um einen massenbesetzten Brennpunkt. Die Gesetze Keplers geben Newton die Kriterien für die empirische Bestätigung der absoluten Bewegung der Erde und damit für den Beweis der Wahrheit des Kopernikanismus. Und Newton sagt ausdrücklich, um dieses Beweises willen habe er sein Hauptwerk geschrieben.

BARBARA BIEŃKOWSKA
University of Warsaw

FROM NEGATION TO ACCEPTANCE

The Reception of the Heliocentric Theory in Polish Schools in the 17th and 18th Centuries

I in company with most Polish historians do understand the term "reception" as the acceptance of certain ideological and scientific values and the manner in which they function, irrespective of what their author originally intended. In the case of the Copernican system, this will mean the process of a gradual penetration, with all its consequences to sciences, methodology and world outlook, of the heliocentric concepts into the scientific perception of the community.

This process lasted in Polish schools continuously for nearly two centuries. Elements of knowledge about the new cosmological system entered the school curricula in different ways and at different times depending on the type of school. Those elements differed and, being differently lectured, aroused reflections and objections. It is, however, possible to select several characteristic stages which (with varying intensity) marked the reception of the Copernican system. At first it was presented as a new system, from the logical point of view absurd and erroneous as inconsistent with the "letter" of the *Bible*, later as a probable hypothesis, then as a perfect hypothesis, and finally as the only true system of the universe.

The heliocentric system was, of course, not included to school curricula in the 16th century. In Polish schools that time, Copernicus was presented as an eminent scientist, his observations were known, and calculations quoted. There is not so far known, however, any lectures from the 16th century in dealing with the conception of the motion of the Earth. From the beginning of the 17th century, this problem was continuously discussed, though in various ways, from negation, through a gradual acceptance, finally to a complete approval at protestant schools from 1722, and at Catholic schools from 1782. Irrespective of the attitude towards the heliocentric theory, it is important that it was discussed in schools all through the 17th and 18th centuries; pupils were acquainted with its principles, arguments for and against were quoted, and they were informed on the Europe-wide dispute over the heliocentric theory. Whatever the attitude towards the actual theory, concerning the motion of the Earth, the memory of Nicolas Copernicus was kept alive through the entire period, he was honoured as a great mathematician, the most eminent graduate of the Cracow Academy, and the pride of Warmia.

The reception of the heliocentric theory in Polish schools depended largely on the Europe-wide scientific and ideological dispute over the heliocentric theory; however, it reflected the basic trends of the dispute under different conditions and in a different scope.

The new theory aroused the interest of only a few scientists in Western Europe in the 16th century; Copernicus was known and highly estimated only as a mathematician, astronomer, observer and elaborator of Tables. The opposition against the heliocentric theory initiated by the leaders of the Reformation was of an ideological character and narrow in scope. The Catholic church, convinced of the hypothetical character of the new theory, did not see in it any immediate danger. The main topic of discussions, in the first half of the 17th century, when more and more scientific arguments supported the new system, were its objective truth and its being contradictory or consistent with the Scriptures. In the second half of the 17th century, the scientific dispute was actually concluded. Till the end of the 18th century, theologians and popularizers of science continued their dispute over the problem as to whether the theory is consistent with the "letter" of the *Bible* and whether the "hypothesis" does not infringe the principle of the literal interpretation of the *Scriptures.* The point at issue was whether the heliocentric theory was to be introduced into popular cosmology, and finally into the school curricula, or whether the existing traditional views were to be manintained.The heliocentric theory finally captured the minds of Europeans not only due to its indisputable scientific arguments but also to the fact that theologians of all Christian denominations accepted the possibility of an allegoric interpretation of the *Bible.* From the second half of the 18th century, the gradually accepted heliocentric system started to fulfill the ideological function then required of astronomy; in its simplicity and perfection it served as an example of the wisdom of the Creator. What was taught in schools was as a matter of fact contributory. It did not show any new scientific or ideological trends, it only conveyed, with some time lag, the achievements of astronomers and mathematicians — supplying more and more new indestructible arguments in support of the heliocentric theory — and philosophers and theologians working on the ideological interpretation of the new system.

It is obvious however, that the level and intensity of conveying this knowledge determined to a great extent the state of education. It cannot be defined by the views of eminent persons, or even by those of a rather small group of scholars. Indirect influence had a much wider range, under Polish conditions mainly through schools.

The majority of the educated gentry and townsmen in Poland formed their religious, political and social outlook at schools, there they also acquired their knowledge about the world "the earth and the heavens" and their model of thinking.

They completed their education at the secondary school and the knowledge they gained there lasted till the end of their life, sometimes updated by newly published books, but rarely completely changed. Even those young people who studied at Polish or foreign higher schools, as a rule chose academies which continued scientific and educational trends with which they had become acquainted at secondary schools. Former pupils kept up their school contacts for many years, since schools hold in the structure of cultural life in Poland a dominating role, especially those remote from bigger urban centres. There was as a rule a theatre run by the school giving several, sometimes as many as a dozen performances or more a year. The schools organized festive inaugurations and windings up of the school year, lectures by professors and "show" days of the pupils. During the entire 17th century and up to the last quarter of the 18th century, schools were the centre of cultural, scientific and even social life of the region.

Thus, materials concerning schools in this period constitute documents on the state of education and knowledge of the community, of significance transcending the interests of a historian dealing with education alone. They are trustworthy historical sources of great interest. Unofficial documents are most important. They include minutes of lectures prepared by teachers, notes, exercises and examination questions for pupils. Manuscript notes, materials not prepared for publication, show the actual contents of lectures more truthfully than materials either published or prepared for publication, much more carefully prepared and always consored by school superiors. Such materials include text books, scientific theses, pupils' parts in displays, inauguration speeches, etc. Printed materials had, however, a wider range of influence. School materials are a very valuable and interesting source very often underestimated and not utilized by historicans of culture and science. In Poland, research in this field is carried on intensively. It is difficult, however, due to the destruction, transferring and dispersing of most of the school libraries during historic calamites, especially the Second World War. A systematic study of extant schools materials should, however, constitute a basis for the study of the general problem of the knowledge and reception of modern scientific ideas in Poland in the 17th and 18th centuries. Research on the reception of the heliocentric theory in Polish schools is a part of wider research work planned in this field.

Links are still closer when separate problems of the complicated process of the reception of the heliocentric system are considered against the background of the dispute between modern natural science (of which the heliocentric theory was one of the main elements), and the traditional system of the doctrines of Aristotle and Thomas Aquinas. In Poland, the philosophy of Descartes became the principal argument in the dispute. That philosophy aroused in the last quarter of the 17th and the beginning of the 18th century the interest of the professors of colleges at Gdańsk and Toruń. Apart from a group of enthusiasts of the French philosopher (Schaeve, Meier, Bormann), there was a group of moderate critics (Sartorius, Sahn, Glosemeyer) and stern opponents (Bohm, Jaenichen, Groddeck). The opposition, however, was not the result of faithfulness to Aristotle's doctrines, but of the acceptance of the empirical method of research in natural sciences, which already then necessitated recognition of the basic achievements of the 18th century natural science.

In Central Poland, the doctrines of Descartes were introduced into the school curricula much later. In the mid-18th century the reformers of Piarist schools, Stanisław Konarski and Antoni Wiśniewski, introduced into the school curricula of philosophy, physics and natural sciences certain elements of the Cartesian philosophy in connection with Newton's physics known as the "recentiorum" philosophy, without, however, including Descartes's philosophy in its pure form. The prevalence of the "recentiorum" philosophy helped to introduce the heliocentric theory into the college curricula, although in Catholic schools it was treated as "hypothetical" till the end of the 18th century. These objections, however, had from the middle of the century a more and more entirely verbal character and finally disappeared completely. *The Apologia of Nicolas Copernicus* delivered by Professor Jan Śniadecki at the Cracow Academy in 1782, containing a full and unconditional approval of the heliocentric theory, may be considered a symbolic end to the dispute over the theory of Nicolas Copernicus in Polish science and education. By that time schools of all types accepted the heliocentric theory.

However, in view of the great number of types of schools the time and manner of abandoning the geocentric image of the world varied considerably in different circles. In this situation, research on the history of the reception of the heliocentric theory in Poland necessitated a study of the process in various types of schools.

* *
*

From the mid-16th century to the end of the 18th century, there were two principal types of schools in Poland — Catholic and Protestant. They included schools of Polish Brothers, Lutheran and Calvinist schools. The small number of mainly elementary schools, Orthodox and Catholic of the Eastern rite, did not play any major role either scientific or cultural. This division is not only justified by the actual division of the community into a Catholic majority and a Protestant minority, but also by considerable differences between their school curricula, the level of education and sources of knowledge. The religious criteria of the division of Polish schools in the period under review are justified by the fact that Polish schools were then entirely religious educational establishments, run mostly by religious orders. Religion defined the character and trends of teaching and education, the educational process and curricula. In this context a religious rather then, say, a social division seems more justified. As a rule, however, colleges run by the Cracow Academy and Protestant schools were attended by townsmen, while the gentry and aristocracy attended schools run by religious orders.

Protestant schools, especially the colleges of university standing at Gdańsk, Toruń and Elbląg, achieving a specially high standard, differed considerably from

Catholic schools, not only as regards, religious aspects. The education and grading system was different, lecturers had completely different qualifications, as a rule using graduates of foreign universities, interested in scientific achievements of great European Protestant cultural centres, such as Königsberg, Rostock, Leyda, Wittemberg, and Altdorf. The curricula of Protestant schools were more flexible both in their range and contents, and changed frequently to include new achievements of natural sciences.

Catholic schools in Poland, however, including the most numerous group – Jesuit colleges based their education on rigid curricula and methods (ratio studiorum), established at the end of the 16th century and only slightly changed by the mid-18th century.

Poland's territory varied in the 17th–18th centuries from about 1 million to about 700,000 square kilometers. The population ranged according to estimates from 10 to 14 million people with 10–20 in habitants per square kilometre. The Northern and Western territories were the most densely populated, and the extensive Eastern territories the most sparsely populated. The same concerned the net-work of schools, of considerable density throughout the country. Roman Catholicism was the official religion of Poland during the whole period under review.

Catholic schools were then attended by the majority of youth in Poland. Those schools, uniform from the ideological and religious viewpoint, were organized in several networks.

The most representative college, enjoying general esteem and appreciation, was the Cracow Academy established in 1364. Till 1780, when the completely reformed academy received the structure of a modern secular higher school, nearly all the professors – excluding those of the Medical Department – were Catholic priests. The Cracow Academy had Departments of Theology, Law, Medicine, and a Humanist and Philosophical Department constituting a preparatory course for studies at the other Departments. The end of the 15th and the first half of the 16th century were the most brilliant period in the development of the Academy. In the following years, the financial situation of the Academy and the social position of the professors deteriorated; the standard of education and the role of the Academy in the development of Polish culture and science also declined. The number of students ranged from one to two thousand. From about the mid-16th century, the Academy became mainly a college for commoners. Both professors and students came mainly from the middle-class. The graduates mostly took posts as secretaries (clerks in municipal offices, doctors (physicians), lawyers, clergymen, and schoolmasters.

From the end of the 16th century, the Cracow Academy started to organize in provincial towns a network of secondary schools administered by the Academy. In the 17th century, there were over 40 such schools. In the 18th century, however, before the establishment of schools by the Commission for National Education, the number fall to about a dozen.

The gentry and aristocracy, the ruling class in Poland during the period under review, sent their sons to colleges run by religious orders, mainly the Jesuits, competing with schools of the Cracow Academy. Universities abroad were generally chosen for further studies.

Jesuit schools developed in Poland from 1565, when the first Jesuit college in Poland was established at Braniewo. Already in 1599 there were 11 such schools in Poland. During the 17th century, the number increased nearly four times, reaching 47 in 1700. In 1740, the number reached 63, and in 1773, when the Jesuit order was dissolved it amounted to 66 schools. Thus the Society of Jesus had the longest number of educational establishments in Poland. Taking into consideration the number of pupils attending each college from a few score to a few hundred, sometimes even over a thousand -- and the fact that the Jesuits carried on their educational activities uninterruptedly from 1565 to 1773, it is easily realized what a great influence the Jesuit schools had on the outlook and culture of the Polish people.

The Piarist order ran a much smaller number of schools in Poland. It played, however, an important role in the history of Polish education. The Piarists, in the 40's of the 18th century, introduced into Polish Catholic schools an up-to-date reform of curricula and methods of teaching.

The order arrived in Poland in 1642, and already in the following year opened schools at Podoliniec and Warsaw. Competing with the Jesuits, they managed to win the esteem and trust of the gentry. By the mid-18th century, they already had 23 schools and continued their educational activities till Poland lost her independence in 1795. The Theatinians and Basilian monks also ran schools for secular youth, their range of influence was, however, very small as compared with the two previously mentioned orders — the Jesuits and the Piarists.

Among Catholic schools should also be mentioned schools established by the Commission for National Education — the first central secular educational authority in Poland, operating from 1773—1794. Schools established by the Commission for National Education were intended primarily to replace Jesuit schools after the dissolution of the order in 1773. They differed, however, very considerably organizationally, didactically and scientifically from the already existing Catholic schools. Mathematics and natural sciences prevailed in the curricula of the new schools instead of the older types oriented towards the humanities — law and what was called moral science. Polish replaced Latin as the language of instruction.

Protestant schools were started in Poland about mid-16th century as a result of extensive religious toleration of that time. Lutheran schools in Lutheran bourgeois centres of Northern Poland were the first established, and represented from the very beginning a high standard.

In 1558 a high-school was established in Gdańsk — actually, a lower-degree awarding university. Similar high-school were established at Toruń in 1568, and at Elbląg in 1590. Apart from these schools, preparing the young for university studies, there was a dense network of Lutheran elementary schools.

Other groups of Polish Protestants established their own schools at their centres.

There were many Calvinist schools in Lithuania, including the well-known secondary school at Kiejdany. The most eminent educational establishment of the Polish Brothers was the Rakow Academy (1635–1638), attended in its years of prosperity by more than a thousand pupils, many from abroad. The Bohemian Brothers opened in 1635 at Leszno Wielkopolskie a college famous for the cooperation spread over many years, of John Amos Komensky. The school flourished until mid-17th century, but declined after the Swedish wars and was finally changed into a Lutheran school.

Humanist and philosophical subjects prevailed in the curricula of all these schools. The teaching of the elements of natural sciences, including astronomy, varied in scope, form, and interpretation.

In general, therefore, natural sciences were in Catholic schools till the mid-18th century part of the course of philosophy, then including logic, metaphysics, and "physics" – comprising knowledge about nature and the universe. The whole course of philosophy was based on the system of Aristotle as interpreted by scholastic writers of the Middle Ages (Albertus the Great, Thomas Aquinas). It contained, although to a rather limited extent, frequently in the form of a polemic, the achievements of modern natural sciences. Even such a course of philosophy was not taught in all Catholic schools. It was covered by the curricula of only the highest grades, achieved by the most keen, and talented pupils. It was thus taught only at the Cracow Academy, the Zamość Academy and a few Jesuit Colleges. Elementary knowledge of astronomy, geography, meteorology, etc. was, however, taught in all Catholic secondary schools for merely practical reasons.

At Protestant schools, especially high schools, "physics" was soon separated from general philosophy. From the beginning of the 17th century, it constituted a separate subject, especially valued in those circles. It is worth noting that most of the graduates of the Gdańsk and Toruń high schools, prepared their examination thesis from physics and writing textbooks on this subject, showing a special interest for astronomy.

Catholic schools in Poland up to the middle of the 18th century, rather carefully observed the Aristotle-scholastic line in lectures on natural science. Protestant schools were much more liberal in this respect. The Aristotelian system was attacked there already at the beginning of the 17th century, and approval was expressed for the views of the anti-Aristotelians from Wittenberg university, especially Johan Sperling. The theses of both students and professors in the 17th century were packed with discussions on all current topics. The fact what in spite of this, the heliocentric system was finally accepted there comparativelly late, was connected with the special role, in the world outlook, of astronomy as yielding knowledge about the heavens, and also in the special ideological aspect of the dispute over the heliocentric system.

Both Catholic and Protestant schools taught astronomy for two reasons – the first merely practical, the second more profound connected with religion and outlook on the world. The first reason concerned ability to use a calendar and cloks, and a basic ability to orientate by means of stars, and weather forecasting. The religious aspect was connected with religious education at schools in those days. Knowledge about the greatness and the spatial distribution of the heavens was intended to prove the wisdom and the goodness of the Creator. "The heavens praise the glory of God" – this expression from the *Bible* was the guiding thought of many school textbooks on astronomy, and even of scientific treatises on astronomy. Both Catholic and Protestants alike recognized the role of astrononomy in forming the religious and world outlook. Moreover the authors of textbooks attributed this role to all natural sciences, declaring that nature as a whole and all its components reflect the Creator. One can then have cognition of God by studying both the written Book of Revelation – the *Bible*, or the unwritten – the "book" of nature. The two "books" cannot, having been created by God, be contradictory. The principle that the *Bible* and the results of research in natural sciences must be fully consistent, accepted both by Catholics and Protestants, recognized as the starting point equally by Jesuit, Lutheran professors and – for instance – Komensky constituted a major obstacle to the development of scientific research and its popularization. Thus in practice the results of the observations and calculations of astronomers, physicists or biologists should supply an image consistent with the description of the creation of the universe and its structure given in the *Bible*. In the course of the progress of knowledge, modern astronomy supplied more and more views controversial with the *Bible*. This especially concerned the heliocentric theory. Many verses of the *Bible* clearly imply that the Earth is immovable and by God fixed once and for ever in its foundations, while the Sun is in motion. In such a controversial situation the only way of preserving the authority of the *Bible*, and at the same time of accepting the results of the discoveries of modern natural sciences was to recognize an allegoric, metaphoric interpretation of the verses of the *Bible* concerning natural phenomena. The traditions of such a metaphoric interpretation of the *Bible* go back in Christian philosophy to St. Augustin. To this tradition Galileo referred in his well known letter of 1615 to Christine, Duchess of Tuscany. He argued that the *Bible* uses a popular language in describing the phenomena of nature, suitable to the concepts and images of common people, in order to make its ideas of faith and morality understandable. Teaching the laws of physics was not the aim of the *Bible* and a literal interpretation of the verses of the *Bible* dealing with the phenomena of nature in a manner undermining scientific statements would be contrary to common sense and tradition. The conception of the allegoric interpretation of the *Bible* represented by Galileo was condemned by the Roman Catholic church in decrees of 1616 and 1633, which were qualified only in the mid-18th century. "The problem of Galileo" thus made it impossible for many years to reconcile within orthodox Catholic doctrine the description in the *Bible* and the discoveries of modern science, and impossible, too, the acceptance of the heliocentric system in Catholic Schools. Protestant clergymen and scientists, however, not restricted by the Roman decree, and furthermore using the decree as an argument in their polemics, took into consideration and worked out in detail, already in the 17th century, the possibility of reconciling the principles of the heliocentric system with the *Bible* interpreted allegorically. In this connection, Protestant schools, in spite of many substantial (Aristotelian physics) and ideological restraints were able much earlier than Catholic schools to overcome the basic difficulty — which was the verbal inconsistency with the *Bible* of the scientific theory considered fully authoritative and credible in the latter part of the 17th century.

The finest traditions of the controversy with and for heliocentrism are represented, of course, by the most famous school of old Poland — the Cracow Academy. Traces of interest taken there in the text of *De revolutionibus* are found as far back as the 16th century, not many years after the death of Copernicus. It must be remembered that Copernicus had been one of its students, and his cult developed there until the end of the period under review.

On the other hand, the Cracow Academy hung on longest in its official lectures to the geocentric theory of the universe, and was the last of all Polish schools to recognize the heliocentric theory. This is one of the eloquent proofs that a sincere and real cult of Copernicus as a great scholar and illustrious student of the Cracow Academy could be accompanied by a slashing criticism of his theory in the minds of Polish scholars of that time.

The cult of Copernicus and the interest taken in his works were inspired by the Cracow professors of astronomy. In 1542 — that is still in the lifetime of the Frombork scholar — Albert Caprinus of Bukowo wrote a foreword to his astrological book of prognoses, in which he highly praised Copernicus as a mathematici of genius. In 1549, another Professor, Hilary of Wiślica, drew up an astronomical calendar, called *Ephemerides*, on the basis of the astronomical tables contained in *De revolutionibus*. A. Birkenmajer has shown that this use of Copernicus's calculations did not need and was not tantamount to acceptance of the theory of a moving earth[1]. In the years 1578—1580, Professor Walenty Fontana explained the tables of Copernicus during his lectures, but even this is not admissible as a proof of the acceptance of heliocentrism[2]. On the basis of the Copernican calculations, Cracow professors carried out in 1563 a group observations of the conjunction of Saturn and Jupiter. Until the end of the 16th century and all through the 17th (which was already a scientific anachronism) Cracow professors used Copernican calculations in their works, demonstrating an accurate acquaintance with the text of *De revolutionibus*... For the significance of the latter work consisted, as A. Birkenmajer has shown, not only in the theory expounded in it but also in the large number of Copernicus's independent observations and calculations which served for a strict determination of the elements of the movement of heavenly bodies, and for the compilation of

---

[1] A. Birkenmajer, *Czy Hilary z Wiślicy był szermierzem systemu heliocentrycznego w Krakowie*? [Was Hilary of Wiślica the Champion of the Heliocentric System in Cracow?], in the quarterly „Kwartalnik Historii Nauki i Techniki", 1959, pp. 419—464.

[2] E. Rybka, *Four Hundred Years of the Copernican Heritage*, Kraków, 1964, pp. 175.

Tables better agreeing with the sky than the old ones. Numerous astronomers of the 16th century and the first half of the 17th attached more weight to the merits of Copernicus as an observer and author of *Tables* than as the author of the heliocentric theory. This was because those *Tables* stood up well to practical examination, while the theory of the movement of the earth defied at once the ages-old scientific authorities recognized by schools and the direct testimony of human senses. During the period under review, the Cracow Academy was pursuing first and foremost practical astronomy. For his *Tables* alone, therefore, Copernicus won at once great recognition, and as a result of the scientific conservatism of the Cracow professors that recognition continued even when a large part of the detailed calculations was already out of date. On the other hand, the lectures on philosophy and natural science held at the Academy itself and its branches – secondary schools situated in smaller towns – contained the geocentric theory taught from a textbook by J. Carpentarius, Mediaeval but still popular in the 17th century. In the historical aspect, therefore, Copernicus could not be identified with the heliocentric theory.

The activity of Professor Jan Brożek (d. 1652) was of great importance for deepening the interest taken in the life and work of Copernicus by the Cracow milieu. It appears from Brożek's private (inedita) notes made in the margins of *De revolutionibus* and of works by Kepler and Galileo, that he was inclined to recognize the heliocentric theory and the theory of the multitude of worlds. All he, like other icholars of those times, wanted, was to back this by further physical and mathematscal arguments. But in his official academic lectures, Brożek remained faithful to geocentrism and Aristotle. The personal interest taken by Brożek in the person of Copernicus was expressed in a scientific expedition to Warmia in 1618, the aim of which was to look for memorabilia and biographic materials connected with the great scholar. The only result of this search was a biography of Copernicus contained in a collection of the lives of Polish writers and scholars published by Szymon Starowolski (1627), and publication of a pseudo-Copernican religious poem called *The Seven Stars* (*Septem sidera*).

Under the strong influence of Brożek, who enjoyed a marked scientific authority, Cracow chroniclers and professors of the latter part of the 17th century frequently extolled Copernicus. At the same time they propagated geocentrism in the version of Tycho or even Ptolemy.

Stanisław Słowakowic, professor of astronomy and physician was also inclined to accept the heliocentric theory in his book attacking the natural science of Aristotle and supporting the modern theory of the origin of comets (*Postliminium cometarum*, Cracow 1681). But Słowakowic was entirely alone among the conservative professors of the Cracow Academy. Many decades passed before those professors began to treat the heliocentric theory as a hypothesis equivalent to the system of Tycho. A veritable turning point as regards the heliocentric theory occurred only in 1761, when Professor Jakub Niegowiecki published a treatise containing observations of the passage of the planet Venus in front of the sun.

In his observations and calculations, Niegowiecki used the heliocentric theory, but he treated it, nevertheless, only as the most probable of all hypotheses, more useful for research than the theory of Tycho. He denied to it, however, the name of an objective physical axiom, and in a special treatise he criticized astronomers accepting the movement of the earth as real.

Thus after many years of complete rejection of the heliocentric theory a new situation arose at the Cracow Academy. Following in the steps of Niegowiecki, many professors admitted that the Copernican theory had remarkable cognitive values, but denied that it reflected reality, arguing that it lacked convincing scientific arguments and was inconsistent with the *Bible*. The recognition of the heliocentric theory as the best existing working hypothesis for the use of astronomers was, of course, a marked step forward, but it was at the same time a glaring anachronism in relation to the then level of science and teaching in Western Europe, or even in other Polish countries. No wonder that this gave rise to sharp protests. In his book *Prodromus Polonus eruditae veritatis*, Berlin—Wrocław 1765, Marcin Świątkowski fought for the removal from teaching curricula of the obsolate Aristotelian and scholastic theories. The book contained a separate chapter devoted to praise of the heliocentric system and its author — Nicolas Copernicus. Świątkowski was a mathematician, professor of the Cracow Academy, a student of Christian Wolf of Halle, a zealous propagator of modern philosophy and science. He was particularly scornful of objections to the heliocentric theory arising from a literal understanding of the *Bible*. Like all other propagators of the ideas of the Enlightenment in Poland, Świątkowski argued that the heliocentric system not only does not stand in contradiction to the *Bible* and theology, but that by its harmony it demonstrates even better the wisdom of the Creator.

Efforts of the propagators of the Enlightenment, particularly those of Hugo Kołłątaj, who in cooperation with progressive professors introduced a thorough reform of the Cracow Academy in the years 1777—1780, created conditions for the introduction of the heliocentric system as the sole recognized cosmological system giving a true picture of the universe. The complete triumph of the theory of heliocentrism in the Cracow Academy was an *Apologia of Nicolas Copernicus*, in a speech made at the Academy in 1782 by the then young scholar, Jan Śniadecki, later a famous mathematician and astronomer.

In his speech, Śniadecki not only extolled the merits and talent of Nicolas Copernicus, but above all presented his theory as the picture of the true system of the sky. The speaker devoted considerable attention to the controversy about the heliocentric system, justly seeing in it a plot of scholastic forces. Śniadecki argued that the clergy had committed a serious error when they attacked the heliocentric theory with arguments from the *Bible*. After all, the Copernicus theory does not undermine religion, it only explains more lucidly and perfectly than other theories the work of the Lord. By placing the whole blame for the fight against the heliocentric theory on the Roman Inquisition, Śniadecki avoided a conflict with the clergy.

Śniadecki's view that it had been only the Copernican theory that fully showed the whole might and wisdom of God had been advanced on numerous occasions by European heliocentrists of the 17th century. It was used by Galileo, it was beautifully expressed by the founder of the Oratoriars, Pierre Bérulle (d. 1629). But in Poland this argument came to be appreciated only in the epoch of the Enlightenment. The men of the Polish Enlightenment repeated after the earlier centuries that the book of nature and the book of the Scriptures are the work of the same God and cannot be mutually contradictory, but they drew different conclusions from this. They no longer thought of subordinating scientific studies to theology, but of interpreting the results of natural studies as proof of the wisdom and might of the Creator. This then necessary model of reasoning opened up possibilities of independent natural investigations without giving rise to conflicts, and even with the approval of theologians.

The triumph of the heliocentric system at the Cracow Academy opened up for this theory a way to lectures at schools connected with the Academy. There too, of course, the acceptance of the heliocentric theory progressed with difficulty, for reasons identical with these applying at the Academy.

Particularly at the Zamość Academy, a school of fine scientific traditions, possessing from 1594 the rights of a higher school but declining in the 18th century, the divergencies of opinion among the professors on the subject of the Copernican theory were still great.

Klemens Poniatowski, a Zamość professor, published in 1768, for example, an outline of modern natural science, in which he conceded high scientific merits to heliocentrism[3]. The author, a propagator of the physics of Newton, asserted that heliocentrism was consistent with the laws of physics and astronomers' observations. Although heliocentrism appeared to stand in contradiction the *Bible*, the latter also, after all speaks of moving the earth from its place (*The Book of Job*' 9,6). There is no doubt that Poniatowski himself did not need arguments of this type, but as a teacher he had to consider the popular habits of thought. And hence propagating a new theory he baked it with the customary biblical arguments. But to the last they did not convince the milieu of the Zamość professors. As late as 1774, printed philosophical theses appeared rejecting the Copernican theory. Even though the author of those theses conceded the consistency of the theory with astronomy, he quoted the letter of the *Bible* and accepted the Tycho system.

* *
*

It was the Jesuit schools in Poland that were the first to turn their attention to the heliocentric theory. A sort of first stage of acquainting students with the Coper-

---

[3] K. Poniatowski, *Summula philosophiae naturalibus in physica phaenomenis exornata*, Zamość 1768.

nican theory was provided by polemical digressions during lectures on the theories of Ptolemy or Tycho Brahe. While refuting the principles of the heliocentric theory, the Jesuit teachers unwrittentionally popularized both the theory itself and the name of its author. It must also be remembered that from the close of the 16th century Copernicus was recognized and trusted in Jesuit schools as the author of accurate calculations.

The Jesuit colleges rejected the theory of the movement of the earth even before it was officially condemned by the Church in 1616. For example, an extent rough manuscript contains the draft of lectures on astronomy according to *The Sphere* by John Holywood held in 1610 in the Jesuit College in Braniewo. The lecturer, Henryk of Gdańsk of whom little is known, expounded in it the geocentric system, but mentioned also the existence of other, different, view with regard to the system of the universe[4]. He wrote, namely, that there were astronomers, including their leader Nicolas Copernicus, who assert that the earth moves in a circular motion. The view, however, could not be accepted for it was denied by scientific reasoning (the Aristotelian physics), religious reasoning (the *Bible*) and the everyday testimony of our senses. Aristotle taught that the elements can have only a single movement along a straight line. The *Bible* clearly stated that the earth is stationary, while the sun and the whole sky are moving. Moreover, we might see with our own eyes that an arrow or a stone cast upwards drops back in the same spot, while all objects on the surface of the earth remain in their place at rest. The Jesuit astronomer left the detailed polemic concerning the theory of the movement of the earth to men more competent in this domain.

The Braniewo lecture is typical of the Jesuit astronomical lectures of the early 17th century. On the basis of Aristotelian physics and the *Bible*, Jesuits propagated the geocentric system, but at the same time treated polemically the views of heliocentrists. The Braniewo MS contains also another element which played a decisive role in the process of the reception of the heliocentric theory — namely, the trend of an allegorical interpretation of those fragments of the *Bible*, which pertained to heavenly phenomena. The teacher from Braniewo dwelled on the meaning of the well-known fragment of the *Book of Genesis* (1,16) which speaks of the creation of two "great lights" (*lumina magna*) — the Sun and the Moon. Astronomers, meanwhile found beyond the shadow of a doubt that the Sun and the Moon are not the greatest lights, since the stars are bigger. The words of the *Bible* must, therefore, be taken to mean that the Sun and the Moon are the greatest not in size but in their effect (non mole sed efficientia), because both in the daytime and at night they illuminate the Earth more than the stars[5].

An even more eloquent example of the tendency towards the allegorical inter-

---

[4] Manuscript, property of Biblioteka Narodowa (National Library). Rps. BN 3468. Henrikus Gedanensis. *Quaestiones astronomicae in sphaeram Joannis de Sacro Bosco in collegio Brunsbergensi Soc. Iesu, Anno 1610*... card 94. v.

[5] MS, BN 3468, p. 115 v.

pretation of the "natural" fragments of the *Bible* is contained in a lecture delivered in 1625 at an obscure Jesuit college in Poland[6]. The Jesuit professor in question held to the geocentric system and criticized the speculations of certain (unspecified) philosophers concerning the possibility of the existence of many worlds (*pluralitas mundorum*). He dwelt at length, though on the results of the studies of Galileo on the movement of the planets and on the sun spots, and distinctly asserted that one must not take literally the fragments of the *Bible* relating to heavenly bodies, for the *Bible* frequently uses terms of little accuracy, unacceptable to science. It uses, for example, the term "heavens" for the air and for the sky itself, and the term "light" for the sun itself and the light of day in general. There is no doubt that it was no great jump from such assertions to finding that the biblical texts concerning the movement of heavenly bodies were of small authority for scientists. Galileo's trial and the Roman decree of 1633 connected with it, strictly banning a non-literal interpretation of all the words of the *Holy Scriptures*, for long put a stop to attempts at a allegorical interpretation of the *Bible* in the Polish Jesuit schools. They arrested considerably, therefore, the process of the reception of heliocentrism in those schools.

Opposition against heliocentrism was maintained in Jesuit philosophy and astronomy lectures from the thirties of the 17th century, until the middle of the 18th. With this difference, that as the years went by, the system was discussed in ever greater detail, and ever more arguments had to be found against it. The arsenal of the anti-heliocentric proofs also kept changing. In the mid-17th century, Polish Jesuit schools imparted to the polemic with the heliocentric theory characteristic traits of scientific criticism. With this aim in view, Jesuit professors usually exploited the arguments of J. B. Riccioli, a renowned Jesuit astronomer and geocentrist of the mid-17th century. The inconsistency of the heliocentric theory with the *Bible* was then barely mentioned, but no emphasis was laid on the problem. No attacks on Copernicus himself are recorded in Polish schools (neither could they be found in Riccioli's work). Invective, if any, was addressed not against Copernicus himself (whose name was protected by his scientific prestige, backed also by the Jesuits themselves and by the foreword of Osiander suggesting the hypothetical character of the Copernican theory), but against contemporary propagators of the heliocentric theory – recruited mostly from Protestant circles.

In order to keep their lectures up-to-date, Jesuit professors had to be ever more painstaking in their polemics with the heliocentric theory. Beginning with the latter part of the 17th century, they already defended not only the principle of the immobility and central position of the earth in the universe, but also, in accordance with instructions from Rome, the principle of a literal understanding of the relevant fragments of the *Bible*. For as physical and astronomical observations and investigations progressed, critics began to lack scientific arguments and the *Bible* remained their sole mainstay. From its scientific track, the discussion began to pass over

---

[6] *Universa Aristotelis philosophia naturalis, 1625*, MS, BN, p. 479.

more visibly to the ideological plane. One may cite here as examples the printed philosophical theses of a 1666 graduate of the Braniewo College. The author launched his main attack against supporters of a metaphorical interpretation of the "anti--heliocentric" fragments of the *Bible*. Since the Roman Curia clung to their literal interpretation, the student in his polemic zeal went so far as to call heliocentrists the enemies of the Pope, even of religion and God. He appealed also, though not primarily, to reason: that one cannot be the solugh of the forests caused by the onward rush of the earth, that all objects on the surface remain at rest, that a bullet fired in an easterly direction hits a flying bird, though it travels, together with the air and the earth, 4 miles in a short moment, while the bullet travels half a mile only. As time passed, arguments of this kind were becoming more and more anachronic, but were invariably repeated in most Jesuit lectures almost until the middle of the 18th century[7].

Similar objections were weighed in the eighties of the 17th century by Adam Kochański, an eminent Jesuit mathematician, well-known in the scientific centres of Europe. He published in 1685 his own reflections on the subject of the heliocentric theory, ever more widely recognized in Europe[8]. Kochański did not accept the heliocentric theory. His counter-arguments were its inconsistency with the *Bible* and lack of convincing scientific proof. But Kochański did not discredit or completely reject the theory on the movement of the earth. He considered that it should be verified experimentally by mathematical and physical methods. If the experiment confirmed the indisputable correctness of the thesis, only then would it be possible to think how to make the text fit the physical truth. The scientist even worked out the project of such an experiment. Kochański's reasoning, though curbed by the Rome decrees of 1616 and 1633 and cautious to a fault in face of the astronomic and physical proofs accumulated by 1685, admitted, however, the possibility of a revision of the theological standpoint in face of obvious scientific truth.

That truth, revealing itself in a manner which defied all denial, compelled a part of the Jesuit professors to relent in the early years of the 18th century.

Some handwritten philosophy lectures from the beginning of the 18th century, delivered at an unidentified Jesuit college in Poland, contained the basic principles of the heliocentric theory and even called it a hypothesis valuable for astronomers[9]. But the Jesuit professors in question came out definitely in favour of the Tycho hypothesis, since it was consistent with the *Bible*. The fact that heliocentrism came to be called the clearest hypothesis explaining the phenomena taking place in the sky was the first step taken by the Jesuit schools towards the acceptance of the heliocentric theory, due to the reception of the elements of a new philosophy and natural history. The development of exact sciences, and above all the publication of Newton's

---

[7] M. T. Sosnowski, *Conclusiones ex universa Aristotelis philosophia*, Braniewo 1666.

[8] *Considerationes et observationes physico-mathematicae circa diurnam Telluris vertiginem. Acta eruditorum*, Lipsiae, 1685, pp. 317–327.

[9] Ms, property of Ossolineum Library, shelf No. Pawl 168, card 152–159.

basic work (1687), removed scientific doubts. The sole obstacle remained the *Bible*, which distinctly spoke of a moving sun and stationary earth.

In this situation, the Jesuit professors of the first half of the 18th century gradually rid themselves of the scientific arguments based on Aristotelian physics. Their place began to be taken by works full of appreciation of heliocentric theory as the best hypothesis, its acceptance banned solely by the Roman interdiction. Thus the Jesuits accepted the scientifically inferior but religiously uncontroversial hypothesis of Tycho.

Kasper Niesiecki, for example, a lecturer in mathematics and related sciences in Kalisz college in 1717, objected to the heliocentric theory solely because it was inconsistent with the *Bible*. He did not offer, on the other hand, any counter arguments based on science or reasoning[10]. A graduate of the Jesuit College in Lwów in 1746, also charged the heliocentric theory only with a formal inconsistency with the *Bible*[11]. Nevertheless, this formal inconsistency was sufficient to prevent the author from recognizing the movement of the earth as truth. The scholar explained even in this connection that Copernicus himself (called "the illustrious astronomer") had presented his theory only hypothetically for a better mathematical explanation of the heavenly phenomena. It was only later that astronomers and writers tried to ascribe to Copernicus a conviction of the truthfulness of his idea, in order to make it easier for themselves to convince their students and readers. One sees in this justification of Copernicus an echo of the all-Polish cult of the great scholar, continuing and developing in defiance of the critics of the heliocentric theory.

Around the middle of the 18th century, Jesuit professors began to depart more and more distinctly from the principles of Aristotelian and scholastic science, and to accept elements of modern philosophy and natural history. The process neither proceeded easily nor in all centres uniformly. The Jesuits were attached to the old conceptions and ideas forming a cohesive philosophical and theological system, a system closely bound up with the traditions of the Catholic Church. Meanwhile, the new science was a product of laymen, mostly of the Protestant persuasion. The Jesuit teachers, most of whom were badly trained scientifically, feared the threat to the foundations of their religion contained in a totally new picture of the universe created by astronomy, physics, chemistry, and biology. But under the pressure of Enlightenment tendencies spreading ever more widely in Poland in the mid-18th century, and in view of the necessity to adapt the curricula of their schools to the changed requirements, the Jesuits had to yield and accept the elements of the new philosophy and science. They adopted, nevertheless, a very far-going eclecticism. They eliminated Cartesianism, contenting themselves with the physics of Newton

---

[10] MS, property of Biblioteka Kórnicka (Kórnik Library), No 1425. About the lectures of Niesiecki: T. Przypkowski, *Astronomia w Kaliszu* [Astronomy in Kalisz], [in:] *Osiemnaście wieków Kalisza*, Kalisz, 1960, pp. 193–194.

[11] *Rozmowa o filozofii przez Michała Drużbackiego napisana* [A Discusion on Philosophy by Michał Drużbacki], Lwów, 1746.

in the philosophical interpretation of the well-known philosopher of the era of the Enlightenment, Christian Wolf. They began at the same time to propagate the new vision of the world as the perfect expression of the benevolence and wisdom of the Creator. This left, of course, no room for clinging to the "letter" of the *Bible*. The Polish Jesuits therefore accepted gradually the principle of the lack of the authority of the *Bible* on scientific questions, a view long since accepted in the Catholic centres of Western Europe.

A portent and a marked example of this new attitude of Polish Jesuits was the 1760 textbook by Benedykt Dobszewicz, containing an exposition of modern philosophy[12]. Speaking of the cosmological systems, the author asserted that the conception of Tycho, although consistent with the *Bible* and explaining heavenly phenomena, does not explain the causes of the movement of the planets. It would have to be surmised that it was a perpetual miracle performed by God, or that the Angels move the planets, which cannot be considered as natural or credible. Thus Dobszewicz mutely espoused the heliocentric theory.

At approximately the same time, another Jesuit professor, S. Szadurski, who taught philosophy in Nowogródek and Warsaw, published in 1761 and 1762 examination theses for his students. They accepted the heliocentric theory as the best of all hypotheses supported by new scientific arguments, especially the discoveries of Newton. The latter's physics thus supported the heliocentric system as well.

The examination theses from subsequent years moved, gradually but consistently, toward a complete approval of the heliocentric hypothesis. It is worthwhile to follow this complicated process on at least a few examples:

In 1763, Stefan Łuskina, professor of mathematics and physics at the Jesuit College in Warsaw, published theses of his students who extolled Copernicus and accepted the heliocentric theory as fully consistent with heavenly phenomena and modern physics. Copernicus was called in them a great Polish astronomer, a native of Royal Prussia, whose theory had been accepted and applied by the greatest scholars, from Kepler and Galileo to Newton. Objections arising from inconsistency with the *Bible* were of no moment if the movement of the earth was accepted as a hypothesis. The students referred also to a work by L. A. Muratori[13] and argued

[12] B. Dobszewicz, *Placita recentiorum philosophiorum explánata*, Wilno, 1760.

[13] D. A. Muratori, *De ingeniorum moderatione in religionis negatio*. Lutetiae Parisiorum, 1714. This work played a considerable role in the history of the European dispute over the heliocentric system. Muratori, a Catholic clergyman, and then a well-known historian and religious writer came out against the literal interpretation of verses of the *Bible* concerning the phenomena of nature. The *Bible*, in his opinion, described the phenomena of nature as they appeared, and not as they actually were. The *Bible* used simple language understandable for common uneducated people, and frequently employed metaphors. Referring to the verses of the *Bible* declaring that the Earth is static Muratori said that they were general observations perceived by the senses, which should not be taken into consideration in scientific research, and beyond that must not be used to challenge the discoveries of astronomers. Muratori then expressed the same opinion as Galileo a hundred years earlier: he was not, however, condemned by Rome.

that many Catholics were of the opinion that the *Bible* did not really contradict the heliocentric theory, since it lacked authority on physical and natural questions.

Theses on general physics defended at an examination of students at the Jesuit Academy in Wilno in 1764 went even further. They explained why an observer moving together with the rotating earth does not feel this movement and ascribes it to other heavenly bodies. This was already not only a recognition of the heliocentric theory, but also a confutation of counter-arguments.

The contents and level of the theses of philosophy graduates in Jesuit schools were never normalized by strict curricula. There existed certain common general tendencies, but smaller or greater deviations were caused by the different scientific preparation and individual views of the various professors. This was why professors of eminent schools in Warsaw and Wilno saw no obstacles to the teaching of heliocentrism, while other schools more distant from cultural centres still felt the old objections. It may be said generally, however, that in the years from 1765 to 1773 (when the Society of Jesus was dissolved) the Jesuit schools in Poland came close to a full recognition of heliocentrism as the most probable, and most convenient hypothesis.

Simultaneously voices began to be heard among Polish Jesuits demanding the removal of the hypothetical barrier to the recognition of the Copernican theory.

There is no doubt that the most momentous and well-considered contribution was made in 1768 by Grzegorz Arakiełowicz, a professor of the Jesuit College in Przemyśl. His publication was a scientific treatise containing his own notes on the subject of the heliocentric theory. It was, therefore, of a higher standard than school lectures which are by the very nature of things more general and superficial[14].

Arakiełowicz considered heliocentrism to be no more than a hypothesis, but one that met all scientific requirements, that is, explained satisfactorily heavenly phenomena and was consistent with the principles of physics. It was, therefore, the most perfect hypothesis, superior to the obsolete theories of Ptolemy and Tycho, and had been accepted, therefore, by all scientists.

Arakiełowicz did not answer unequivocally the question of whether heliocentrism was inconsistent with the *Bible* or not, feeling no doubt dependent on the authorities of the Society and the Church. He asserted generally, however, questing the earlier opinions of Gassendi and Muratori, that the *Bible* was incompetent when it came to scientific disputes, and its fragments referring to heaven and nature could not be taken literally, nor used as arguments either in favour or against in scientific discussions.

Arakiełowicz's publication, though restricted no doubt in the formulation of the final conclusions concerning heliocentrism, taken together with the views of

---

[14] G. Arakiełowicz, *De mundi systemate dissertatio cosmologica qua de Copernicani systematis cum philosophia Sacrisque praesertim litteris quaestio discutitur*, Przemyśl, 1768.

other Jesuit professors, constituted an important stage in the controversy around the heliocentric theory in Poland. Textbooks and school lectures of that time contained information on Copernicus's theory as the best hypothesis. The Ptolemaic system was shelved, and the theory of Tycho Brahe began to be disqualified. Thus it was an intermediate stage in the reception of heliocentrism, which was indispensable for the subsequent leap from the prejudice of the forties to the complete recognition of the objective truth of heliocentrism in the seventies and eighties of the 18th century.

The subsequent period was marked by an increasingly comprehensive and affirmative introduction of the heliocentric theory into the Jesuit school curricula. A textbook by L. Hoszowski, for example, published in Lwów in 1766, enumerated heavenly bodies in the Copernican order. Without specially discussing this system, the author only pointed out that it results from the theory of gravitation.

The Copernican planetary system was taught by a geography manual written by a Jesuit father, Karol Wyrwicz, and published in Warsaw in 1768. A collection of examination questions and answers for students of the Warsaw College appeared in printed form in 1771. The section called *Physica particularis* contained also information on the theory of Nicolas Copernicus. It was said, inter alia, that the development of astronomy first pushed out the Ptolemaic system, and then discredited the theory of Tycho, "while almost all men recognize today the system of the Pole, Copernicus, as nearer to the truth than the others, and fully consistent with astronomical observations".

This shows that in the period immediately before the dissolution of the Society, the leading Jesuit schools were already teaching rudiments of the heliocentric theory. Scholars, on the other hand, were still weighing deeply the scientific values of heliocentrism and its contradiction of the "letter" of the *Bible*.

A graduate of the Jesuit College in Wilno, Jan Konarski, for example, published in 1769 philosophical theses in which he used scientific arguments in favour of a (universal, as he put it) recognition of the heliocentric theory. The Jesuit scholar accepted the principles of the new physics, but criticized the materialistic interpretation of natural phenomena, represented over the ages by Epicur, Toland and Covard. He himself, however, rejected constant intervention by supernatural forces, guiding himself by Newton's principle that the causes of natural phenomena, are what they really are, and that these causes are sufficient for explaining those phenomena; identical causes produce identical results which it is possible to determine. The discoveries of Kepler, Newton and Bradley confirming the movement of the earth proved that the Copernican hypothesis was the best of them all. The movement of the earth explained all heavenly phenomena so well and satisfactorily that it was unnecessary to search for other causes of these phenomena. Konarski consistently called heliocentrism a hypothesis, but nowhere did he even hint that it was inconsistent with the *Bible*.

On the other hand, the inconsistency was dealt with especially by a Jesuit author of a physics lecture at the beginning of the seventies of the 18th century[15]. He wrote inter alia as follows:

The *Holy Scripture* speaks of the phenomena of nature as the ordinary man understands them. This is why the causes and the origin of a given phenomenon can be explained solely by men conversant with physics and mathematics. The *Holy Scripture* speaks of a phenomenon as of a historical fact, the explanation of the manner in which that physical phenomenon occurred belongs to mathematics. The problem was very skilfully dealt with by Gassendi: God reveals himself to us in two ways, in the world of material nature by the experiments of mathematicians, in things revealed through revelation. And just as we were right in saying that a mathematician who might try to solve questions of faith on the basis of geometry transcends his competence, so, on the other hand, theologians and preachers ignorant of geometry and optics and pronouncing verdicts on natural phenomena exceed their powers. And were a man so dull as not to understand our arguments and so childish as to regard as impossible the fact that official interpreters of the *Holy Scriptures* do not understand some of its details, he should be advised after Kepler to leave astronomers in peace and go home, or even having condemned the astronomers' views should he wish to do so, go back to his own work, and having lifted heavenwards the eyes with which he looks at the Sun praise the Lord, our Creator with all his heart. He may be sure that he will do no less for God's glory than the astronomer who by the grace of God sees better, and can and wants to worship God for what he discovers.

With this one might end the remarks on the history of heliocentrism in the Jesuit schools in Poland. For this quotation distinctly shows that toward the end of their pedagogical activity the Jesuit schools overcame the religious objection with regard to the movement of the earth. Victory went to the tendency of Galileo and Muratori to treat the *Bible* as a text unauthoritative on questions of natural history.

As we have tried to demonstrate, Jesuit professors began to take an interest in the heliocentric theory in the early 17th century — that is, as soon as this theory became one of the most important and most controversial problems of science. At first, this interest was critical, and negative, but it popularized the main principles of the Copernican theory among several generations of students of Jesuit schools. As a result of a general drop in the standard of Jesuit schools in Poland in the latter part of the 17th and the early 18th century, the attitude toward the heliocentric theory began to change too late, together with the rejection of the obsolete scholastic learning and the acceptance of the fundamental elements of modern natural history, particularly the physics of Newton. The real achievement of the Jesuit schools in the reception of the heliocentric theory consisted in the recognition of the Copernican theory as the best of all hypotheses and the rejection of the prejudice against it arising from its formal inconsistency with the *Bible*. This achievement became the basis for the recognition of the heliocentric theory as an undeniable scientific truth in the schools of the Commission of National Education.

* *
*

The heliocentric theory was accepted at Polish Piarist schools in a different way and time. There are as yet no materials on the reception of the theory of Coper-

[15] MS in possession of the Ossolineum Library, No 2929/I. This piece was quoted in a Polish

nicus in those schools before 1746. This date has been accepted conventionally by W. Smoleński[16] and other scholars as the turning point in the history of Piarist schools in Poland. In the preceeding period, most probably Aristotelian doctrine, including the geocentric system of the universe, were taught in natural sciences. This opinion, however, still needs to be chacked, after a detailed study of historical sources.

Thus far, the publication of these discussing the main principles of the so-called recentiorum philosophy by a pupil of the Warsaw professor Antoni Wiśniewski, in 1746, was considered the turning point. The theses discussed principles concerning philosophy proper, as well as chemistry, physics, and natural sciences (biology). It was a complete revelation in the circles of Polish Catholic schools, none of which would allow a public discussion on and acceptance of the principles of a philosophy openly opposing the officially accepted, Aristotelian system. A Piarist, Antoni Wiśniewski, the actual author and propagator of the above views, was one of the pioneers of the Enlightenment trends in Poland. Together with Stanisław Konarski, he played an eminent role in the reform of Piarist schools and in introducing elements of modern natural science into the school curricula. The tenth item of the theses referred to dealt with the motion of the Earth. It was formulated as a reflection following an exercise for pupils during which they displayed what was called an armillary sphere showing the mutual position of the planets and the Sun, according to the Copernican system. The explanation stated that the adherents of the Copernican system had not managed to prove that the Earth at different seasons of the year, either approached or receded visibly to the senses, from a point in the universe. According to the senses, it seemed always to be at the same distance from the heavens. The Piarist thesis, however promptly retorted to this objection, most commonly used as an argument against the heliocentric theory. It held that the orbit of the Earth during its seasonal cycle is only a tiny speck compared with the space sprinkled with stars.

Continuing this trend, other pupils of Wiśniewski published in 1752 other theses, this time on modern philosophy and physics. They already polemized with the supporters of scholasticism from the Augustinian and Dominican orders. Items 106 and 107 were devoted to the heliocentric theory. They stated that the Ptolemaic and Tychonic systems had caused chaos in astronomy. At present, continued the theses, the only acceptable system was the hypothesis of Copernicus, so satisfactorily and easily explaining all the phenomena of the universe. The authors stressed that their choice had not been influenced by desire to honour the Polish nation, from which Copernicus originated, but rested solely on the scientific value of the hypothesis and its great credibility.

---

translation by S. Bednarski [in:] *Upadek i odrodzenie szkół jezuickich w Polsce* [The Decline and Revival of Jesuit Schools in Poland], Kraków, 1933, pp. 334–335.

[16] W. Smoleński, *Przewrót umysłowy w Polsce wieku XVIII* [The Intelectual Revolution in Poland in the 18th Century], Warszawa, 1923.

It is worth emphasizing that the Piarist theses of 1740 and 1752 made no reference to the inconsistencies of the heliocentric theory with the *Bible*, while, as we have seen, Jesuit works analyzed this problem extensively. The Piarists at that time already lectured the heliocentric theory, in a sense hedging the Rome decree, however, by defining the heliocentric system as a hypotheses, credible and the best. By comparison with other schools, the Piarists from the leading, reformed Warsaw college very early undertook the decisive step in the reception of the Copernican theory – from a complete negation to lecturing it as the best hypothesis. The Jesuits arrived at that point much later and following much more prolonged disputes. Such an early and firm acceptance of the heliocentric theory was the result of rejection of the Aristotelian system and scholasticism in favour of the system of recentiorum philosophy.

This concerns, of course, mainly the Warsaw Collegium Nobilium, which under Konarski and Wiśniewski became a real school of Polish Enlightenment. In provincial Piarist colleges, progress was neither so quick nor so unambiguous. In 1759 was published a textbook envisaged for those schools, containing instruction on the geocentric system.

There is a unique print of 1760 containing philosophical theses by a pupil of the Piarist Warsaw Collegium Nobilium, Kazimierz Biernacki. In content these did not differ from the theses of 1752, alreay referred to. It was then most probably an accepted principle repeated every year. Of major importance, therefore, is the conclusion that year by year opinions were repeated that Copernicus' theory was the most satisfactory and most probable hypothesis, which should, as being true, be accepted in the motherland of its creator, many foreign scientists having already considered it true.

In examination theses in Wilno (Vilnius) in 1762, there are also questions concerning the heliocentric system. Since the papers contained questions on the theory of the movement of the Earth, the problem must have been in the school curricula. The questions were as follows:

1. What is the principle of the system of Filolaos or the Copernican system, and does it explain all the phenomena of the universe? 2. What arguments based on the *Bible*, physics and mathematics are brought forward by the opponents of this system, and what are the answers to them? 3. What are the arguments in favour of stating, that the Earth is a planet? 4. Are there any other systems of the universe which would have the same astronomic and physical qualities as the Copernican system? The manner in which the questions were put makes it obvious that the author favoured the heliocentric system. It may certainly, therefore, be stated that the pupils of Piarist schools in Wilno, like those in Warsaw, were in 1762 systematically taught the principles of the heliocentric theory – more, they were prepared for polemics in its defence.

The process of accepting the heliocentric theory at Piarist schools was thus actually concluded at the beginning of the 16's of the 18th century. The fact that differences existed in some Piarist colleges cannot change the general opinion that Piarist schools were the first of all Polish Catholic schools definitely to introduce the heliocentric system into the school curricula. They also contributed greatly to the popularization of the heliocentric theory, publicly coming out in its favour from the mid-18th century, in popular calendars and journals, and organizing public demonstrations and discussions.

* *
*

The second half of the 18th century was marked by a considerable enlivening of scientific activities at Catholic schools in Poland.

A manuscript of school lectures on modern physics, dated 1757-1758, though without any note on the school it came from[17], may serve as an interesting example of this general trend. It certainly came from Catholic circles, and frequent quotations from a textbook on physics by a member of the Theatin order, G. M. della Torre, indicate that the lectures may have been held at the College of the Theatin order in Warsaw.

Those lectures on physics were based on the Great French Encyclopaedia and the textbook by della Torre. They accepted the Copernican theory as the most credible of the known theories, and refuted objections as to its inconsistency with the *Bible*. After the already mentioned Piarist theses of 1746 and 1752, which, however avoided discussion of the problem of inconsistency with the *Bible*, it in the second early document proving the acceptance of the heliocentric theory in school circles. It is at the same time an attempt to abandon the traditional, literal understanding of the *Bible*. The item concerning the heliocentric theory is so characteristic that it is worth quoting in full (in translation from Latin):

> ... Nicolas Copernicus, a Pole, the Canon of Warmia, born at Toruń on the 19th of February 1473 discovered (reperit) the structure of the universe, in the centre of which is a static Sun rotating only around its axis, around the Sun at various distances from it, orbit planets, the first being Mercury and the last Saturn. The system of Copernicus has been accepted by all the most famous universities and we consider it as the most credible and justified due to the following arguments: 1. The Copernicus' system explains all the phenomena of the universe and the movements of the heavenly bodies; 2. In the Copernicus system, constant stars are motionless, while in other systems such a rapid motion is attributed to them in such a great space that it is more probable that the Earth, more speck in the mass of stars, rotates at a much slower speed and in a smaller space; 3. In systems assuming motion of the Sun, two opposite motions are designated for the Sun and the planets — the motion specific to each of them from the West to the East, and the motion common to all, caused by primum mobile, from the East to the West. Two such opposite motions have never been observed in nature; 4. Other systems assume the motion of all planets except the Earth, which in weight is similar to them and is apparently as stable as they are.
>
> Arguments against the motion of the Earth are easily met: 1. If the Earth orbited, water would flow out of the wells. This argument can be unnihilated by stating the existence of the pressure of

[17] MS in possession of the Biblioteka PAN (Library of the Polish Academy of Sciences) in Kraków, shelf No 3869.

air surrounding the Earth and the water in wells; 2. If the Earth rotated, a stone thrown perpendicularly upwards would never fall on the place it was thrown from. One may reply to this that the rotation of the Earth makes the air over its surface rotate too, and the stone thrown up rotates together with the air and falls down perpendicularly. 3. Known verses for the *Bible* deny the rotation of the Earth, especially the words of Joshua, and on the basis of these verses the Rome Congregation of Cardinals banned in 1616 and 1633 teaching about the rotation of the Earth; this argument, again, may be refuted by explaining that in the *Bible*, as regards purely physical matters, not involving any dogma of faith, descriptions of physical problems are based only on perception and common ideas.

This excerpt undoubtedly proves that the author was convinced of the truth of the heliocentric theory, and with a great art and facility solved the problem, difficult for a Pole in those days, of reconciling this theory with the *Bible*.

The heliocentric theory was lectured consequently and according to programme in the first Polish lay school, the Warsaw School for the Nobility (Szkoła Rycerska) established in 1765. The up-to-date system of teaching introduced in that school envisaged a curricula covering all known achievements or mathematics and natural sciences, including of course the Copernican theory. The director of the School, Adam Kazimierz Czartoryski, laid special stress on the popularization of the heliocentric theory among the pupils. To this end he purchased in Britain at his own expense a "planetary machine" well known later in Warsaw, showing the orbiting of planets around the Sun[18].

By the 70's of the 17th century, all types of Catholic schools in Poland had finally overcome than reluctance towards the heliocentric theory. Therefore, schools established by the Commission for National Education established in 1773 was able from the outset to take the last step and lecture on the Copernican system as the true system of the universe.

When introducing natural history propositions into school curricula, the new educational authority was aware of their impact on world outlook and the possibility of opposition. This was why an instruction issued for teachers in 1774 recomended lectures on physics and biology without polemical and controversial accents, without defending or attacking individual views. The lectures were to consist in a concise presentation of propositions scientifically most probable. This is why the manuals and examination answers of the Commission schools do not contain any polemics or arguments either for or against — used so often in the monastic schools.

A concise physics manual published by J. M. Hube in 1783 does not evaluate any cosmological theories[19]. Writing of the theory of the motion of the Earth, the author did not even mention the name of Copernicus (perhaps precisely to avoid

---

[18] K. Mrozowska, *Szkoła Rycerska Stanisława Augusta Poniatowskiego (1765–1794)* [The Stanisław August Poniatowski College for the Nobility (1765–1794)], Warszawa, 1961, pp. 86, 222.

[19] J. M. Hube, *Wstęp do fizyki dla szkół narodowych* [Introduction to Physics for Public Schools], Warszawa, 1783, pp. 66–67. See J. Lubieniecka, *Przedmioty matematyczno-przyrodnicze w programie Towarzystwa do Ksiąg Elementarnych* [Subjects Dealing with Mathematics and Natural Sciences in the Programme of the Society for Elementary Text Books], [in:] Rozprawy z dziejów oświaty, vol. II, 1959, pp. 76–77, 83–84.

polemics). As regards the motion of the Earth J. M. Hube wrote as follows: "It is not to be supposed, however, that the heavens are in fact revolving... It is quite possible that the Earth revolves round its axis from West to East unceasingly and without stopping, and we do not perceive its revolutions, and it seems to us, therefore, as if the Sun and all stars revolved round the Earth from West to East continuously". The meaning of this fragment is quite obvious.

The examination questions printed in 1784 for elementary schools in Warsaw contain the following formulation: "The Sun seems so to run its daily round as if the entire heaven were revolving round its axis, which might be done, however, by the revolving of the Earth". For the elementary school standard, this is a good explanation of the seeming movement of the Sun. The formulation, couched in the conditional mood, leaves no doubt as to the intention of its author. Another examination question concerned the law of general gravitation causing the movement of all planets around the Sun.

A second manual on physics written by Hube appeared in 1792[20]. By comparison with the earlier one, many more problems were dealt with and better documented. Six consecutive pages, for example, were devoted to the problem of the Earth. The author accepted the heliocentric theory without any reservation whatever. Writing about scientists who had contributed most to the discovery and proof of the motion of the Earth, Hube mentioned Copernicus first and Bradley second.

This shows, then, that the last physics manual published before the partition of Poland spoke of the heliocentric theory without reservation or limitation, without mention of the notorious "hypotheses" or inconsistency with the *Bible*. This was eloquent proof of the triumph of enlightened thinking in the educational system. In the same spirit of the Polish Enlightenment did Hube interpret the achievements of mathematical and natural sciences. He emphasized very strongly that the new picture of the world worked out by scholars may be considered the best proof of the wisdom and goodness of God.

* *
*

The road of Polish protestant schools from rejection to acceptance of heliocentrism was shorter but equally difficult and important for the history of the national culture. These schools turned their attention to the heliocentric theory as early as did the Jesuit schools. But, more important it had in these schools its supporters and propagators, throughout 17th century, and it was fully and finally accepted in them several decades earlier (1722), than in the most progressive of Catholic schools in this respect the Piarist schools.

In examining the attitude of the Protestant schools to Copernicus and the helio-

---

[20] J. M. Hube, *Fizyka dla szkól narodowych* [Physica for Public Schools], part I – *Mechanika pierwszy raz wydana* [Mechanics Published for the First Time], Kraków, 1792, pp. 478–483, 536.

centric theory, we are compelled in the absence of source material to confine ourselves above all to the three most important high schools in Gdańsk, Toruń, and Elbląg. These schools were staffed by the best professors, frequently renowned scholars, and by the very nature of things they showed the greatest interest in the heliocentric theory, because only in those schools did the curriculum include a full course of philosophy, covering also natural history and astronomy.

In the latter part of the 16th century, the dissenter schools in Poland, like are other European schools, taught geocentric theory. This was inevitable, since only a small handful of scholars — astronomers and physicists — took an interest in the heliocentric theory at that time.

Pedagogues among the Polish Brothers took a keen interest in the heliocentric theory in the early 17th century. Joachim Stegmann, for example, the mathematics professor at the Arian school founded in Raków in 1602 is said to have been the first man in Europe to introduce in his lectures, in the years 1634—1638, the discoveries of Copernicus, Kepler and Galileo[21]. Martin Ruar, Rector of the Raków school in 1620—1621 corresponded later with M. Mersenne on Gassendi's work *De apparente magnitudine Solis* (1642). He conceded in one of his letters that he recognized the daily motion of the Earth, convinced by the arguments of Copernicus, Kepler, Galileo and Gassendi. Ruar criticized, of course, the tendency to submit the problem of the motion of the Earth to the judgement of the Roman Catholic Church which was completely incompetent in scientific matters[22]. The thoughts expressed in Ruar's correspondence were his private opinions, and we do not know whether they had been expressed earlier in his didactic activity.

The geocentric system was taught and propagated in manuals at the end of the 16th and the beginning of the 17th century by Bartholomew Keckermann (d. 1609), an eminent professor at the high school in Gdańsk. In his astronomy manual[23] Keckermann, writing of the geocentric system, mentioned Copernicus as an illustrious fellow countryman and world-famous scholar. Keckermann praised the detailed calculations of Copernicus and was indignant at J. C. Scaliger who had criticized the distances from the Earth to the Sun established by Copernicus. This statement by the Gdańsk professor is the first known document of the local cult of Copernicus in the Gdańsk school milieu.

All this produced a situation similar to that prevailing among scientists in Cracow: Copernicus was praised for his astronomical calculations and represented a source of pride even before attention was turned to the heliocentric theory. But the development of this starting situation followed different directions in the two places.

---

[21] Z. Ogonowski, *Arianie polscy* [Polish Brothers], Warszawa 1952, p. 102.

[22] L. Chmaj, *Wstęp* do: J. L. Wolzogen, *Uwagi do Medytacji metafizycznych Rene Descartesa* [Introduction to L. J. Wolzogen, Remarks on the Methaphysical Reflections of Rene Descartes], translated by L. Joachimowicz, Warszawa, 1959, pp. XXI.

[23] B. Keckermann, *Systema astronomiae compendiosum*, Hanoviae, 1611, pp. 77, 85, 99, 156, 188, 227.

In a treatise not intended for direct school use, Keckermann showed deep interest not only in the person of Copernicus but also in the theory of the motion of the Earth[24].

Keckermann accepted on basis of the authority of the *Bible* and Aristotle the immobility of the Earth as a scientific truth, as also its central position in the universe. This was already a firm statement within the controversy on the heliocentric theory, a controversy whose existence was fully appreciated by the Gdańsk scholar. Guided by sympathy for Copernicus (which he distinctly emphasized), a compatriot and a great scholar, Keckermann attempted a specific defence of him. He tried to convince his readers that Copernicus, a man of great genius and tremendous store of knowledge, could not have considered the motion of the Earth as the truth but only as a working hypothesis of assistance in his calculations. Those who charged against Copernicus that his theory was absurd should remember that Ptolemy's hypotheses had been also considered absurd. Copernicus said that the Earth had three movements, which was contrary to the view of Aristotle to the offer that everybody can have but a single movement. But Ptolemy also sinned against this; he granted two movements — for example, to the Moon and Mercury.

Keckermann concluded his reflections on the subject of the Copernican theory as follows: "Everything indicates that the movement of the Earth as assumed by the greatly famed Copernicus must not be regarded as a realistic term but as an astronomical working term. As a result of it, Copernicus won the recognition of the most illustrious scholars, because starting from wrong premises he proved better that something was true than Ptolemy and Alfonsian astronomers who had started from correct premises". Keckermann remained silent on the system of Tycho Brahe, pleading lack of competence and leaving the evaluation of his hypothesis to astronomers.

The entire statement of Keckermann should without any doubt be treated as an attempt at defending Copernicus against the opponents of the heliocentric theory. From the modern point of view, this appears harmful because it deprived Copernicus of his most valuable discovery. But in those times such a defence was able to protect the astronomer from the condemnation of theologians of different denominations and Aristotelian schoolars, who in fact included Keckermann himself. The declaration that the heliocentric theory was only one of the astronomical hypotheses enabled it to be referred to and commented upon (which, even without its being accepted contributed to its propagation) without discriminating against Copernicus as a scientist. Keckermann's activity should be considered, therefore, in the light of the actual situation as objectively conducive to the acceptance of the heliocentric in its early stage, and initiating at the same time an ideologically and educationally important local cult of Copernicus.

---

[24] B. Keckermann, *Contemplatio gemina, prior ex generali physica de loco altera ex speciali de Terrae motu*, Hanoviae 1958, 1607.

The next major step in the process of acceptance of the Copernican theory in the protestant schools was made by Peter Krüger. He opened in fact a new phase in the history of the reception of the heliocentric theory by publishing in the printed form the fundamental principles of the heliocentric system with their partial approval.

Peter Krüger (1580–1639), Keckermann's successor in the chair of mathematics and astronomy at the high school in Gdańsk, a student of Kepler and a professor of Hevelius, was one of the most outstanding Polish astronomers of the 17th century, enjoying great scientific and pedagogical prestige. Through the evolution of his astronomical views, he reached in time the full Copernican heliocentric theory, which he expressed in his *Prognostics for 1631*.

Krüger was educated on Aristotelian physics, and slowly and with great difficulty changed his views. As far back as 1615, he had a dispute with his pupil, J. Gerhard, in which he regarded the Earth as immobile around its axis also circling the Sun[25].

The publication of the 1615 dispute was as far as we know the first statement printed in Poland broadly explaining and partly also propagating the heliocentric theory as expounded by Copernicus. The text of the dispute was contained in 65 successive theses for the description of the "nature" of the heaven and the Earth, understood according to the principles of Aristotelian physics. Subsequently, the author discussed in turn the systems of Ptolemy, Copernicus, and Tycho. The presentation of the Copernican system was accompanied by a drawing of the heliocentric system and by the two most frequent charges against the heliocentric theory: incompatibility with Aristotelian physics and inconsistency with the *Bible*. Particularly interesting was thesis 64 which was a summing up of the argumentation. Translated from Latin its conclusion runs as follows: "One cannot do other than regard this residence of ours – that is, the Earth – as immobile around its axis. But contrary to the third thesis of Tycho (preserving his system of the universe and astronomical calculations), together with the Pythagoreans, Heraclides and Ekfantos, one must recognize the Earth as moving in a circle in a way which makes the stars seen from it appear to rise and to set".

This is a somewhat fantastic combination: the Pythagorean idea of the Earth circling around a central fire (explaining solely the day-and-night phenomena) and the up-to-date system of Tycho. But this seemed to save the principle of Aristotelian physics to the effect that everybody can have but a single movement. Although Krüger did not yet accept the full Copernican heliocentrism with its three movements of the Earth, he made a big step towards such acceptance. The concessions in favour of Aristotelianism and the system of Tycho are fully understandable when one takes into account the fact that it was after all a dispute about school programme, treated at the same time as a demonstration for the outside world, and that it occurred in 1615 when professors could hardly depart completely from the almighty authority of Aristotle. Even so, the very fact that a professor could suggest such a dispute

---

[25] *Disputatio de hypothetico systemate caeli*, Gdańsk, 1615.

at all and obtain the permission of the school board and the borough council to print, it shows the very great interest taken in Gdańsk in modern natural history, especially astronomy. No public announcement of a thesis admitting any movement of the Earth whatever was possible in any Catholic Polish school in 1615.

Krüger took a serious interest in the theory of the movement of the Earth and began about 1623 a thorough study of *De revolutionibus*. A copy of it which was once in his possession and kept now at the Copernicus Museum in Frombork bears many marginal notes in the owner's handwriting, showing his approval of the heliocentric theory[26].

In 1931, Krüger published a collection of prognostics for the individual heavenly bodies for the years 1617—1631[27]. In the 1631 prognostic, he wrote of the heliocentric theory, devoting the whole of Chapter IV to deliberation as to whether the Earth revolves or the Sun. According to Krüger, the observations of Galileo, confirmed the theoretical theses of Copernicus. The scholar perceived that the acceptance of the heliocentric theory opened up tremendous new prospects for astronomy. On the basis of the heliocentric theory one may assume the existence of many other worlds and possibly confirm these assumptions by observations. Krüger fully realized the consequences of such a view for world outlook, since he closed his remarks with: "I dare not say here what might be the result of the assumption of a multitude of worlds". This significant formulation says much for the scientific thoroughness of Krüger and his spiritual distraction; his own deeply rooted convictions, sense of responsibility and partly perhaps also his post as professor in a Lutheran religious school forbade the publication of conclusions challenging the authority of the *Bible* and religious views on the divinity of a single heaven and the uniqueness of a chosen Earth. Small wonder then that in his lectures and manuals Krüger, though saying a great deal about Copernicus, accepted as binding the system of the heaven according to Tycho.

Still in Krüger's lifetime, another professor of mathematics in Gdańsk, Peter Lossius, held with one of his pupils a dispute about the "heaven", in which he criticized the heliocentric theory[28]. The criticism was based on the typical arguments about inconsistency with the *Bible* and the testimony of the senses, without even any attempt at a scientific discussion. Krüger's views did not take root for the time being at the high school in Gdańsk. But they did not remain without an echo in the long run, because one of Krüger's pupils, for example, Benjamin Engelcke, who

---

[26] T. Przypkowski, *Notatki astronomiczne Piotra Krügera, nauczyciela Jana Heweliusza, na egzemplarzu „De revolutionibus" Mikołaja Kopernika* [The Astronomical Notes of P. Krüger, the Teacher of J. Hevelius, on His Copy of the *De revolutionibus* by Nicolaus Copernicus], [in:] Sprawozdanie z czynności i Posiedzeń PAN, 1949, No 10, pp. 607—609.

[27] The works were published under the title *Cupediae astronomicae Crugerianae*, Wrocław, 1631.

[28] *Disputatio physica de caelo*, Gdańsk, 1636. Items XVII and XIX concerned the heliocentric system.

travelled to Italy in 1632 met Galileo, discussed with him the question of the movement of the Earth and received from him a copy of the Italian original of the work *A Dialogue on the Two Systems of the World.* Engelcke subsequently took the book to Vienna and persuaded an Austrian scholar, Matthew Bernegger, to translate it into Latin. Together with some minor works of Galileo and Foscarini, the Latin translation appeared in Strassbourg in 1636. Krüger's ideas were consistently developed by his pupil, the Gdańsk astronomer John Heweliusz (1611–1687).

Krüger was succeeded in the chair of mathematics and astronomy in the Gdańsk high school by a physician, Wawrence Eichstadt (d. 1660). He held lectures on astronomy on the basis of manuals written by Keckermann and Krüger, which means that he did not as yet teach heliocentrism. But there are grounds for supposing that Eichstadt himself recognized the motion of the Earth, and based on the heliocentric theory his calculations of the position of planets in various years — that is, what are called prognostics the calculation of which was among the duties of the professor of astronomy at the high school in Gdańsk. His pupils regarded Eichstadt as a heliocentrist. The same was held by his friends writing after the death of this very popular Gdańsk pedagogue. Moreover, the private note book of one of his pupils contains a sketch of the heliocentric system[29]. Also in existence is a printed notice announcing the astronomy lectures which Eichstadt intended to held in 1648. The notice bore all the marks of an advertisement, and it is not known, therefore, whether the plans were real and actually carried out. Eichstadt pointed in his notice to the usefulness of the astronomical science — the possibility of determining the geographical coordinates, familiarity with the calendar, orientation according to the stars. He announced lectures on the ancient science on epicycles, excentrics, etc., and "... instead of the movement of the stars with a tremendous velocity from the East to the West. I shall describe the daily revolutions of the Earth from the West to the East... I shall also explain the planets surrounding Jupiter and discovered by Galileo... The pupils will hear how changes in new stars and comets abolish the Aristotelian immutability of the world... In its proper place in the theory of the planets, I shall explain in what manner one should use the subtle hypotheses of Copernicus, a great pride of astronomy, and what to think of the application of the magnetic theory of William Gilbert..." If Eichstadt really delivered all these lectures, that would mean that as early as 1648 there were courses on the principles of modern astronomy, together with heliocentrism, though the latter was treated as a "subtle hypothesis". Even if professor did not succeed in the implementation of such an ambitious programme, it is significant that the plans were so formulated.

In Eichstadt's time, there were in Gdańsk no school textbooks accepting heliocentrism. There is no doubt, nevertheless, that the school milieu felt sympathy for Copernicus, as well as understanding and acceptance of the heliocentric theory

---

[29] MS in the possession of the Polish Academy of Sciences Library in Gdańsk. Shelf No 673 card 32v.

though such were not demonstrated too openly at school. These tendencies are expressed in the activity of John Heweliusz, a famous Gdańsk astronomer and a friend of Eichstadt. Heweliusz was of the opinion that the theory about the movement of the Earth excellently explained all heavenly phenomena, and was in no way contrary to common sense, and that it was accepted for this reason by nearly all outstanding mathematicians[30]. But even if the heliocentric theory used to be explained at the Gdańsk high school in mid-17th century, these lectures were of a sporadic nature depending on the individual views of the different professors.

Conservative professors deeply attached to the Aristotelian system, and fanatic Lutheran clergy perceiving in this theory a complete incompatibility with the *Bible*, constituted the obstacle to a full introduction of the heliocentric theory in the curricula and school textbooks in Gdańsk, and Toruń. For example, Daniel Lagus, a professor of philosophy in Gdańsk, published in 1650 a collection of disputes[31], in which he discussed at length the physics of Aristotle. A school textbook of astronomy[32] also appeared in print in Gdańsk in 1651. The textbook contained just a lecture on the geocentric theory, without any mention, even a polemical one, of the heliocentric system. But this conservative trend in the school life of Northern Poland was disturbed more and more frequently by enlightened professors. Two treatises, for example, very important in the history of the acceptance of the heliocentric theory in Poland, were published in Elbląg in 1652/53. The publications are connected with the name of Henry Nicolai (d. 1660), a professor of philosophy at the high school in Elbląg. Nicolai took an interest in the progress of modern science and declared himself as an opponent of the authority of Aristotle. He supported the Polish Brothers and was constantly attacked by Lutheran theologians. The first of the treatises is a typical school dispute signed by the professor and a pupil[33]. The second treatise, on the other hand, is signed only with the name of Nicolai and is clearly intended to be a scientific treatise[34]. The first publication is called in the title "constructive" (*adstructoria* — that is, gathering proofs of the motion of the Earth). The second is called by the author "destructive" (*destructoria ac refutatoria*), since it contains arguments against the motion of the Earth and the replies to these charges

---

[30] J. Heweliusz, *Selenographia sive Lunae descriptio*, Gdańsk, 1647, pp. 163–164.

[31] D. Lagus, *Theoria astronophica mathematico-physica e praelectionibus publicis dedecade disputationum repetita*, Gdańsk, 1650.

[32] N. Kauffman, *Cosmographia sive descriptio caeli et Terrae in circulos*, Gdańsk, 1651.

[33] *De Quotidiana Telluris revolutione... disputatio in gymnasio Elbingensi proposita praeside Henrico Nicolai... respondentis vices garente Fausto a Raciborsko Morstinio Equite Polono*, Elbląg, 1652. There is a copy in the possession of the Polish Academy of Sciences Library in Gdańsk, No Ab 23.8. See T. Przypkowski, *Zainteresowania matematyczno-astronomiczne Braci Polskich* [The Interest of the Polish Brothers in Mathematics and Astronomy], [in:] *Studia nad arianami* pod red. L. Chmaja [Studies on the Polish Brothers edited by L. Chmaj], Warszawa, 1959, pp. 407–410.

[34] *De quotidiana Telluris revolutione. Exercitatio peculiaris philosophica, mathematica et theologica*, Elbląg, 1653. There is a copy in the Polish Academy of Sciences Library, Gdańsk, shelf No Fa 123 adl. 8.

collected by Nicolai. Both treatises taken together constituted a kind of anthology of arguments for and against the motion of the Earth, a concise encyclopaedia of knowledge on this subject. Nicolai was no astronomer, but neither did he treat heliocentrism as a purely astronomical problem. Both treatises show that their author was a widely read and erudite man, both are written with a polemic passion and ideological dedication stemming from an understanding of the consequences in methodology and world outlook of the controversy about the heliocentric theory.

Prepared under the guidance of Nicolai, the Morsztyn dispute began with the enumeration of over 50 authors, beginning with Melanchton, who rejected the theory on the motion of the Earth for religious reasons. This was followed by a long list of scientists and philosophers, beginning with Plato and the Pythagoreans, who conceded the Earth some sort of movement. The place of honour among the latter was accorded to Copernicus, although Osjander wrongly denied him, according to the authors, full conviction of the truth of his system. Supporters of the theory on the motion of the Earth included the Gdańsk professors – Krüger and Eichstadt[35]. Further on, Morsztyn analyzed the differences of views among the supporters of the theory on the motion of the Earth. Following in Copernicus's steps, some conceded movement only to the Earth, and immobility to a centrally positioned Sun and the stars. Others, called "semi-Copernicans", considered that the Earth, the Sun and the stars all move.

The authors of the dispute stood by this last opinion. Thus Morsztyn gives the following conception of the cosmological system: the Earth performs a natural revolution around its axis and is in the centre of a stellar sphere. The Moon and the Sun revolve around the Earth in an orbit. The five planets revolve not around the Earth but around the Sun. This peculiar combination of the Copernican system with the system of Tycho, rightly called by contemporaries the semi-Copernican system ("systema semi copernicanum"), was the work of a Danish astronomer, J. Ch. Longomontanus (d. 1647), a pupil and creative continuator of the work of Tycho Brahe. The Longomontanus system, accepting from the Copernican theory only one day-and-night movement of the Earth, made great concession sto Aristotelian philosophy and the *Bible*.

Nicolai's treatise is a continuation and supplementation of the dispute of Morsztyn. It is confined to refuting earlier arguments against any motion of the Earth, irrespective of its being a day-and-night or yearly motion. That is why from the point of view of the acceptance of the heliocentric theory in Poland, Nicolai's treatise is more important than the Morsztyn dispute. Perhaps most interesting was the foreword to the treatise. Nicolai wrote in it that the main opponents of the theory on the motion of the Earth are uneducated people believing only in the testi-

[35] Morsztyn quoted the sentence by Krüger concerning the motion of celestial bodies: "The day is made up of the motion of the Earth with an addition of the movement of the Sun, the month is made up of the movement of the Moon, and the year and its four seasons of the motion of the Sun".

mony of the senses and the authority of the *Bible*. According to Nicolai, the authority of the *Bible* on problems of nature may be countered with essential arguments, while the voices of ignoramuses based on sensory perceptions alone may be disregarded in this respect. Both assertions were expanded by Nicolai in the text of the treatise. Refuting "biblical" arguments against the motion of the Earth, the author said that the *Bible* used pictures understandable for simple people, unacquainted with astronomy, and that was why the *Bible* could not be an authority in the field of astronomy and physics. Continuing in this trend, Nicolai added that the opinion of theologians could not be decisive on mathematical questions.

The same page of the treatise[36] contains another sentence very essential methodologically, namely that the authority of Aristotle on mathematical questions was non existent, and little attention need be paid to it in questions of physics.

Nicolai was, therefore, probably the first Polish scholar to combine elements of criticism of the authority of the *Bible* with elements of criticism of Aristotelianism in questions of natural sciences. It is not difficult to understand, therefore, that it was the milieu of protestant schools in Poland that was the first to create conditions for an early acceptance of the new science on nature, including also the Copernican theory, or at least its fundamental assumption of the motion of the Earth.

Protestant circles in Western Europe had already popularized the idea of the alegoric interpretation of the *Bible*. It found, therefore, numerous supporters among Polish Protestants, if only as a form of opposition to Catholics. It was much more difficult to overcome obstacles resulting from the deeply rooted views of the physics of Aristotle. This was why that semi-Copernican system with its considerable concessions, alone all to Aristotelianism, became in this milieu an intermidiate stage in the reception of the heliocentric theory. In Polish Catholic schools, the objections to the heliocentric theory assumed different proportions: Aristotelianism was rejected earlier than the literal interpretation of the *Bible*, and the semi-Copernican stage became unnecessary.

The conviction of the existence of a motion of the Earth, though not always in conjunction with the entire Copernican system, was even stronger among professors of Protestant high schools in the latter part of the 17th century. For example, the theses of a pupil of a Gdańsk professor, Frederic Büthner, published in 1660, contained the assertion that right lay with those astronomers who saw the causes of the phenomena of night and day in the motion of the Earth around its axis. Büthner was most probably still a semi-heliocentrist, but two professors of the high school in Toruń – Henry Schaeve and John Meier – proved themselves to be consistent heliocentrists.

Taking over the post of Rector of the Toruń high school in 1660, Schaeve made a speech in which he warmly extolled the pride of Toruń – Nicolas Copernicus. Ever since that time, praise of Toruń in the Rectors' inauguration

[36] *De quotidiana Telluris revolutione...*, p. 63.

speeches was always combined with praise of Copernicus. The custom had also, of course, its obvious propagating and popularizing, and even scientific, repercussions.

In 1661, Schaeve published six pupils' disputes held under his guidance at the Toruń high school. The disputes pertained to various fields of science, with special reference to the latest achievements of modern mathematical and natural sciences[37]. The fourth dispute was devoted to astronomical problems. In it Schaeve expressed full support for the views of Copernicus and Galileo, while severely criticizing the Aristotelian and scholastic science about the heavens.

The dispute was divided into 12 questions, dealing with all the most important astronomical discoveries of the 16th and 17th centuries. The Copernican system, enriched by subsequent discoveries, was recognized in it as the only scientifically authoritative system.

When compared with the publications of Krüger and Nicolai, the above dispute marks further essential progress in the reception and popularization of the heliocentric theory. For the pupils of Schaeve were already assimilating the principles of the entire Copernican system enriched by the observations of Galileo and the arguments of Descartes. The novelty of the dispute lay also in its opposition to astrology, since it ridiculed, for example, such views as that the natural heavenly phenomena (such as the eclipse of the Moon) may have an effect on human destinies. But despite the value of statements of this kind, one must not forget that they were only sporadic, made possible to the most progressive professors by the differentiation of curricula in Protestant schools, and a comparatively greater elasticity of programmes and scientific tolerance. For dissenter schools still continued to use textbooks teaching the geocentric system, while conservative professors still gave their pupils for preparation theses based on geocentric principles[38].

As far as is known, the first documented lectures on the principles of the heliocentric system in Poland were introduced in 1676 by a professor of the high school in Toruń — John Meier. The perfectly preserved complete MS of the lecture enables us to follow the sources and the reasoning of the author. The lecture shows that Meier accepted the most important principles of the Copernican theory: the central position of the Sun, the course of the planets (including the Earth) around the Sun, the order of the planets from Mercury to Saturn. He accepted also the astronomical calculations of Copernicus, later brought up to date by the calculations of other astronomers, including also Kircher and Riccioli.

Meier based the entire conception of the lecture on Galileo's *Dialogue on the Two Most Important Systems of the World*. As a concession to Aristotelian philosophy, Meier has in fact only a single movement — a progressing annual movement around

---

[37] H. Schaeve, *Dissertations pansophicae ad modum Ianuae aureae Comenii sex*, Toruń, 1661.

[38] In 1684, for instance, Christopher Hartknoch, later a historian of Pomerania, published at Toruń a physics thesis arguing the immobility of the Earth. He used arguments both of reason and the authority of the *Bible*.

the Sun. Meier drew correct cosmological conclusions from the heliocentric theory: he accepted the possibility of the existence of a multitude of worlds (*pluralitas mundorum*) with systems similar to the Solar system. In support of his suppositions, he included a relevant, lengthy quotation from the *Principia philosophiae* of Descartes. For a professor holding very firm religious views (as may be seen from the MSS of his lectures on theology), the heliocentric theory was just another proof of the greatness of the Creator. Meier wrote in his astronomical lecture, ended with a comprahensive justification of the acceptance of the heliocentric theory (mainly on the basis of Galileo's arguments): "This globe of amazing size, though greater have been created by the might of God, freely suspended in the air, has been revolving unceasingly for ages"[39].

The last two and a half decades of the 17th century passed in Protestant schools in an atmosphere of increasing interest taken in modern science (including also heliocentrism) and an ever greater and more strongly built up cult of Copernicus.

In 1688, for example, a professor of the high school in Toruń, Paul Peter, gave a public lecture about the stars. The lecture is known solely from a printed invitation card which emphasized that the lecturer would speak of "the incomparable Torunian, Nicolas Copernicus". The name of Copernicus served thus as a publicity argument, no doubt an effective one. Three years later (1691) the same Professor Peter staged at the school theatre a play ridiculing astrology and alchemy, and extolling useful sciences, such as astronomy, geography, natural history, etc. In the play, astronomy was represented by Copernicus (his part was played by a Toruń pupil, Daniel Schiller). The figure of Copernicus symbolized all the qualities of astronomy. This was not, therefore, any dispute between heliocentrism and geocentrism. Copernicus was presented as the greatest scholar in the finest field of science.

Another proof of the interest taken by the Protestant milieu in the heliocentric theory and of excellent orientation in the most modern studies were three disputes held in 1702 by pupils of the high school in Gdańsk. They presented with great precision arguments both for and against the heliocentric theory, but did not take sides themselves.

It may be said, however, with complete certainty, that geocentrism was still taught in Toruń at the beginning of the 18th century and this knowledge was exacted during examinations. Known, for example, is a 1704 publication containing examination theses and questions for several years back[40]. Physics theses from Toruń contained in 1701 one proclaiming that: "The Earth constitutes the immobile centre of the universe". As arguments used were the identical distance at all points from the

---

[39] The lecture is entitled: *De ordine, magnitudine et motu maiorum mundi partium*, MS, property of National Library, shelf No BN 3190/IV.

[40] *Cursus disputatorius theses ex logica, physica et metaphysica CCCCVIII complectens praeside M. Mariino Bertlefio a supremae classis auditoribus in gymnasio Thoruniensi hactenus confectus*, Toruń, 1704, pp. 68–69, the thesis: *Terra est centrum universi immobile, and arguments in favour of it.*

Earth to heaven, the *Bible*, the testimony of the senses which feel the immobility of the Earth, and the lack of a convincing proof of any motion of the Earth. This arsenal of geocentric arguments rolled out at the beginning of the 18th century marks no doubt a great regression by comparison with the previously quoted pro-heliocentric arguments. It constitutes, however, just one more proof that the Protestant schools of that time did not have any stable, binding modern physics curriculum. Every professor expounded simply his own views and knowledge. What was important was not so much the sporadic statements for or against as the general trend. And this was no doubt more progressive than in the Catholic schools of the same period. In 1715, for example, a Toruń professor, F. Bormann, held with one of this pupils a comprehensive dispute on the subject of the theory of a multitude of worlds. Bormann accepted as the basis the Copernican heliocentric theory and the Cartesian theory of the multitude of worlds. This hypothesis, too, was contrary to the *Bible*: to certain parts of the *Old Testament* and the whole idea of the *New Testament*. In the 1715 dispute, Bormann was considering solely the possibility of reconciling the two theories with the *Bible*. Two years later, however, the Toruń professor published his own theses concerning the multitude of worlds, in which he made a distinct division between the competence and authority of the *Bible* and scientific questions. It may be supposed that this stand took a stronger hold in the Polish Protestant school system and opened the way to the introduction of heliocentrism and other achievements of modern natural sciences in the curricula and textbooks.

This new and at the same time final stage of the reception of the heliocentric theory is connected with the name of John Arndt, a graduate and later professor of the high school in Gdańsk, who published a textbook on elements of astronomy, teaching without any reservations (and without any reference to opposition systems) the heliocentric theory[41]. This was as far as we know the first textbook on principles of the heliocentric theory intended for the Polish Protestant schools, and at the same time the first textbook of this type in the Polish lands. Arndt reached his heliocentric principles slowly. As a graduate of the high school in Gdańsk, he published in 1707 theses on mathematics and physics. One of them concerned the theory of the motion of the Earth. Arndt admitted in it that heliocentrism had great scientific qualities and was not at all contrary to common sense. It was contrary, on the other hand, to the *Bible* and this made its acceptance difficult, for in his opinion the text of the *Holy Scriptures* could not be interpreted alegorically or its incompetence on scientific questions be admitted. As a mature man, Arndt retained his scientific convictions, but changed his views as regards the interpretation of the *Bible*. His textbook published in 1722 not only taught the heliocentric theory and other axioms of modern astronomy, but also criticized all the earlier objections to the heliocentric theory.

[41] J. Arndt, *Collegium astrognosticum in quo doctrina de caelo huiusque corporibus et systemate mundi secundum oculum, artem et ratiocinationem perspicuis aphorismis scholiisque exponitur*, Królewiec, 1722.

He taught, among other things, that the *Bible* described the heavens in the way a man sees it with his eyes. Later astronomers, Nicholas Copernicus among them, used scientific methods (*ars*) and reason (*ratiocinatio*) and gave a true picture of the heavens. The *Bible* could not, therefore, be considered, when it came to scientific questions.

Arndt's textbook really closes the period of controversies about heliocentrism in the Polish Protestant schools. As a printed book, published in a large edition and, moreover, produced in the form of a manual, it was by the very nature of things to have a stronger and deeper impact than the ephemerial theses of various professors and their pupils, or even a systematic but sporadic lecture of J. Meier. In later years, the Polish Protestant schools used an astronomy and physics textbook of Chrystian Wolff, teaching the heliocentric theory without any reservations whatever. In 1730, lectures on the heliocentric theory had already a long tradition and were delivered by a specialized physicist[42]. Nobody was surprised, therefore, in Elbląg in 1744 by a brochure of the professor of mathematics in the local high school, Jakub Woit, which contained explanations for a model of the heliocentric system placed on public view there. This was but anothter and at the same time extremely eloquent proof of the full acceptance of the heliocentric theory by the Polish Protestant schools.

* *
*

The controversy about heliocentrism in the Protestant schools ended in full recognition of the new theory at approximately the same time when Catholic schools in Poland had not even reached the decisive phase of the fight. The earlier triumph of the Copernican system in the Protestant milieus was due in large measure to the greater liberalism of the more enlightened part of Protestant theologians on questions of allegorical interpretation of the *Bible*, a higher level of the teaching of mathematical and natural subjects, the cult of Copernicus, remarkably expanded and nurtured in the northern lands of Poland.

In the process of the acceptance of the heliocentric theory, Protestant schools had to overcome the same opposition as the Catholic schools: to overcome the conviction of the truth of the Aristotelian physical laws and the necessity of compability of scientific theories with the *Bible*. The Protestant professors, scientifically traditional, like the Catholics, but unhampered by the Roman Decree, had a longer fight with the objections of Aristotelian physics than with the literal interpretation of the *Bible*. This was why the semi-Copernican system with its concessions in Aristotelian philosophy became in this milieu an indispensable stage and a starting point for the acceptance of the heliocentric theory. In Catholic circles, on the other hand, objections connected with physics were removed comparatively earlier than ideolog-

---

[42] See, for instance, *Catalogus lectionum et operarum publicorum in Athenaeo Gedanensi cursu annuo expediendarum*, Gdańsk, 1730.

ical objections, though later, too, than in Protestant circles. This was why the semi-Copernican system did not play any role here, and the entire fight went on for the "hypothesis". None the less, complete triumph for the heliocentric theory became possible in both milieus only after Aristotelian philosophy had been broken through and the recognition of the *Bible* as competent on scientific questions set aside. The essence of the controversy was identical, the only difference lay in the centres of gravity and the historical rhytm of the transformations taking place.

BIBLIOGRAPHY

Bednarski S., *Upadek i odrodzenie szkół jezuickich w Polsce* [The Decline and Revival of Jesuit Schools in Poland], Kraków, 1933.

Bieńkowska B., *Kopernik i heliocentryzm w polskiej kulturze umysłowej do końca XVIII wieku* [Copernicus and Heliocentrism in Polish Culture till the End of the 18th Century], Wrocław, 1971, [in:] *Studia Copernicana*, vol. III.

Birkenmajer A., *Osiągnięcia duchowieństwa polskiego w zakresie nauk matematycznych i przyrodniczych* [Achievements of the Polish Clergy in Mathematics and Natural Sciences], [in:] Roczniki Filozoficzne Towarzystwa Naukowego KUL, vol. XII, 1964, No 3, pp. 31–43.

Brożek J., *Wybór pism* [Selected works], vol. I, edited by H. Barycz, Warszawa, 1956.

Chmaj L., *Kartezjanizm w Polsce w XVII i XVIII wieku* [Cartesian Ideas in Poland in the 17th and 18th Centuries], [in:] Myśl Filozoficzna, 1956, No 5, pp. 67–102.

Gdańskie Gimnazjum Akademickie. Księga pamiątkowa [The Gdańsk High School. Memorial Book], Gdynia, 1959.

Lipko S., *Kopernikanizm w szkołach polskich XVIII wieku* [The Copernican Theory in Polish Schools in the 18th Century], [in:] Komentarze Fromborskie, Olsztyn, 1965.

Lubieniecka J., *Przedmioty matematyczno-przyrodnicze w programie Towarzystwa Ksiąg Elementarnych* [Mathematics and Natural Sciences in the Curricula of the Society for Elementary Text Books], [in:] Rozprawy z Dziejów Oświaty, vol. II, 1959.

Nadolski B., *Walka o myśl Kopernika i jej losy w Polsce* [The Struggle for the Idea of Copernicus and its history in Poland], [in:] *Wkład Polaków do nauki. Nauki ścisłe*, Warszawa, 1967.

Polkowski I., *Kopernikijana czyli materiały do pism i życia Mikołaja Kopernika* [Materials on the Works and Life of Nicolas Copernicus], vol. I–III, Gniezno, 1873–1875.

Przypkowski T., *Dzieje myśli kopernikańskiej* [The History of the Copernican Theory], Warszawa, 1954.

Przypkowski T., *Z dziejów heliocentryzmu w Polsce* [On the History of the Heliocentric System in Poland], [in:] Myśl Filozoficzna, 1953, No I, pp. 176–190.

Rybka E., *Four Hundred Years of the Copernican Heritage*, Kraków, 1964.

Smoleński W., *Przewrót umysłowy w Polsce wieku XVIII* [The Intellectual Revolution in Poland in the 18th Century], second edition, Warszawa, 1923.

Suchodolski B., *Nauka polska w okresie Oświecenia* [Polish Science in the Enlightenment], Warszawa, 1953.

KRISTIAN PEDER MOESGAARD
University of Aarhus

HOW COPERNICANISM TOOK ROOT IN DENMARK AND NORWAY

I. EARLY CONFRONTATIONS

Dybvad's Commentary on Copernican spherical astronomy (1569)

During the period 1578–1590 when at Hveen Tycho Brahe laid the foundations of modern observational astronomy, the chair of mathematics, including astronomy, at the University of Copenhagen was held by *Jørgen Christoffersen Dybvad* (Dibvadius; died 1612), who for several years had studied at Wittenberg and Leipzig. As early as 1575 he obtained an extraordinary chair of theology, natural philosophy, and astronomy, crowning his career in 1590 by taking over the chair of theology. Being a quarrelsome and unsociable character, he was dismissed in 1607.

Dybvad is the earliest Danish author known to have dealt explicitly with Copernican astronomy. His teachers had been Caspar Peucer and Sebastian Theodoricus in Wittenberg [*27*: fol. A 6v, E 1v][1], and in 1569, as an outcome of his studies, he prepared a book with the interesting title: *Short Comments on Copernicus' 2nd Book which by unmistakable arguments prove the truth of the doctrine of the first motion and show the composition of the tables* [*27*].

Its dedicatory preface presents Copernicus' 2nd Book as perfectly illustrating the necessity of founding scientific studies on geometry. "For you can scarcely understand one single chapter unless by using considerations and methods from the doctrine of triangles which alone unveils and exhibits all the secrets of astronomy". And astronomical science as such presupposes three stages, viz. observation, hypotheses, and the explanation of the latter by the aid of geometry. It is stated that the 2nd Book of the *De revolutionibus* falls in two parts, namely the chapters 1–13 in which Copernicus, in due succession and by using the most manifest method, explains and proves the whole doctrine of the first motion, and the 14th

[1] See the combined *Bibliography* to the essays *Copernican influence on Tycho Brahe* and *How Copernicanism took root in Denmark and Norway*, p. 145–151.

chapter on the procedure of determining stellar positions from observations. Dybvad, however, has decided to pass over the latter subject, following Ptolemy who placed it after his lunar theory.

"Since, then, this book is concise and extremely well-ordered, too, it deserves to be read and studied most thoroughly by everybody who aims at a sound foundation of his knowledge of astronomy" [*27*: fol. A 3r — A 4r]. And a little later Dybvad declares that he has commented upon this useful and valuable book for the benefit of those who did themselves never study the monuments of the old astronomers [*27*: fol. A 4v].

Dybvad's commentary follows the said 13 chapters of the *De revolutionibus*, explaining and proving each proposition with reference to the figures from the *De revolutionibus*, and illustrating each theorem by means of numerical examples.

So far Dybvad has proved to be a clever student, having learned spherical astronomy by a careful study of the *De revolutionibus*, but nothing has been said as to whether he does or does not actually accept the motion of the Earth; he may well have avoided to decide upon this question since he had chosen for his subject the very least controversial topic which had, for the very same reason, been dealt with rather briefly by Copernicus himself [*60*: II, 64].

However, it is clear from the first chapter, *On the Circles and their Names*, that originally Dybvad intended to accept not only Copernicus' description, but also his explanation of the heavenly motions. "It is, moreover, at this point useful to have in mind the author's treatment — *De revolutionibus* I. 11 — of the triple motion of the Earth in order to comprehend more properly his definitions; for some circles, e. g. the equator and the two tropics, are brought about by the daily motion of the *center* of the Earth, whereas others, e. g. the ecliptic, are described by its annual motion. [...] The equator is described by the motion of the *center* of the Earth when this center has, by its own motion, been carried to the intersections between the equator and the ecliptic. [...] But when the *center* of the Earth has reached the first point of Cancer it describes by its daily revolution the Tropic of Cancer" [*27*: fol. A 7; my italics].

Evidently Dybvad visualized a sphere around the Sun upon which he attempted to make the Earth trace the equator and the tropics by its daily rotation, just as it traces the ecliptic by its annual motion. Having thus failed to grasp the idea of those circles as projections upon the very distant heavenly sphere, he met with surprising troubles of comprehension which caused him to give up the Copernican point of view in a rather indifferent way, namely by noticing that the equator, the tropics, and the ecliptic may, too, be described with reference to the motion of the Sun, and without the Earth to be movable. "And we shall adopt this doctrine below because it is the most ancient as well as the most agreeable one" [*27*: fol. B 1r].

One decade later Dybvad published textbooks within the fields of geometry, astronomy [*28*; *29*], and meteorology. On one occasion he emphasizes Copernicus' improvement of Ptolemy's lunar model which involved a far too great variation of the apparent diameter of the Moon [*28*: fol. A 6v], but otherwise he adheres to the traditional picture of the Universe. This is especially clear from his dissertation *On the Aethereal Region of the Universe* [*30*].

Fincke, a cautious mathematician about 1600

In 1590 *Anders Krag* (1553–1600) succeeded Dybvad at the mathematical chair. However, the next year already he was in turn followed by *Thomas Fincke* (1561–1656), whereas he himself took the chair of physics. Both of them had studied for several years abroad, had obtained their doctor's degree of medicine, and were greatly influenced by Petrus Ramus.

Already in 1581 at Strassburg Fincke had published an ephemeride computed on the basis of the Prutenic Tables [*34*]. On the whole it appears that these tables came into common use during the 1580's. As it could be expected Morsing's *Astrological Diary* published at Uraniburg in 1586 contains comparisons with Copernican and Prutenic parameters [*23*]. But an almanac for the year 1585 by *Erik Christoffersen Dybvad* [Dibvadius] also brings, under the 11th of March, a definite reference to the Prutenic Tables [*26*].

As a result of his mathematical studies Fincke in 1583 published his main work, *Geometria Rotundi*, which brought some new trigonometrical formulas. It is remarkable for its ordering of subjects according to the precepts of Ramus in contrast to the more cumbersome arrangement of Euclid's geometry. Its dedicatory letter to King Frederick the 2nd explicitly mentions Copernicus among the mathematicians of astronomy whom Fincke had studied intensively, and whose works form the basis of his own book [*35*: fol. 4r; cf. *36*: fol. A 1v]. In his preface to the reader Fincke, addressing himself to Thomas Digges, expressed his doubt as to whether tho calculations of Geber, Regiomontanus and Copernicus lead more compendiously and briefly to the positions of the stars than Ptolemy's and Reinhold's, which Fincke himself has used [*35*: fol. 3r]. He concludes his tenth book, on the calculation of plane triangles, by mentioning his original intention of comparing the calculations of Regiomontanus, Ptolemy and Copernicus to his own in order to decide which were the most compendious. However, he has renounced this plan in order not to confuse young students or to prevent them from reading the monuments of the immortal masters themselves [*35*: 295]. The *Geometria Rotundi* brings specific references to Copernicus regarding his plane geometry [*35*: 61; 63, 275, 291], his table of signs [*35*: 126, 134] and his spherical geometry [*35*: 338, 367, 392]. Fincke's own trigonometrical Tables are based on Rheticus [*35*: 137–269].

From another reference it is clear that Fincke had also studied Copernicus' theory for the variation of the solar eccentricity [*35*: 326]. In his next major work, *Horoscopographia* (1591) he, following Reinhold, records and uses Copernicus' theory for the variation of the obliquity of the ecliptic, although he wants to leave it to the astronomers to decide to what extent this theory squares with the facts [*36*:

fol. C2r; cf. *35*: 361; *38*: fol. A2]. And his *Theses on Months and Years* (1602) mentions the Copernican variation of the length of the tropical year [*39*: fol. A2—A3].

So far Fincke has shown himself well-versed in astronomical subjects and calculations although he was not a professional astronomer. However, he cannot refrain from reminding the astronomers to be cautious as to hypotheses. Thus his *Theses on astronomical hypotheses* (1592) establishes that the heavenly bodies revolve, fastened to their spheres and traditionally arranged around the stationary, central Earth, and thence the author continues: "But according to quite sane philosophers the heavens appear to be continuous; the heavenly bodies are lively and active, and the above-mentioned harmonious world order has already long ago commenced to be mistrusted, not merely by Copernicus, but also by other prominent men. In so far as we, by means of such principles labelled as hypotheses, smooth the path for astronomical measuring we must be extremely cautious not to violate the truth of physics" [*37*: fol. A1v—A2r]. Fincke may well by his "other prominent men" have aimed at Tycho and his breaking down of the heavenly spheres [cf. *119*: 240—241], but it is manifest that he did not accept Copernican cosmology either. Perhaps he wants to take up the attitude of an expectant towards all innovations within this field.

From 1603 onwards Fincke was a professor of medicine and a highly influential person at the university. In 1626 he is said to have published a *Methodical Treatise on the Doctrine of the Sphere*, but I have not succeeded in finding this work [*40*].

Riber, Tycho's Copernican pupil?

From 1603 to 1607 the chair of mathematics was held for the first time by one of Tycho's assistants, *Christian Hansen Riber* (Ripensis; 1567—1642), who later on became a bishop at Aalborg.

During his mathematical professorship Riber wrote four small academic dissertations with several allusions to the variation of the obliquity of the ecliptic, which he evidently accepts as real, and attributing to this obliquity a minimum value of 23;28° at the time of Copernicus [*90*: fol. B1, cf. fol. A3v, B2r; *91*: fol. B1r, cf. fol. A3v; *92*: fol. A2r, B2r].

His *Theses on the Definition and the Division of the Sphere* treats of the motion of the ethereal world by exposing the discordance between Ptolemy and Copernicus on this question. Riber faithfully adduces a few familiar arguments of scientific nature to both sides, as well as quotations from the Holy Scripture against the motion of the Earth together with their counterpart depending on a non-literal interpretation [*89*: fol. A3v—B1r]. Thereupon he passes to the order of the heavenly spheres and orbits. He reproduces the Ptolemaic ordering, then Copernicus' arrangement, adding: "And thus he demonstrates that the phenomena are saved by the harmony of the triple motion of the Earth". And finally Riber records that more recent astronomers have devised another hypothesis of the Universe which, of of course, appear identical to the Tychonian system. He mentions the alleged superiority of this arrangement compared to Copernicus' audacious placing and moving of the Earth as well as to the Ptolemaic lavishness of concentric and eccentric orbits. He states that Copernicus had himself been aware of the last-mentioned inconvenience "and that was exactly the reason why he devised one single circle (which he calls the great because of its numerous advantages) to revolve the Earth once a year, thus liberating nature from the penetration and the redundance of the orbits, and from other preternatural motions with which the Ancient world picture is encumbered. So by means of this circle he strove to aid that very nature which is simple and loves unity" [*89*: fol. B1r–B2r].

Riber's following section, On the Earth, quotes the arguments put forward by Ptolemy to prove the Earth to be situated at the center of the Universe, and he continues: "But although the truth of these arguments may consistently be proved according to the Ptolemaic hypothesis they do not, however, destroy Copernicus' system. Within *his* arrangement of the orbits they do consequently not prevent the Earth from revolving around the central Sun, either". Thence Riber refuses to say antyhing definite on the motion of the Earth, referring to the arguments brought forward above, and concerning the daily motion he concludes that it must necessarily be attributed either to the heavens or to the Earth [*89*: fol. B2v–B3r; cf. *90*: fol. A2r].

All this is not very definite, and being theses for disputation perhaps not even so intended. However, it is remarkable that Riber does not mention Tycho by name at all, so I find it hard to avoid the conclusion that he was for some reason or other in opposition to Tycho, and after all more of a Copernican than he dared openly admit. He was succeeded by Longomontanus whose compromise, agreeing in most essentials with Tycho's ideas, will be the subject of a later section [cf. p. 126].

II. CASPAR BARTHOLIN'S TYCHO-ARISTOTELIAN PHILOSOPHY

Aslaksen formulates a Tychonian philosophy and theology (1597)

As made clear in the foregoing chapter the Professors of mathematics at Copenhagen took up a sceptical or even opposing attitude towards Tycho's innovations within astronomy. The natural philosophers, on the other hand, fully accepted Tycho's ideas. So they were enabled for another half century to do with moderate alterations of the Aristotelian physics and cosmology instead of facing an open confrontation with the possible result of radical change or even total rejection.

Krag was the first and most conspicuous Ramean logician and natural philosopher in Denmark. He took up the attitude of an open-minded intermediary, e. g. between Paracelsus and Galen in the field of medicine. His authorship contains nothing directly relevant to the development of Copernicanism. However, one of his students [*63*: fol. A1r], *Kort Aslaksen of Bergen* (Aslacus; 1564–1624), had for three years been Tycho's assistant at Hveen and had afterwards studied for six years at numerous universities all over Europe. On his return in 1600 he became a professor at Copenhagen, and from 1607 professor of theology. Influenced by Johannes Piscator of Herborn and imbued with Calvinistic ideas he was met with considerable resistance from the Danish clergy, but owing to his diplomatic character he managed to avoid the fate of his predecessor Dybvad [cf. p. 117].

In 1597 Aslaksen published a treatise, *On the Nature of the Triple Heavens*, dedicated to Tycho Brahe. In this work he aims at reconciling the Holy Scripture with the results of natural philosophy and astronomy with a special reference to refraction and to the non-existence of impenetrable heavenly spheres. Thus his book formulates what may be called "Tychonian" philosophy and theology which, however, does not include any elaborate debate with Copernicus. Tycho had already seen to that. So apart from a couple of trivial references [*7*: 42, 45, 110] Copernicus is mentioned only in the 23rd chapter *On the Immobility of the Sidereal Heavens.*

Note that this heading does not make Aslaksen a Copernican as suggested by V. Brun [*134*: 72]. According to Aslaksen the heavenly bodies do perform a daily revolution, but here he deals with the heavenly stuff containing these bodies, and he argues its immobility from its extreme rareness and thinness. Thereafter he considers the alternative solution that the heavens should perform a daily revolution carrying with them the heavenly bodies. But he deems this opinion absurd because the thin heavenly material is no more able to move the dense and solid stars than the aereal atmosphere can turn the globe of the Earth around its axis. And just because the great Copernicus had realized the latter impossibility he left behind the heavens and let the Earth move [*7*: 166—168]. It should be noticed that this argument presumes that Copernicus had already tacitly done away with the material heavenly spheres, a probable assumption, indeed, but at variance with the general opinion of the time [cf. p. 43].

The influence of Aslaksen depends mainly on the fact that the Tycho-Aristotelian cosmology which he originated was accepted and reformulated by Caspar Bartholin who more than anybody else came to stamp the natural philosophy in seventeenth century Denmark. But quotations from Aslaksen are found in several places, e. g. in a corollary to a dissertation by *Hans Arnoldsen de Fine* (1579—1637). "Against Aristotle and following the opinion of the most excellent and illustrious Aslaksen" he affirms that coming-to-be and passing-away take place in the celestial world [*41*: fol. B3v — B4r]. Otherwise de Fine's series of dissertations is principally Aristotelian and irrelevant to the development of Copernicanism.

This is the case, too, with a good deal of the academic as well as the popular scientific literature of the time. It would be idle to mention separate works belonging to this even flowing undercurrent of Aristotelianism which, in relation to our subject, only indicates that their authors were not Copernicans. Indeed they may best be characterized by quoting the motto of *Carl Jørgensen Bang*: "Nothing new under the Sun" [*8*: fol. A1r; cf. fol. F1]. Niels Nielsen mentions a work by *Hans Lang* [*65*], with an interesting anti-Aristotelian title, which I have, however, not succeeded to find.

Caspar Bartholin's *Systema Physicum* (1628)

With Thomas Fincke's son in law, *Caspar Bartholin* (1585—1629) a family entered the University of Copenhagen which was to produce several distinguished members and thus came to dominate medicine and natural philosophy at the university for more than a century. Being an exceptionally gifted person Caspar Bartholin acquired the knowledge of his time within several disciplines all in the first decade of the seventeenth century. In 1611 he became a professor of eloquence at Copenhagen, in 1613 of medicine, and 1624 of theology. Besides an extensive authorship within these fields he produced, during the first quarter of the century, a series of writings on natural philosophy, published several times in various combinations, and finally in 1628 in a joint edition entitled *Systema Physicum* [*11*].

This *Physical System* in due order covers all the subjects of Aristotelian natural philosophy from the first principles of natural bodies to the rational soul of man. No real originality is found in Bartholin, but owing to his clearness of expression and his ability of exposing and judging between different views his works became highly influential for many decades to come. Excerpts were published for use in Danish schools as late as in 1676 [*12*].

As the general ordering of subjects within Bartholin's natural philosophy is Aristotelian in reaction to the Ramism of the foregoing decades, a great part of its contents is Aristotelian, too, which will be fully confirmed from the author's discussion on Copernican ideas. However, it would be unfair not to mention his willingness to accept new scientific results and let them influence his philosophy.

Concerning the Milky Way he thus prefers the explanation of Galilei — with whom Bartholin had studied — that it is made up of innumerable individual stars. He defends at length this position against Tycho's view that the Milky Way should be a diffused celestial substance formed by residual primordial light, as well as against Longomontanus' attempt of reconciling the opinions of Tycho and Galilei. By the way Bartholin expresses the wish that Tycho had been in possession of the Galilean telescope [*11*: fol. P4r — P8r]. Bartholin's treatment of comets is particularly illustrative of his being a man of compromise between Aristotle and the new science of his own time. In the astronomical section he fully accepts the nonexistence of solid heavenly spheres, thus continuing Aslaksen's work of building a Tychonian natural philosophy [*11*: fol. P1v—P4r, T3v—T4v, Aa2v—Bb3r]. Accordingly he sets forth and defends, especially against Keckermann, Tycho's opinion of comets and new stars [*11*: fol. Y4v—Z5r], touching incidentally Tycho's wondering at Copernicus' poor and crude instruments which had caused many a failure in the description of various motions [*11*: fol. Z2v—Z3r]. That, however, does not prevent Bartholin from distinguishing — in his meteorological chapter on comets — between, one on hand, *celestial* comets, including new stars and dealt with previously, and on the other hand, *elementary* comets. The latter originate in the superior part of the atmosphere and are treated along purely Aristotelian lines [*11*: fol. Zz6r—Aaa7v].

Caspar Bartholin's Tycho-Aristotelian refutation of Copernicus (1619)

However, concerning the world systems Bartholin was not at all inclined to envisage any compromise. That he did not accept the motion of the Earth is clear already from a dissertation of 1605 [*9*: fol. B1r]. Dealing in his *Systema Physicum* with the order of the planets he rattles off various old and new arrangements. He declares that he will himself maintain the opinion of the most ancient philosophers and of Ptolemy, although he admits that Copernicus, by following Aristarchos of Samos, has demonstrated very neatly the celestial phenomena [*11*: fol. R2v—R3r].

And finally Bartholin shows his hand clearly by his book *On Earth, Air and Fire* [*10*] in two sections on the site and on the state of rest of the Earth. By traditional arguments he supports Plato's and Aristotle's opinion of the central Earth in contrast to Pythagoras' less probable doctrine that fire, being the most worthy element, occupies the centre, around which the dark earthern star is revolving. Aristarchos made a real astronomical theory by assuming the Sun to be the central body, and Copernicus did save phenomena properly enough along the same lines. "But being physicists we shall not forthwith admit the astronomers' hypothetical assumptions which are sometimes plainly false" [*11*: fol. Ff3v—Ff4r].

The section on the stationary Earth opens by asserting that the Earth is resting perpetually immovable in the middle of the world. Thereafter the most excellent mathematician Nicolaus Copernicus is introduced as a link of the chain of opponents to this view, from Heracleides of Pontus to William Gilbert. Then Bartholin supports his thesis by a series of arguments.

Firstly he adduces some scriptural passages. The objection that the language of the Scripture should have been adapted to the apprehension of common man is met by Bartholin who among other things asserts that the human intellect could easier grasp the motion of some trifling body than that of the numerous and immense stars — an astounding context, indeed, for the "Copernican" argument of economy.

Secondly Bartholin asks why the Earth should move since it has been shown in his astronomy that the stars do so?

Thirdly he attacks the problem indirectly by assuming the Earth not to be at rest, making thereafter, by Aristotelian argumentation, nonsense of all conceivable forms of its supposed motion, be it rectilinear or circular and, in the case of circular motion, be it natural, violent or preternatural. As for the Aristotelian axiom of an unequivocal relation between the simple motions and the simple bodies it had been restricted by Copernicus to regard circular motion primarily and simple bodies only as long as they remain in their natural places [cf. *60*: II. 20]. Bartholin points to the obvious difficidulty of attributing to a body any motion when it is already in its place.

Fourthly Bartholin assumes a circularly revolving Earth, and he maintains that things thrown vertically upwards could in that case not return vertically to their starting point. Copernicus had made an attempt of explaining how the lower part of the atmosphere partakes in the revolution of the Earth either (a) because of its being mixed with earth and water and accordingly of the same nature, or (b) because the air has acquired such a motion from its being contiguous to the perpetually revolving earth [*60*: II. 19].

Bartholin in turn charges Copernicus with (1) wavering between two conjectures, (2) arguing, in his former guess, the matter in question, viz. that rotation should be a natural motion of the Earth, (3) neglecting that the falling body while performing two partial motions must necessarily lose (!) some velocity of both components and thus cannot follow a vertical trajectory, and (4) attributing a mixed motion to a simple falling body. — I can make sense of the "lose" above only be relating it to the upwards path of a thrown body, but that is contrary to Bartholin's formulation — Bartholin goes on assuming that Copernicus had himslef realized the fourth absurdity and accordingly (a) attached circular motion to universals and rectilinear motion to their parts which, however, only makes up another absurdity, viz. that parts have a nature different from that of their whole, and (b) made circular motion natural for a simple body as long as it remains a unity and stays in its natural place, while rectilinear motion is an additional motion of parts which have been removed from their natural place and thus do not entirely possess their real nature.

"Rise from the dead, Stagyrite; and learn a marvellous equivalence. You divide motion into natural and violent, the former again into rectilinear and circular. To Copernicus natural motion equals the circular, and rectilinear the violent". Thus a stone should fall contrary to its nature and in consequence it might tend upwards with Daedalus. Moreover he maintains that one subject simultaneously performs circular and rectilinear motions, i. e. natural and violent, and hence contrary motions, which cannot but intellectually be separated. "Behold loads of absurdities! by reciting them I have refuted them".

Fifthly Bartholin charges Copernicus with a petitio principii when he, answering Ptolemy's fear that a revolving Earth should disintegrate, maintains the rotation as springing from nature. In addition Bartholin adduces a dubious analogy between the Earth and the sea. In spite of its supposed natural motion the latter drifts ships to and fro, while we never perceive any similar effect on land.

So Copernicus, according to Bartholin, failed to refute the arguments brought forward by the ancients to support the immobility of the Earth, but he was even far less successful in establishing positively his own case. For (1) his argument drawn from the spherical shape of the Earth applies equally well to the heavenly bodies or even to the heavens themselves, (2), his considerations on the relativity of motion prove nothing, since the possibility of a motion does not imply its existence, and (3) his assertion that motion belongs to the contents rather than to the container agrees better with Bartholin's own idea of movable stars contained within the immovable heavens.

In conclusion Bartholin concedes that the doctrine of the movable Earth is a convenient hypothesis to astronomers although it is really false and not to be trusted in physics. In logic similarly true conclusions may sometimes be drawn from false premisses. To support his conclusion Bartholin finally quotes the preface of the *De revolutionibus*. It is impossible to infer from his wording whether he were aware that Osiander, and not Copernicus, did write this preface [*11*: fol. Ff6r — Gg2r].

The influence of Bartholin on natural philosophy may be traced in numerous contemporary and later academic writings, Danish as well as Swedish [*132*: 169—175]. However, in so far as they do not add further arguments, they do not deserve to be mentioned separately, either.

Of course Bartholin's discussion proved nothing but the incompatibility of the Copernican cosmology with the basic ideas of Aristotelian physics; it does not prove which point of view is the true. But apart from easily discernible sophisms it must be admitted that he has with merciless acuteness unveiled the weakness of Copernicus' embodying his new cosmology in an Aristotelian vocabulary.

III. LONGOMONTANUS' ATTEMPT AT SYNTHESIS

The *Astronomia Danica* (1622)

According to Dreyer "the Tychonic system did not retard the adoption of the Copernican one, but acted as a stepping-stone to the latter from the Ptolemean" [*119*: 181]. This is certainly true as a statement of the development of astronomy proper, especially as materialized in the work of Kepler. But one must not forget that Kepler adopted the Copernican system in opposition to Tycho's wishes, and the natural philosophy outlined above and perceived as another result of Tycho's work calls upon the directly opposite judgement. Moreover the formulation of a comprehensive theoretical astronomy most strictly adhering to Tycho's spirit and intentions, to wit Longomontanus' *Astronomia Danica*, reduces Dreyer's statement to be true only in the restricted sense that the Copernican system was taught and made known as an attractive hypothesis, but not adopted as the true world picture behind the heavenly phenomena.

A peasant's son, *Christian Sørensen Longomontanus* (1562—1647), spent nearly ten years of learning from and assisting the astronomer of noble birth Tycho Brahe, from 1589 to 1597 at Hveen and again at Prague about 1600. Thus he became a professional astronomer himself and by his *Astronomia Danica* [*75*] came to realize Tycho's plans of restoring theoretical astronomy on a Tychonian foundation. Secondly he founded an educational tradition of astronomy at the University of Copenhagen where he was a professor from 1605 to his death — from 1607 professor of mathematics and after 1621 of astronomy, mathemata superiora, which was that year made a separate discipline to be distinguished from ordinary mathematics, mathemata inferiora. Longomontanus planned the observatory at the Round Tower of Copenhagen, and his *Introduction to the Astronomical Theatre* contains precepts for the education of astronomers [*77*: fol. C1r — C4r]. In a dissertation from 1636 he proposed a veritable programme for studying the mathematical disciplines [*76*, cf. *70*], but the way of transmission down through the different levels of education was long. For example, in 1656 *Jørgen Eilersen* (Hilarius; 1616—1686, since 1672 professor of mathematics) published, for use in the schools, a spherical doctrine which maintains that the opinions of Ptolemy and the Alphonsines had hitherto been the most current. He does not at all mention Copernicus or Tycho [*50*: fol. A2v].

According to Neugebauer the spell of tradition within astronomy was broken with Tycho and Kepler. "The very style in which these men write is totally different from the classical prototype. Never has a more significant title been given to an astronomical work than to Kepler's book on Mars: »*Astronomia Nova*«". [*129*: 206]. By way of comparison it must be admitted that Longomontanus' *Astronomia Danica* is not "new". It is modelled essentially after Ptolemy's *Almagest* and Copernicus' *De revolutionibus* [*75*: 2, 146, 459]. Longomontanus refused to accept Kepler's elliptical planetary models [*75*: 343; cf. *71*: fol. B2v and *73*], since he faithfully adhered to Tycho's basic principles including that of uniform circular partial motion. He preferred the "prostaphairesis method", invented by Tycho and Paul Wittich, to Neper's logarithmic calculation [*75*: 7—15], and he considered the telescope of very limited use within observational astronomy [*77*: fol G4v].

That, however, did not prevent Longomontanus from considering his own work an innovation, formulated "in Tycho's spirit" and intended to form the "perennial" theoretical outcome of Tycho's worke [see . g. *75*: 181]. We may briefly characterize his achievement as contrasted to Kelper's in the following way. Kepler built up a complete model for the motions of Mars from Tycho's observations only and so to speak "infinitesimally". Longomontanus, on the other hand, did not posses Tycho's diaries of observation, and so was unable to use a similar procedure. But he imagined that he would have arrived at the same models in case he had been possessed of these observations [*75*: 146]. For Longomontanus, therefore, Tycho's results merely constituted one truly reliable set of positions of the heavely bodies, and Tycho's work on refraction, parallax, and the use of instruments enabled him to attach reasonable corrections to the results of earlier astronomers.

But even thus improved these results had only a corrective value. And no trustworthy dynamics of celestial motions being at hand [cf. p. 54] Longomontanus has to search for independent criteria elsewhere. Hence he is happy to find "perfect" numbers in the solar model [*75*: 186] as well as in the theory of the precession of the equinoxes [*75*: 221], and to extrapolate some basic heavenly motions to "neat" initial positions at the time when the world was, according to the mosaic tradition, created [*75*: 187, 196—197].

Strange as this sort of science must necessarily appear to modern minds I shall only add that Rheticus' predictions from Copernicus' long term motions proceed quite similarly [*93*: 121—122; cf. *128*: 136], and I am not at all convinced that

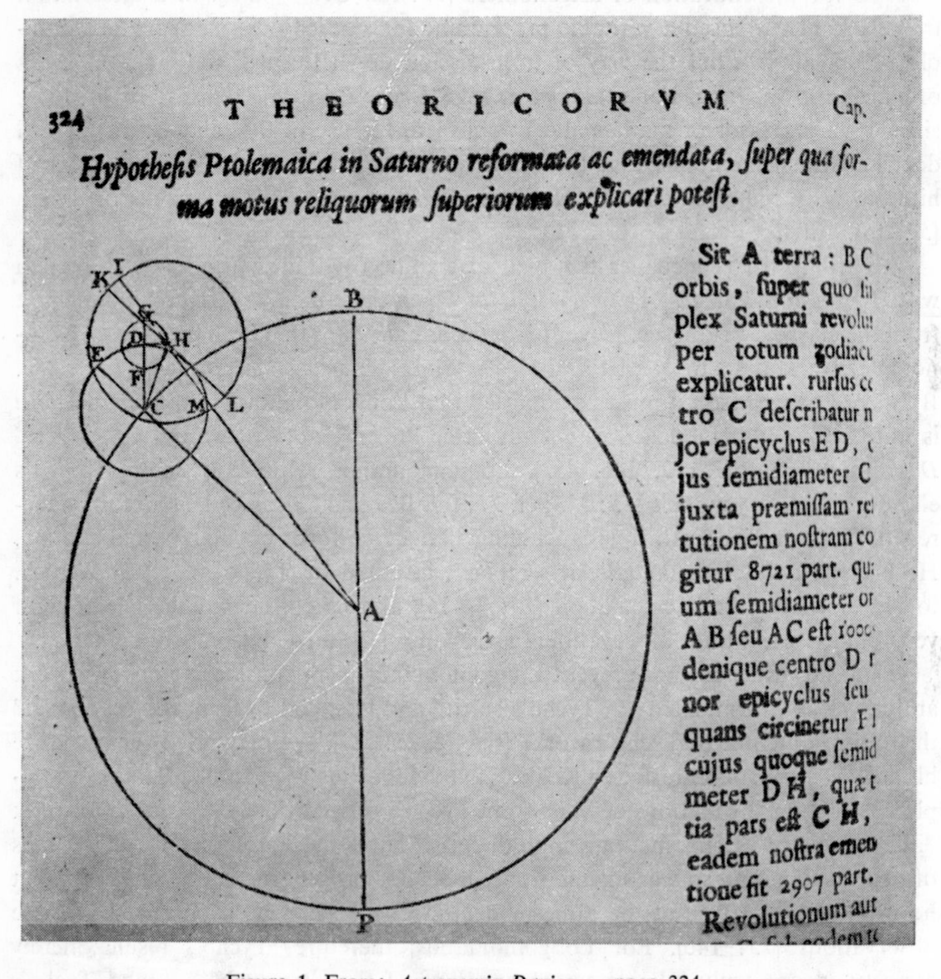

324 THEORICORVM Cap.

Hypotheſis Ptolemaica in Saturno reformata ac emendata, ſuper qua forma motus reliquorum ſuperiorum explicari poteſt.

Sit A terra : B C orbis, ſuper quo ſi plex Saturni revolu per totum zodiac explicatur. rurſus ce tro C deſcribatur n jor epicyclus E D, ju s ſemidiameter C juxta præmiſſam re tutionem noſtram co gitur 8721 part. qu um ſemidiameter or A B ſeu A C eſt 1000 denique centro D r nor epicyclus ſeu quans circinetur F I cujus quoque ſemid meter D H, quæ t tia pars eſt C H, eadem noſtra emen tione fit 2907 part. Revolutionum aut

Figure 1. From: *Astronomia Danica*..., page 324.

Longomontanus deviated from "Tycho's spirit" in his proof that the heavenly motions really are harmoniously arranged and governed by divine laws or, put the other way round, that the records of the Holy Scripture might be affirmed by sufficiently thorough scientific investigations.

As in Tycho one finds in the *Astronomia Danica* a wealth of references to Copernican parameters, but here merely for the sake of comparison. Ptolemaic and Copernican astronomy had finally, after Tycho, become numerically equivalent in that the parameters set forth by both authors were equally uncertain. Accordingly the old Ptolemaic, the admirable Copernican, and the modern Tychonian world system as well as the three authors' formally different models for the individual heavenly bodies had at last become equivalent, too, in the sense that the true Tychonian

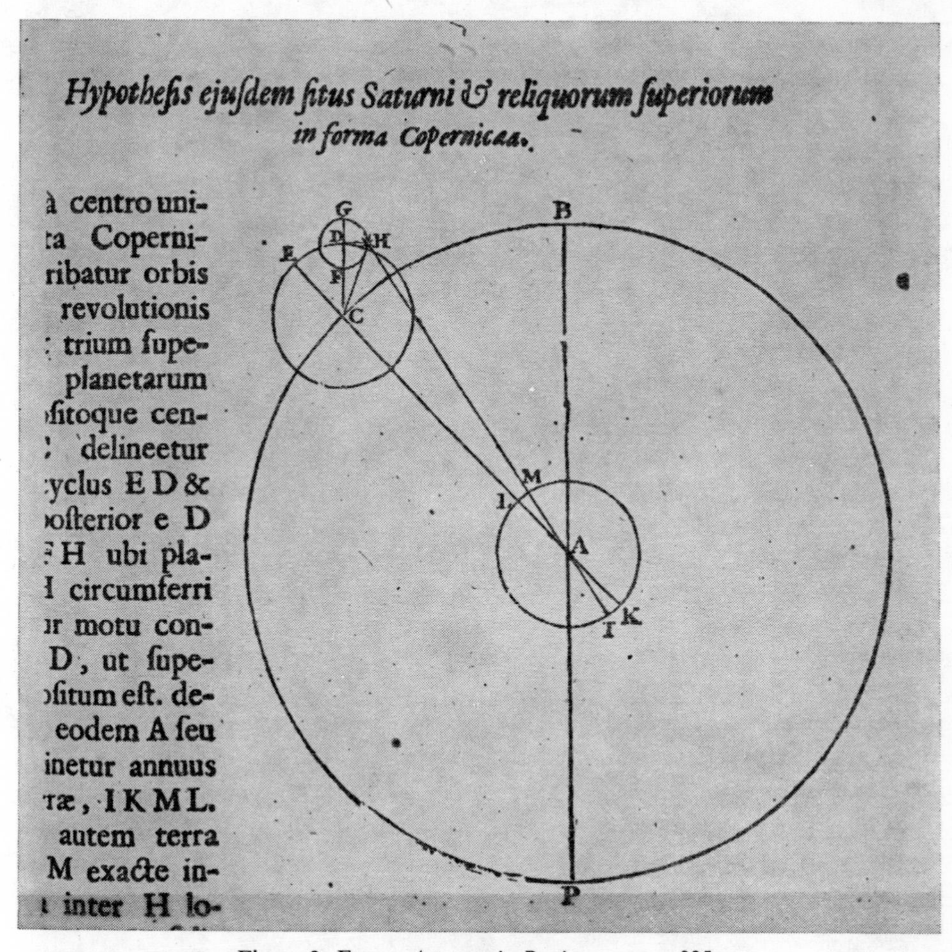

Figure 2. From: *Astronomia Danica...*, page 325.

parameters could equally well be pressed into any of the three frames. And to make his point clear is exactly one main end of the second part of the *Astronomia Danica* [*75*: 146–147, 149–150].

This programme includes a detailed confrontation with the equant-circle of Ptolemy's planetary models which are in consequence transformed, firstly into an eccentro-epicyclic model in the manner of Copernicus and al Zarkali, and secondly into a model with two epicyclets as proposed by Tycho and — not mentioned by

Longomontanus — by Copernicus in his *Commentariolus* [*93*]. Longomontanus considers this bi-epicyclic planetary model to be in agreement with internal causes of the continuous and harmonious, well-ordered motions in the heavens, and hence

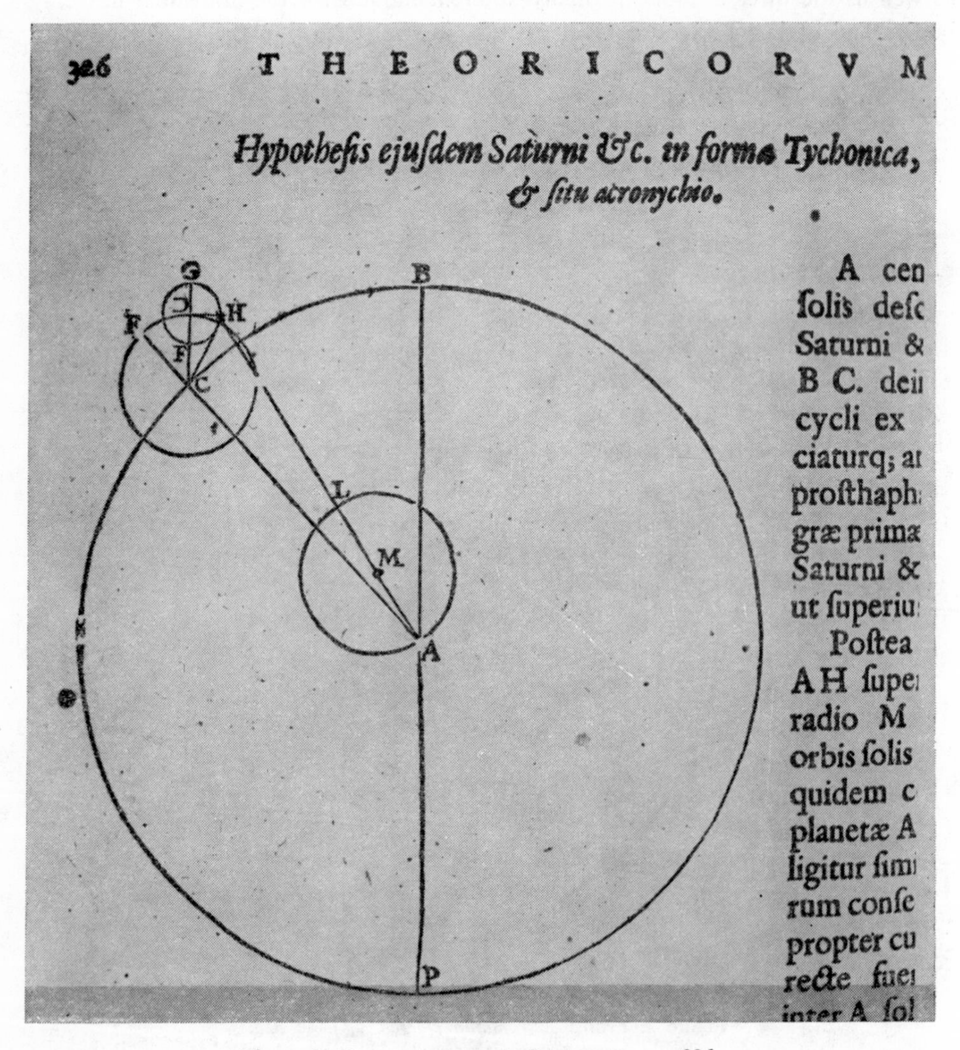

Figure 3. From: *Astronomia Danica…*, page 326.

it came nearer to truth than Copernicus' eccentro-epicyclic model [*75*: 163—169]. So all planetary models of the *Astronomia danica* are furnished with two epicyclets irrespective of their being labelled as Ptolemaic, Copernican or Tychonian. And Longomontanus' three versions of e. g. the model for the motions of Saturn differ only as to the placing of the partial motion with the period one year [*75*: 324—326, see Figures 1—3].

Longomontanus' Tycho-Copernican world system

The chapter of *Astronomia Danica* "On the triple arrangement of the world system" [*75*: 151–163] brings a thorough representation and discussion of the Ptolemaic, the Copernican, and the Tychonian systems. The transition from the Ptolemaic to the Copernican system takes the form of a brief, but enthusiastic exposition of what multitude of phenomena Copernicus could explain by letting the Earth perform a triple motion. After sketching this wonderful invention Longomontanus goes into details concerning its advantages, especially stressing its usefulness, economy and harmony. And he concludes this section with a poem of applause about this highly economic explanation of nature.

Thence he goes on: "So far the great Copernicus' model of the Universe, including his assertion that the Earth performs a triple motion. And it must be admitted that everything is very ingeniously devised to save the celestial phenomena and entrenched moreover by means of physical arguments, inserted into the 1st book of the *De Revolutionibus*, to the effect that it may but with difficulty be destroyed. Nevertheless, there are among others two arguments, concerning mainly the annual motion of the Earth, which apparently deprive the system of its credibility".

Longomontanus' first counterargument regards the discrepancy between the Copernican system and Genesis. Heavens and Earth were first created and, in consequence, ought to occupy the primary places within the Universe, i. e. circumference and centre respectively.

His second argument is directed against the incredibly great distance and the vast size of the fixed stars resulting from an annual motion of the Earth which "by one stroke ruins any well-established symmetry between the parts of the Universe". This argument of Tycho's is amplified by numerical calculations of hypothetica stellar distances and sizes.

Thirdly Longomontanus, as an argument against the librations of the pole of the Earth, adds that the latitudes of the stars appear to vary rather than their declinations. This point is further elaborated in his chapter on the precession of the equinoxes [*75*: 217–221]. Longomontanus' own solution does away with the offensive librations by making the slowly moving axis of the Earth responsible only for the uniform mean precession. And it accounts for the variable rate of precession and for the variations of the obliquity of the ecliptic by a secondary precessional motion of the solar orbit itself.

It is remarkable that Longomontanus does not at all mention Tycho's mathematical proof of the truth of his own system [cf. p. 45].

For these reasons Longomontanus prefers the Tychonian system which, if any, perhaps represents the true arrangement of the world. Yet, following David Origanus (1558–1628), he accepts the daily rotation of the Earth because so many swift apparent motions may thus be explained by the assumption of a single rather slow motion. In case somebody wants to attribute to the stars a real daily motion he has to cling fanatically to the wording of a few scriptural places [cf. *72*: fol. A3r].

Longomontanus' choice of a rotating Earth appears to spring from considerations concerning the equation of time which in turn originate in his investigation of the lunar motions. Thus he ends a discussion on these topics in his *Introduction to the Astronomical Theatre* by the following statement: "Nor do we shrink from maintaining that the Earth rotates about its own centre. And much earlier than Origanus in his Ephemerides we have once at Hveen discussed with Tycho the reason why this is necessary" [*77*: fol. E3r—E4v; cf. *75*: 181—182, and *72*: fol. B4r]. He regards the problem irrelevant to spherical astronomy [*72*, fol. A3r; cf. fol. A4r] as well as to astrology [*74*: fol. A4]. His wish, however, to make all celestial motions take place in planes of great-circles forms another motive for his choice, since a rotating Earth does away with the motions of fixed stars along small-circles [*72*: fol. A3v, B3v—4r].

Already in 1609 Longomontanus denied the possibility of restoring astronomy without using hypotheses as proposed by Petrus Ramus [*70*: fol. B4v]. In his *Astronomia Danica* he concludes the chapter on the world systems by characterizing the Ptolemaic, the Copernican, and the Tychonian system as explaining the annual partial motion by a mechanism placed "outside", "at", and "around" the centre of the Universe respectively [Figures 1—3]. And from this apparent exhaustive classification he infers that a fourth world system is unthinkable [*75*: 163]. Accordingly he had, in his preface, argued that other systems might easily be reduced to one of these three [75: 147]. However, the three possible systems have in principle equal right, and the reader is as free as the author to choose the solution he finds most probable [*75*: 149—150].

Copernicus at the focus of a theoretical astronomical discussion

After the two preliminary chapters quoted above, the second part of the *Astronomia danica* attacks the main question of contemporary and — as convincingly argued by Ravetz [*131*] — much earlier theoretical astronomy, to wit that of establishing reference frames. Longomontanus states that the order of treating the Sun, the Moon and the fixed stars represents an important problem. Copernicus started with the doctrine of the fixed stars and the precession of the equinoxes while Ptolemy's starting point was the theories of the Sun and the Moon. "However, after exposing the matter to a through logical analysis and estimate we did choose to agree with Ptolemy upon the investigation, but with Copernicus upon the order of representation". [*75*: 169].

An inserted commentary of some 27 pages [*75*: 169—196], including a detailed restoration of Copernicus' solar observations [*75*: 179—181], produces, indeed, the main parameters of the theories of the Sun, the Moon, and the fiexd stars, treated in the said order. The following chapters deal with the motions of these same bodies in the order of the *De revolutionibus*, i.e. the fixed stars, the Sun and the Moon. Finally, the 2nd book treats of the remaining planets "as part of the furniture" [cf. *131*: 48].

This curious plan reflects Longomontanus' wavering between his primary pattern, viz. the *De revolutionibus*, and "Tycho's spirit". The said inserted commentary finds its parallel in the *Introduction to the Astronomical Theatre* which brings, in brief outline, instructions to the observer concerning the right ordering, the means, and the methods to lay a firm foundation for investigating the heavely motions. Here the proposed order is once more: Sun, Moon, fixed stars [*77*: fol. D1r—E1v].

A whole group of contemporary writings mainly concentrate upon questions of formulating a fundamental theory of observational astronomy. And several Danish authors besides Longomontanus entered the discussion, viz. *Peder Bartholin Kierul* (1586—1642) with his *Apology for Tycho* [*59*] against Martin Hortensius and Philip Lansberg [*68*], *Jørgen From* (1605—1651; professor of astronomy since 1647) with his *Astronomical Dissertation* [*43*] and *Response* [*44*] defending Longomontanus' *Introduction to the Astronomical Theatre* against Johs. Baptiste Morin [*81, 82*], and *Erik Olufsen Torm* (*1607—1667*; professor of mathematics 1636—1645) with his *Mechanical Disquisition* [*103, 104*] against "the most illustrious mathematicians of this century: Schickard, Galileo, Keckermann, Morin, Humius, Landsberg, Hortensius and others" [*103*: 2]. These writings include numerous references to details from Copernicus' astronomy of which the adequate treatment would, however, take us into technicalities quite outside the scope of the present essay. But their bare occurrence shows, quite apart from cosmological questions, that Copernicus' methods and numerical improvements in describing calestial phenomena were still by several professional astronomers considered up to date in the sense that they might be used for restoring astronomy no less than Tycho's results. By the Tychonians these astronomers are in turn accused of being mere theorists whose speculations may well be true, but cannot possibly be transformed into useful observational practice. The problem concerning the right order of treating the Sun, Moon, and the stars is dealt with at some length by From [*43*: fol. E4v—F1v; *44*: 35—39]. To the said group of polemic writings may be added a dissertation by *Christian Stenbuch* (1625—1665) directed against some of Scipio Claramontius' assertions [*99*]. But this work does not treat of subjects directly relevant to the discussion on reference-frames and world systems.

Longomontanus' Tycho-Copernican cosmology

To round off the above picture of Longomontanus one must emphasize his full acceptance of Copernicus' physical arguments [cf. p. 131] and his avoiding of counterarguments drawn from Aristotle. This point may be clarified by taking into account his dissertation *On the Prerequisites* from 1611 [*71*]. This brief cosmological treatise on the nature of the material and on the major bodies of the Universe was republished verbatim in the *Astronomia danica* [*75*: 35—41], and in his *Introduction to the Astronomical Theatre* Longomontanus repeats some of its main theorems [*77* fol. G1r—G2v].

It claims to rest on the most recent experiences of astronomical science and maintains the existence of a universal luminiferous medium spread everywhere from the interior of the Earth to the outermost limit of the Universe. This medium forms the place for a number of spherical universal bodies among which the heavenly bodies probably have their own aëreal atmosphere as well as the Earth. Everybody is held together by its own gravity and is moving according to divine precepts without any aid from material spheres or axes.

The Copernican features of this cosmology are easily discernible. Being anything but an Aristotelian Longomontanus definitely denies the existence of any boundary or essential difference between an elementary sublunar world and the celestial region. In brief he grasped the spirit of Copernican physics and was indifferent to Copernicus' wording, while Bartholin spent his efforts in vain verbal altercations. As an astronomer Longomontanus did not feel the need for compromising with Aristotle and the Bible to the same extent as did his physical and theological colleagues. Thus it appears quite legitimate to attach the label Tycho-Copernican to his astronomy.

Longomontanus gave everybody a free hand in the choice of world system, but his own choice was really in several respects opposed to the Tycho-Aristotelian cosmology formulated by Aslaksen and Bartholin. Yet to my knowledge no open conflict did ever arise between Tycho's astronomical and philosophical heirs. But after all Longomontanus' cosmology sketched above may well have facilitated the adoption of Cartesian ideas during the last half of the 17th century.

IV. THE RISE OF CARTESIANISM

Astronomers between Longomontanus and Rømer

Kierul who had studied with Longomontanus [*72*: fol. A1r], in the last section of his *Apology for Tycho* entitled "On the motion of the Earth along the ecliptic, and also on the triple Lansbergian heavens" [*59*: fol. N2v—O3v], defends Longomontanus' version of the Tychonian system by using mainly the same arguments as Longomontanus, but adducing further evidence from the Holy Scripture.

A Norwegian student, *Trygve Dieterichsen*, in a dissertation from 1638, at Rostock defends Origanus' system against well-known Ptolemaic arguments [*31*: fol. A3, B1] and against a comprehensive series of Scriptural evidence [*31*: fol. B2r—B3v]. The disseration quotes Tycho but bears no sign of being particularly influenced by Longomontanus.

In *Torm* I find but a cosmological curiosity added to his *Mechanical Disquisition* by the respondent: "In case the firmament were solid and ellipsoidal and the Sun and the Earth were placed at its foci the latter would be consumed on the spot by a conflagration". This hypothesis is not discussed further [*104*: 41].

Longomontanus' immediate successor as a professor of astronomy, *From*, does not in his writings express himself very definitely on the question of the world systems. On the whole he appears open-minded and critical, but an ironical character, too, and cautious in making his choice. Thus he, in his dissertation *On certain principles of the spherical doctrine*, concludes the 40th thesis on the daily motion by handing over to those "granted with the power of discerning coppers from mock-money" the judgement whether the adherents of a rotating Erath should be considered tricked by giddiness, and hence defenders of the most vain things [*45*: fol. B4]. It is beyond doubt, however, that From was anxiously searching for a true and natural system of the bodies within the Universe. In this respect he probably felt disappointed in Kepler's elliptical planetary orbits, while he seems to have put more than ordinary confidence in Schickard, and especially in certain inventions by de Baulne (Florimund de Beaune?) [*43*: fol. D2v; cf. *44*: 4].

From was in turn succeeded by *Wilhelm Lange* (1624—1682) who held the chair of astronomy until his death, but being since 1661 a judge, too, he had to leave his academic duties to substitutes. His publications deal mainly with time-reckoning. After repeating Longomontanus' arguments for the daily rotation of the Earth as well as against its annual motion he makes without hesitating Longomontanus' theoretical astronomy the foundation of his highly learned work on the subject [*66*: 14—17; cf. 49]. Lange moreover brings, in a series of dissertations some critical comments on Copernicus' solar model [*67*: 13, 26—37; cf. 41, 50].

During the period 1660—1676 *Rasmus Bartholin* (1625—1698; after 1657, a professor of geometry, and in 1658 of medicine also) acted as a substitute for Lange. In 1663 he published an *Astronomical consideration on the great conjunction* (cf. p. 33]. This *consideration* uses Tychonian planetary, models, and its calculations are founded upon the tables of the *Astronomia Danica* [cf. *17*: 17, 34]. R. Bartholin even follows Longomontanus in passing over "the somewhat easier logarithmic calculation in order to give a good example to the young students and not accustom them to compendious methods until they had tried the longer road and had often safely reached the prescribed goal by following the royal way" [*16*: 4, cf. *17*: 19, 24]. And had it not been for a critical discussion on astrology [*16*: 32—39] inserted after the calculations, Bartholin would, in his *consideration*, have found no opportunity at all of quoting Kepler.

However, R. Bartholin's next astronomical work, his excellent little book *On the Comets of 1664 and 1665* reveals that cosmology in Denmark is now on its way from Aristotle [*17*: 4] towards Descartes [*17*: 57—69, 87], and accordingly towards Copernicanism, too. This if affirmed in his collection of essays on natural philosophy, *Academical Questions on the Wonders of Nature*, particularly in the essays *On Cartesian Physics* (1664) and *On Physical Hypotheses* (1669) [*18*: 69—81; 101—118], which deal, in fact, with philosophy of science rather than with natural philosophy.

"Since changing times demand fresh studies and habits, too, and many things occur, appreciated by former centuries, which a subsequent era has condemned..." — This very opening sentence of the former essay calls attention to the establishment of the absolute monarchy in Denmark in 1660, and it heralds a new scientific era with campaings of human reason against pseudo-learning based upon authority and deeply rooted customs. After having by a series of hypothetical questions exposed to doubt the entire foundation of theoretical astronomy R. Bartholin concludes: "From the fact that Aristarchos, or Copernicus or René Descartes has successfully established one hypothesis it does not follow, indeed, that another one of equal virtue may not be imagined. But that much follows that you have to rely on that invention, or else to conceive another equally exquisite one. And whoever carefully considers the details will acknowledge the latter to be difficult, nay, far too arduous for mortal men" [*18*: 73—76; cf. 114—116].

By 1655 at latest the King of Denmark had bought Tycho's original observations from Kepler's son, and in 1664 R. Bartholin was entrusted with the task of their publishing [*19*: X, v—vii]. At the time Ole Rømer studied en Copenhagen and lived as an assistant in Bartholin's house. This brilliant young student spent four years in preparing the Tychonian observations for the press. He has himself, in a letter to Leibniz from 1703, testified to the decisive influence of this work on his own later achievements within astronomy [*54*: II. 159]. Thus Tycho had posthumously got still another Danish pupil.

Nicolaus Mercator (Kauffman; circa 1620—1687) was active mostly at London and Paris, where he might well have worked together with Rømer at the fountains of Versailles. Although he only spent a few years about the middle of the century at the University of Copenhagen, his *Institutiones Astronomicae* [*80*] deserves to be mentioned here, because this work was probably for some time used as the standard fundamental textbook of theoretical astronomy and thus filled the gap between the *Astronomia Danica* and Horrebow's *Opera mathematico-physica.* At least it forms an essential prerequisite of the last-mentioned author's exposition of the planetary models [*54*: IV, § 10].

This work shows Mercator as a convinced Copernican. Thus he brings, in an exemplary chapter *On the true world system*, a unified account of the Copernican and the Tychonian systems, of which the latter is delineated by dotted circles in the figure. After adducing a number of arguments in favour of Copernicus' solution, the last one depending on Kepler's third law, he concludes: "And that is why now-a-days most astronomers appreciate the Copernican world system as by far the most probable of all" [*80*: 114—120].

Wavering natural philosophers

In this dissertation of 1639, *On the elements*, Jakob Fincke (1592—1663; professor, since 1623 of mathematics, since 1635 of physics) follows C. Bartholin and maintains that the Earth is stationary [*33*: fol. D4r] at the center of the Universe

[*33*: fol. C4v; D3v; cf. *32*: fol. B2v]. But it is perhaps possible to trace a nagging doubt when he concludes his "far too superficial" account on this very copious subject by adding to the doctrine of the stationary Earth a prayer that God will, after this rambling life, permit him "to rest safely at the bosom of the mother of everything until the time of eternal and true rest" [*33*: fol. D4r].

Both of two dissertations on the sidereal heavens and the heavenly bodies from January 1651 by *Johannes Clausen* (Orthunganus; died 1673) and *Gisle Thorlakson* (1631–1684) in most essentials follow Bartholin without revealing any conflict with Longomontanus' rotating Earth [*102*: fol. A2v–3r] or non-existence of a boundary between the terrestrial and the heavenly regions of the Universe [*85*: fol. A3v]. But the mode of expression against the Copernicans has been considerably softened since Bartholin's time, and neither acception nor total rejection of their opinion is implied. Thus Thorlakson confines himself to put the question why the Copernicans demand from their opponent a proof that the Sun is situated at the center of the Universe, while they themselves cannot furnish any proof of the central position of the Earth. And their charging the Tychonians with the assumption of two centers of motion within the Universe misses the mark, too, because they themselves make the Earth a secondary center, viz. of the lunar motions [*102*: fol. A4r]. But after all a corollary to the dissertation says that the Earth cannot be considered a planet [*102*: fol. B2r].

Thorlakson mentions – to my knowledge for the first time in Danish scientific literature – Descartes' vortices of the heavenly bodies [*102*: fol. A2v]. But the 1660's should see a Copernican "heretic" before the Cartesian Cosmology gained a footing in Denmark.

The fate of a 1660 Copernican heretic

In 1661 *Claus Nielsen Lesle* published, among other small treatises, a discussion *On the mobility of the Earth* [*69*: fol. A2v–B2r]. Lesle affirms that the Earth does move, and he at once expresses his fear that by defending that point of view he will lift too heavy a burden. Thereafter he reproduces briefly, but in due order, the five argumentations from C. Bartholin's chapter on the stationary Earth [cf. p. 124].

Against Bartholin's quotations from the Scriptures Lesle maintains that they only prove that the Earth will not age and be destroyed; it is, however, quite free to perform a local motion within certain boundaries. Secondly nobody has, neither from experiences nor from the Scripture, shown that the stars really do move. Against Bartholin's third and fourth arguments Lesle asserts that the rectilinear motions of the elements are due to their seeking their own like, and are brought about by some external cause which does not hinder their performing simultaneously a circular motion together with the entire globe of the Earth. And fifthly: "Ptolemy's fear (that a rotating Earth should dissipate) is vain, and since *he* attacks us by laughing we shall pay him back in the same coin and get the laugh of him".

After having defeated Bartholin, Lesle argues, from the Genesis, that the stars do not move since they are fastened to the immovable firmament. And finally he plays his trump card, proving that the Earth does really move since it forms the vessel which, being moved itself causes the motion of its contents, i. e. the tides of the oceans. — This erroneous theory of the tides evidently originates from Galileo's *Dialogo dei massimi sistemi* [*46*: VII. 442 ff.] Neither Bartholin, nor Galileo are mentioned by name in Lesle's discussion.

During the first half of the seventeenth century the Danish clergy had grown more and more orthodox under the leading bishops *Hans Poulsen Resen* (1561—1638) and *Jesper Brochmand* (1585—1652) to the effect e. g. that the works of the highly esteemed theologian *Holger Rosenkrantz* (1574—1642) were banned and remained unpublished. So Bartholin's *Systema physicum* may well have been considered by theologians an invariable authority like the Bible and Luther's catechism. For the centenary of Luther's formulation of his theses, Resen in 1617 published his "*Lutherus triumphans*" [*88*], a title reflecting better than many words the prevailing theological spirit of the time.

In consequence Lesle's open attack on Bartholin had to be taken as heresy, and accordingly it was at once met by an extensive and highly learned theological rejoinder [*106*: fol. A2r—D3r] written by the Zealandian vicar *Hans Hansen Windekilde* (Windefontanus; 1625—1711). Apart from subtle philological and theological interpretations Windekilde brings but a couple of references to Sperling [*98*] and to Tycho. Lesle's theory of the tides is brushed aside as an attempt of proving something unknown from another equally unknown thing, whereupon Windekilde explains the tides as a combined effect of lunar influence and resistance of the different shores [*106*: fol. D1r—D3r].

From 1651 to 1659 Lesle had been a vicar in Scania [*125*: V. 39]. I have found no direct relation between Lesle's leaving the said clergical post, his heresy, and his later miserable life as a physician at Aarhus where in 1685 he was finally imprisoned. However, he was anything but popular among the local vicars [*127*: II. 249]. And Jakob Fincke's caution in expressing his doubt [cf. p. 137] as well as From's ironical style [cf. p. 135] may well indicate, too, that one had better follow the leading theological authorities concerning cosmological matters.

A century of Cartesian physics from the 1660's

The Cartesian influence on R. Bartholin in the 1660's has been touched upon above [cf. p. 135]. And by 1688 it became possible reading about Copernican concepts in the Danish language. For in that year *Johan Brunsmand* (1637—1707) published a calendar, approved by Rømer, discussing at some length whether the daily motion belongs to the Sun or to the Earth [*21*: 210—213]. Brunsmand also mentions the enormously great distances of the fixed stars proposed by the Copernicans [*21*: 331]. However, he rejects the Copernican views, mainly from Scriptural evidence, and it was not until 1748 that the time came for a Danish translation of le Bovier de Fontenelle's Cartesian and Copernican *Conversations with a Lady on the Plurality of Worlds* [*42*].

As for the academic literature numerous writings could reveal how during a whole century Cartesianism dominated the natural philosophy taught at the University and in the schools. The similar development in Sweden has been traced and excellently described by H. Sandblad [*132*: 79 ff.]. Let it suffice for our purpose to indicate a few selected textbooks illustrating the development.

C. Bartholin's grandson, *Caspar Thomesen Bartholin* (1655–1738; professor of physics 1674–1733, and since 1680 of medicine, too), late in the 1680's prepared his *Summa* [*14*] destined to replace the elementary writings drawn from the authorship of his grandfather [cf. p. 123]. This *Summa* rejects the Ptolemaic system, briefly sets forth the Copernican and the Tychonian systems, and mentions the Longomontanian compromise between the two without preferring definitely any of the three last-mentioned systems [*14*: 10–12]. A few years earlier C. T. Bartholin had given a quite similar account on the world systems in a series of private physical dissertations, where for disputation he adds the thesis: "There is no reason why we should maintain the Earth, or the Sun to be the center of the Universe" [*13*: VII].

Thus natural philosophy had now, even on the elementary level, given up its orthodoxy and had in a way adopted the freedom of chosing among different cosmological explanations proposed at the beginning of the century by Longomontanus. C. T. Bartholin's *Summa* was used until late in the 1770's [*112*: 267].

Its parallel on the university level was the *Specimen philosophiae naturalis* offering a more elaborate version of the above-mentioned dissertations and revealing C. T. Bartholin's cautious inclination towards the Copernican supposition as the simpler one and hence easier to understand. It is remarkable that he mentions "Thales (!) and others" as advocates of the daily rotation of the Earth [*15*: 69–76].

C. T. Bartholin's immediate successor, *Georg Detharding* (1671–1747; professor at Rostock 1697, and since 1733 of medicine and physics at Copenhagen), prepared his own textbooks of natural philosophy, but he continued, regarding the world systems, Bartholin's neutral exposition with a clear inclination towards the Copernican solution [*24*: 66–72; cf. *25*: 25–27].

When during the period 1747–53 the ageing professor of astronomy, *Peder Horrebow*, held the chair of physics, too, he felt the need for revising the praiseworthy *Specimen*, then 56 years old, from which he had himself in his youth acquired his knowledge of natural philosophy. And after having in vain tried to comment upon the *Specimen* and to change its text as well as its ordering of subjects he finally produced a completely new text-book, the *Elementa philosophiæ naturalis* [*56*: 5].

Of course, Horrebow emphasizes the astronomical aspects of natural philosophy, but he adhered to the Cartesian cosmology as well as C. T. Bartholin. "Those who hastily rejected Descartes' vortices did unawares throw out the baby with the bath water". [*56*: 16]. According to Horrebow the real enemies of the establishment of a true natural philosophy were the Newtonians whose exaggerated employment of attraction means an explanation of the phenomena of nature by words instead of by reasons, even if Newton himself had, in his Optics, made his reservations [*56*: 55–56]. Already in 1718 Horrebow expressed his attitude towards Newtonian physics by a remark on the variation of the lunar parallax: "The benevolent reader may ask Newton himself for the causes, but my physics of the heavens will, God willing, teach the causes of the causes" [*57*: 42].

So when about 1750 Newtonian physics nevertheless gained a footing in Denmark, it was due only to *Jens Kraft* (1720–1765; professor of mathematics since 1746) and his teaching of Newtonian mechanics at the Academy of Sorø. In 1747 Jens Kraft gave a lecture on Newton's and Descartes' systems to the Learned and Scientific Society of Copenhagen [*61*]. He prepared a series of philosophical textbooks following in most respect his own former teacher Christian Wolf, but for the physical aspect of this world he has "directly opposed to the great multitude of contrary opinions dared to establish the Newtonian hypothesis while rejecting the Cartesian one, and that the more confidently because I take it to be fully demonstrated that the latter is impossible" [*62*: preface].

Descartes had, on one hand, freed the learned society from orthodox Aristotelian authority, but on the other hand the Cartesian cosmology had deprived the astronomers of their freedom to chose at will their hypotheses, or rather it favoured the Copernican hypotheses more than presupposed by the Longomontanian compromise earlier uncritically accepted. After all most academics, except some of the theologians, were probably in their hearts Copernicans. But no stellar parallax had so far been found, and consequently no world system had been *proved* true. My next chapter will make it clear that Rømer believed and that his pupil Peder Horrebow openly declared to have found an annual parallax of the fixed stars. But before Horrebow Copernican sympathies were often expressed rather ambiguously [*58*, *97*, *87*].

V. THE TRIUMPHANT COPERNICUS

Rømer's Tychonian planetary machine (1697)

With *Ole Rømer* (1644–1710) Denmark had another astronomer comparable in genius with Tycho. Like his great predecessor he advanced observational astronomy in an epoch-making way, so that Bessel could praise his ideas within that area as being far in advance of their time [Astr. Nach. No 49, 1823, col. 14]. That Rømer had been influenced by Tycho's works was mentioned above [cf. p. 136], and it is well-known that he spent the years 1672–1681 at Paris, where he associated with the leading scientists of the recently founded Académie Royale des Sciences de Paris of which he became a prominent member.

On his return to Copenhagen he took the chair of astronomy at the University, furnished the observatory at the Astronomical Tower with new instruments, and moreover established two private observatories, viz. about 1690 the *Domestic* Observatory in his home at St. Kannikestraede, and in 1704 the *Tuscular* Observatory near the Bartholins' countryhouse some 15 kms west of the Round Tower. Becoming entrusted with a variety of administrative and engineering duties, Rømer had but little time left for the astronomy, and his printed astronomical works consist of a few small papers. In addition his instruments and the major part of his observational records were consumed by the fatal conflagration of Copenhagen in 1728. We therefore have to found a study of his astronomical achievements mainly upon the *Mathematico-physical works* of his pupil and successor, P. N. Horrebow, who devoted his life to spread the knowledge of the astronomical inventions of his teacher and to restore theoretical astronomy on a Rømerian basis.

Peder Nielsen Horrebow (1679–1764), a poor fisherman's son, at the age of 17 entered the grammar school at Aalborg; privately he studied Longomontanus' *Astronomia Danica*, and he made observations with simple wooden instruments. During the period 1704–06 he studied with Rømer, working as an assistant in the Rømer observatories. After that the earned his living as a private tutor and since 1711 as an accountant. In 1714, finally, he took the astronomical chair at Copenhagen.

Rømer accepted the third law of Kepler's on planetary motions [*96*: 161], and he used, in his observatories, the *Rudolphine Tables* [*54*: IV § 3]. This is no wonder, but the fact that, to my knowledge, nobody in Denmark had done so before is remarkable and shows the strength of the foregoing Tycho-Longomontanus tradition. It is worth noticing, too, that Kepler's planetary models were not until 1713 made directly the subject of an academic dissertation [*94*: cf. *95*].

Rømer was a Cartesian and particularly influenced by Huygens, so he was a Copernican, too. It is tempting to add: of course he was; but to Rømer these things were not a matter of course. He concludes, in his *Adversaria*, some critical considerations concerning the vortex theory of planetary motion by adding: "And I acquiesce in these reflections, or rather I withdraw from the immense sea into the harbour" [*96*: 55].

His wonderful mechanical devices reproducing the motions of the planets according to the Copernican system became widely renowned, and copies were prepared for the monarchs of France, Siam and China. But when, in 1697, a copy had to be mounted at the Round Tower of Copenhagen it was adapted to the Tychonian system [*54*: III, 143 ff.]. Of course this may be a gesture in honour of Tycho, but it may also reflect a theological resistance as the model should serve educational purposes. But for all that it cannot be excluded that Rømer was himself "acquiescing" until he dared rely on his determination of stellar parallax, which, perhaps, never happened.

Rømer's search for an annual stellar parallax (1692–1710)

For Rømer's scientific mind demanded proofs before he would express himself definitely; regarding the annual motion of the Earth only the discovery of a stellar parallax could provide the necessary proof. The above-mentioned letter from 1703 to Leibniz [cf. p. 136] contains Rømer's review of his own achievements within theoretical astronomy; among other things he maintains to have been convinced for ten years that the fixed stars were subjected to parallactic displacements so that he might well soon publish his results regarding this subject. A little later he asks Leibniz for whatever information, from books or from letters, concerning the parallax of the fixed stars [*54*: II. 160–162].

In 1728 P. N. Horrebow found among Rømer's posthumous papers a sheet entitled: *The Movable Earth, or the Parallax of the Annual Orbit from Observations of Sirius and Lyra, Carried out at Copenhagen in the Years 1692 and 1693.* Horrebow published and commented upon this masterly written essay in his *Basis Astronomoe* [*54*: III. 61–68; a Danish version is found in *123*: 83 ff.]. It fully reveals Rømer's continuous and persevering preoccupation with the problem of the stellar parallax which evidently formed his principle motive for establishing this *Domestic* meridian circle, and perhaps, his *Tusculan* observatory too. He maintains to have found, for the said two stars separated by about 180 degrees, a semestrial variation of the difference between their right ascensions which shows the double sum of their parallaxes to be about 0;1°. – He dared not put confidence in variations of stellar declinations mainly because of the influence of refraction.

It is hard, not to say impossible, to avoid the conclusion that Rømer did himself believe to have shown that the Earth performs an annual motion around the Sun. His last hesitation concerns the supposed different working of his clocks during day and night hours respectively, but he asserts that he has made even that doubt disappear. Yet his scruples, whatever they were, prevented Rømer from publishing his paper. – In 1853 C. A. F. Peters from an analysis of the Rømerian observations concludes that the indisputably produced effect *is* probably due to the influence of variations of temperature upon the clocks as well as upon the positions of the instruments of observation [*130*: 15–18].

Peder Horrebow's determination of the stellar parallax (1714–27)

However, P. N. Horrebow was less cautious than his master. Regarding stellar parallax he had in 1714 arrived at a conclusion similar to Rømer's [*54*: III. 249]. In 1717 he had an improved and larger copy of Rømer's planetary machine at the astronomical Tower so changed that it could be switched over between the Tychonian and the Copernican system of which "only the latter is true" [*54*: III. 148]. And a few months before finding the Rømer MS quoted above, he had published his *Copernicus Triumphans*, or *Epistolary Treatise on the Parallax of the Annual Orbit* to the heir to the Danish throne, later King Christian the VIth, on the occasion of his birthday on the 30th of November 1727 [*54*: III. 241–290].

Having duly referred the origin of this work to talks with Rømer and to a note from Rømer's diary of observations Horrebow discusses Flamstead's unsuccessful attempt of proving the Earth to be moving because of variations of the altitude of the pole star. From Rømer's *Domestic* observations during the period 1701 Febr. — 1704 Oct., and his *Tusculan* observations during the period 1705 March — 1709 Sept. he thereafter demonstrates the existence of semestrial variations of the difference between the right ascensions of some selected pairs of fixed stars. Thence he challenges the anti-Copernicans to explain this phenomenon in a reasonable way without moving the Earth, and the excellent French, English, and Russian astronomers to investigate the matter in order to confirm his result.

From a supposed stellar parallax of 0;0,15° and on the condition that all the fixed stars in question are as great as the Sun and equally distant from it, the following chapters deduce the distance and the apparent magnitude of the fixed stars and the orbital velocities of the Earth and Saturn in the true Copernican system as compared to the unreasonably high velocities of the Sun, Saturn, and the fixed stars demanded by the spurious Ptolemaic system. The tenth chapter brings Horrebow's biting attack on the catholic clergy which evidently wants to defend perpetually the Ptolemaic nonsense. The eleventh chapter contains a detailed philological and theological discussion of the Scriptural texts which Riccioli had, in his *Almagestum Novum*, adduced against the Copernicans.

As for the protestants Horrebow ventures to say that the more sane theologians had never been concerned at the progress of Copernican astronomy; some did even write to support it; and he remembers that a highly learned and distinguished Copenhagen theologian many years ago had declared that he did not find the Copernican system harmful in any respect, but should gladly accept it, provided only that some proofs were obtained. And Horrebow himself had for 21 years, i. e. since in 1706 he published his dissertation on the precession of the equinoxes [*53*: 7 ff.], advocated and taught the Copernican system without having for that reason been blamed by any Danish clergyman.

Horrebow concludes his tract by asserting that his demonstration has finally raised the Copernican world system from the sphere of hypotheses to that of truth which you may teach publicly and privately without any obligation to silence the squabble of cantankerous persons. Finally he explains how to apply the found correction of parallax to stellar observations in order to establish a star catalogue with heliocentric coordinates. Having, however, realized that the fixed stars are unlikely to be found at equal distances of the Sun Horrebow later on dropped this correction as in most cases insensible compared to observational errors [*54*: IV § 12].

In brief Horrebow, by his astronomical authorship, succeeded in building an astronomical temple of which his main works form the *Key*, the *Foundation*, the *Court*, and the *Sanctuary* [*54*]. And Copernicus was — like Resen's Luther — a century earlier [cf. p. 138] — the triumphant reformator to receive the homage of the astronomical flock. Horrebow's proof was based upon a pseudo-effect, but nobody was aware of that, so it did not matter. Denmark became at last Copernicanized, and Peder Nielsen Horrebow did it.

Christian Horrebow's re-determination of the stellar parallax (1742–46)

The *Copernicus Triumphans* was kindly reviewed by the *Acta Eruditorum* [1729: 65–69]; the *Basis Astronomiæ* quotes some letters of approval from foreign astronomical colleagues [*54*: III. 73–79], and from letters exchanged between Horrebow and Johann Lulofs we learn that the latter did even translate and comment upon the book [*54*: III. 291–304]. However, the reliability of Horrebow's work was questioned by Eustachio Manfredi [*79*: 612–618], so Christian Wolf, in his *Elementa astronomiæ*, concludes that no annual motion of the Earth had so far been demonstrated from any determination of stellar parallax [*107*: III. 602–603; cf. V. 123–125]. So even Horrebow's own sons began to doubt the trustworthiness of their father [*55*: 109; *51*: 4].

Hence Horrebow's son, *Christian Horrebow* (1718–1776; professor of astronomy, designated in 1743 and ordinary from 1764), together with his brother, *Andreas Horrebow* (c. 1720–1745), and *Niels Jensen Gundestrup* (1713–1779) in 1742–1743 and 1746 made new observations, and they arrived at much the same conclusion as P. N. Horrebow in 1714, although their resulting value of parallax appeared to be a little smaller [*51*].

In 1728 Bradley had, however, already advanced his theory of aberration which, together with Rømer's demonstration of the progressive motion of light, provided further strong evidence to the advantage of the Copernicans. Bradley denied that any sensible parallax in the fixed stars had yet appeared [Phil. Trans. 1728, *35*, 660], but even apart from that his theory was likely to outshine entirely P. N. Horrebow's "proof" that the Earth did move. So Horrebow maintains that the finds Bradley's theory difficult to understand [*55*, Danish version: 111]. Consequently, Horrebow established his own theory of aberration, or — in Horrebow's terminology — of *anaclasis*, explaining the phenomenon as being due to some sort of refraction at the outer limit of the vortex in which he placed the Earth excentrically. His treatise on the subject opens with a poem of congratulation to the triumphant Copernicus by another of Horrebow's sons, *Niels Horrebow* (1712–1760) [*55*: 112].

Eventually, around 1750 Christian Horrebow had to acknowledge that most astronomers simply did not accept any demonstration of the stellar parallaxes [*52*: 133–134], so we find him busy working out a new micrometer method for determining these parallaxes [*52*: 140–142]. However, all the said activities did not include any serious querying of the annual motion of the Earth; the discussion was about technicalities of observational astronomy, and in 1747 Chr. Horrebow could note that no important astronomer had found it worth the trouble to pay attention to Nicolai Müller's voluminous dissertation from 1734 entitled *On the unquestionable motion of the Sun and the firm rest of the Earth* [*51*: 12; *83*].

So it would be off the point to go into further details of the subject; the Copernican system had now become a commonplace precondition within astronomy.

But outside astronomical circles anti-Copernicans were of course still found. E. Spang-Hanssen has sketched appropriately the general spirit at the University of Copenhagen during the first half of the 18th century within the scientific disciplines, and especially traced the influence of P. N. Horrebow's astronomy upon the authorship of the outstanding Danish writer of comedies, *Ludvig Holberg* (1684–1754). Thence he convincingly argues that not only the common man, but also a great many people of culture did not at all accept the new scientific ideas of the time and in particular not the Copernican world system [*124*]. And my bibliography exemplifies later anti-Copernicanism [*5*, *86*, *111*].

BIBLIOGRAPHY

List of Sources

1 Aasheim, Arnoldus Nicolaus, *De systemate Copernicano sive vero disputatio* (Hafniæ, 1767). – A rather thorough dissertation containing several useful references.

2 Agerholm, Christianus, *De semidiametro lunæ dissertatiuncula mathematica* (Hafniæ, 1715), – Agerholm was a pupil of P. N. Horrebow's. He does away with the dilemma concerning the alleged stand-still of the Sun by proposing that the event referred to in Josua's book should regard the rotation of the Sun around its own axis.

3 –, *De parallaxi orbis annui* (Hafniæ, 1716). – I have not succeeded in finding this dissertation; but it is mentioned by Nielsen [*112*: 6].

4 Alburgo-Danus, Nicolaus Nicolai, *Disputatio physiologica de elemento terræ* (Witebergæ, 1622). – This dissertation echoes Caspar Bartholin's chapter on the Earth [cf. p. 124].

5 Alethophilus, Timotheus, Pseudonym for Peder Gutfeld (1727–1797). Orthophilus, Philolaus, Pseudonym for Otto Christensen Riese (1697–1779). – Both partakers in the controversy which covers more than 500 pages were vicars.

5a *Alethophili Brev til Philolaum Orthophilum angaaende Jordens Stilstand eller Ubevægelighed* (Aalborg, 1768). – The letter is dated 1765, Nov. 19th; it claims to answer an earlier letter from Riese which probably did not appear in print.

5b *Ortophili Copernikanske Skrivelse til Alethophilum.* – I have found no date for this letter; it claims to answer *a*. Hence I identify it with *Orthophili Giensvar* although the latter is mentioned by Nielsen as a separate writing [*112*: 177].

5c *Alethophili Tychonianske Svar og imodsatte Beviiser. Første Part.* – Dated 1766, Oct. 14th.

5d *Alethophili Giensvar og imodsatte Beviiser. Anden Part.* – Dated 1767, March 26th.

5e *Philolai Orthophili Undersøgning, hvor vidt Alethophili Tanker om Jordens Ubevægelighed i den første Part af hans Responsum findes at være grundet* (Aalborg, 1768). – Dated 1767, June 1st.

5f *Philolai Orthophili Undersøgning... i den anden Part ...* (Aalborg, 1768). – Dated 1768, May 28th. The letter opens with a preface by Prof. Peder Rasch. I have used a joint edition of the letters *a* – *f*. Some of these were published earlier in various combinations and sometimes with slightly different titles, too.

5g *Alethophili fornødne Anmærkninger imod Orthophili fortsatte Undersøgning* (Haderslev, 1768). – Dated 1768, July 19th.

6 Anonymous, *Orbis Terrarum in Centrum Mundi restitutus atque a Vertigine Copernicana liberatus* (MS at the Royal Library of Copenhagen: Kallsk Sml. No 354.4°). – From a cryptogram [fol. 19v] I find the date 1715 for the composition of this MS. The tract makes up a neat, but – from an astronomical point of view – naïve piece of geometry directed against the motion of the Earth.

7 Aslachus, Cunradus Bergensis, *De natura Cœli triplicis libelli tres. Quorum I. de Cœlo Aëreo, II. de Cœlo Sidereo, III. de Cœlo Perpetuo E sacrarum litterarum et præstantium Philosophorum thesauris concinnati* (Sigenae Nassoviorum, 1597).

8 Bangius, Carolus Gaeorgius, *Compendium naturalis scientiæ; ex octo libris physicæ auscultationis Aristotelis, et aliis Philosophis decerptum ...* (Witebergæ, 1599).

9 Bartholinus, Casparus, *Exercitatio physica de natura* (Witebergae, 1605).

10 –, *De terra aëre et igni Institutio Physica succincta: Cum præmissa elementorum theoria generali* (Hafniæ, Rostockii, 1619; Gryphiswaldi, 1624), – Inserted into *11*: fol. Ee-Ji.

11 –, *Systema Physicum Ex Autoris genuinis ... libris ... coagmentatum* (Hafniæ, 1628).

12 –, *Physica præcepta ex Caspari Bartholini scriptis potissimum excerpta et digesta pro tyronibus* (Hafniæ, 1656, 1676).

13 Bartholinus, Casparus Thom. F., *Collegii Physici Privati Disputationes I–VIII* (Hafniæ, 1684–85).

14 –, *Summa philosophiæ naturalis. Ad recentiorum mentem accomodata, seu Explicatio Præceptorum Physicorum, quæ ex scriptis b. m. Avi Casp. Bartholini hactenus in Scholis tradita sunt* (Hafniæ, 1689). – Other editions 1688, 1706; annotated English edition, London, 1754.

15 –, *Speciminis philosophiæ naturalis Novissimis Rationibus et experimentis illustratæ disputationes I–V* (Hafniæ, 1690–92).

16 Bartholinus, Erasmius, *Consideratio Astronomica Conjunctionis magnæ Saturni et Jovis præsentis Anni 1663* (Hafniæ, 1663).

17 –, *De Cometis Anni 1664 et 1665 opusculum. Ex observationibus Hauniæ habitis adornatum* (Hafniæ, 1665). – Reviewed in *Phil. Trans.* 1669, *4*, p. 1071.

18 –, *De naturæ mirabilibus quæstiones academicæ* (Hafniæ, 1674).

19 Brahe, Tycho, *Opera Omnia*, vols. I–XV (Hauniæ, 1913–1929). – Editor I. L. E. Dreyer.

20 –, *Description of his Instruments and Scientific Work* (København, 1946). – A translation of *Astronomiæ Instauratæ Mechanica* by Hans Ræder, Elis Strömgren, and Bengt Strömgren. A similar translation into Danish is found in *Nordisk Astronomisk Tidsskrift*, 1933, 1946–1955.

21 Brunsmand, J., *Et Almindeligt og Stedsevarende Kalender, Hvor udi visis Hvad dertil hører... ved Gammel og Ny Stiil, Samt Solens og Maanens Gang og Lysning, Saavelsom andet mere* (Kiøbenhavn, 1688).

22 Bröndlund, Laurentius, *Anonymi autoris Novam inquisitionem in sententiam Copernicanam præsenti Schediasmate paucis excussam Publicæ censuræ subjiciet* (Hafniæ, 1728). – The said anonymous tract is *Inquisitio Nova in Sententiam Copernicanam ...* (Analecta ex omni meliorum literarum genere ... que ... evulgat Societas Caritatis et Scientiarum, vol. I, Lipsiæ, 1725), pp. 129–158. Its author is indicated by the letters V. E. L. D., and – in the preface to *Analecta...*, vol. II, 1730 – called "collega noster". He proposes a modified Tychonian world system with several centres of the planetary motions. His paper contains useful further references relevant to the history of Copernicanism.

23 Cimber, Elias Olai, *Diarivm astrologicvm et metheorologicum anni a nato Christo 1586; Et de Cometa qvodam rotvndo ... consideratio Astrologica* (Vranibvrgi, 1586). – Part of this work is edited by Dreyer in *19*: IV. 398 ff.

24 Dethardingius, Georgius, *Fundamenta scientiæ naturalis quibus in rebus naturalibus ... hactenus detecta ... exponuntur ...* (Havniæ, 1735; 2nd edition 1740).

25 –, *Epitome erotematica Physices, in vsvm jvventvtis scholasticæ* (Havniæ, 1746).

26 Dibvadius, Ericus Christophorus, *Almanach oc Practica paa det Aar ... MDLXXXV. Tilhobe-beregnid ved (E. C. D.). Oc der hoss læris ... at gjøre et slags solskiffuer ...* (Hafniæ, 1584 (?)).

27 Dibvadius, Georgius Christophorus, *Commentarii breves in secundum librum Copernicii, in quibus argumentis infallibilibus demonstratur veritas doctrinæ de primo motu, et ostenditur Tabularum compositio* (Witebergæ, 1569).

28 —, *Propositiones aliqvot mathematicæ, in eorvm vtilitatem, qui ... Matheseos sunt studiosi, collectæ et publicatæ* (Hafniæ, 1577).

29 —, *Ratio Componendi præcipvos, primi motvs, canones ... explicata* (Hafniæ, 1577).

30 —, *Theses de regione mvndi ætherea ...* (Hafniæ, 1581).

31 Dieterichsen, Trugillus Asloja-Norweg: *Disputatio Theorematica adstruens Motum Νυχϑημεϱινον a telluris circumgyratione dependere* (Rostochii, 1638). — Sub præsidio Petri Laurembergi.

32 Finckius, Jacobus: *Disputatio physica de Elementis tum in Genere tum Specie consideratis* (Argentorati, 1612).

33 —, *Theses philosophicæ de Elementis* (Hafniæ, 1639).

34 Finckius, Thomas, *Ephemeris coelestium motuum Anni 1582, supputata ex Tabulis Prutenicis* (Argentorati, 1582).

35 —, *Geometriæ Rotvndi Libri XIIII. Ad Fridericvm Secundum ...* (Basileæ, 1583; 2nd edition 1591).

36 —, *Horoscopographia sive De inveniendo stellarum sitv astrologia. In qua tabulæ declinationum, ascensionum rectarum ... Ad Henricvm Ranzovivm, Vicarium Regium* (Slesvici, 1591).

37 —, *Thees de hypothesibvs astronomicis, dimensionis mvndi ac primi motus circulis* (Hafniæ, 1592).

38 —, *Theses de diebvs ac noctibus* (Hafniæ, 1601).

39 —, *Theses de mensibus et annis* (Hafniæ, 1602).

40 —, *Methodica tractatio doctrinae sphæricae* (Coburgi, 1626) — see *115*: vol. I, part 1, no 2816.

41 de Fine, Johannes Arnoldi, *Enneas dissertationum physiologicarum... Disputatio prima physiologica, de definito et genere physicæ* (Giessæ, 1609).

42 (Fontenelle, B.), *Samtaler Om Meer end een Verden, Imellem et Fruentimmer og en lærd Mand. Af det nyeste Franske Oplag... Med Professor Gottschedens og egne nye Anmærkninger forsynede af Friderich Christian Eilschow* (Kiøbenhavn, 1748; new issue 1764).

43 Frommius, Georgius, *Dissertatio Astronomica De Mediis qvibusdam ad Astronomiam restituendam necessariis pro Introductione in Theatrum Astronomicum Hafniense... Christiani Longomontani ... cum ... Johanne Bapt. Morino ... instituta* (Hafniæ, 1642).

44 —, *Responsio ad ... Johannis Bapt. Morini ... defensionem astronomiæ restitutæ* (Hafniæ, 1645).

45 —, *Exercitationum sphæricarum prima, de principiis quibusdam doctrinæ sphæricæ* (Hafniæ, 1648). — The proverb quoted on p. 135 of our text is found in Horats, *Ep.* 1, 7, 23.

46 Galilei, Galileo, *Opere*, vols. I–XX (Firenze, 1929–1939).

47 Galschiöt, Johannes, *Dissertationis de differentia systematum Ptolemaici, Copernicani et Tychonis Brahe particulæ I–II* (Hafniæ, 1774–75). — A rather trivial reproduction of the main features of the said three world systems.

48 Haltorius, Thorlefus, *Schediasma mathematicum de Aplane* (Hafniæ, 1707). — A rather elaborate dissertation in which *Thorleif Haldorsen* (died 1714) follows C. T. Bartholin. He argues that the Universe is indefinite in extension and considers the immense magnitude of the Copernican Universe far more reasonable than the incredible swiftness of the stars assumed by Ptolemy.

49 —, *Schediasma de Sole Retrogrado Es. XXXVIII v. 8* (Havniæ, 1710). — Haldorsen defends the Copernican system against the Scriptural argument concerning the alleged retrogradation of the Sun. And he refers to:

49a Henr. Nicolai: *Dissert. pec. de Revol. Terræ.*

49b anonymus quidam: *Epist. de Terræ motu; edita ultrajecti A. 1651.* I have not identified any of the two writings.

50 Hilarius, Georgius: *Præcepta doctrinæ sphæricæ Brevissima ... ad captum incipientium accomodata* (Hafniæ, 1656).

51 Horrebow, Christian, *De parallaxi fixarum annua ex rectascensionibus, quam post Roemerum et parentem ex propriis observationibus demonstrat* (Hafniæ, 1747). — Published earlier as three separate dissertations.

52 —, *Afhandling om Fixstjernernes Distance fra Jorden* ("Videnskabernes Selskabs Skrifter", 1751—54, Siette Deel) pp. 129—152. — A Latin version was published separately as a dissertation in 1755.

53 Horrebowius, Petrus, *Tentamen academicum III pertractans æqvinoctiorum præcessionem* (Hafniæ, 1706).

54 —, *Opera Mathematico-physica*, vols. I—III (Havniæ, 1740—41). The main separate writings within this joint edition are the following:

54a *Clavis Astronomiæ, sive astronomiæ pars physica* (1st unfinished edition 1725). — Reviewed in Acta Eruditorum, 1726, p. 506 ff.

54b *Basis Astronomiæ sive astronomiæ pars mechanica... cum Triduo Roemeri...* (1st ed. 1735).

54c *Copernicus Triumphans, sive de parallaxi orbis annui tractatus epistolaris* (1st ed. 1727).

54d *Atrium astronomiæ, sive de elementis astronomiæ ex observationibus* (1st ed. 1732).

54e *Adytum Astronomiæ sive Astronomiæ pars theoretica.* — MS at the Royal Library of Copenhagen (ADD 104.4°), never published, but intended to form the IVth volume of the *Opera*.

55 —, *Anaclastice, in parte priori constituto vorticulo Telluris... pro adserendo Copernico Triumphante* ("Scripta a Societate Hafniensi", pars tertia, 1747), pp. 107—180. — A shorter Danish version is found in "Videnskabernes Selskabs Skrifter", 1747, Tredie Deel, pp. 107—120.

56 —, *Elementa philosophiæ naturalis* (Havniæ, 1748).

57 —, *Nova Theoria Lunae, cvm tentamine restitutionis Motvvm Lvnarivm* (Bibliotheca Novissima Observationvm ac Recensionvm, *Sectio I*, 1718), pp. 34—46.

58 Judichaerus, Olaus, *Ψεύδοκρισις sensuum et ἀναγκαιοκρησις hypothesium sive disputatio mathematico physica prior Ostendens fallaciam sensuum in judicando et Hypothesium necessitatem atque utilitatem* (Hafniæ, 1684). — *Ole Judichær* (1661—1729) was a pupil of Rømer's. In this most lively written dissertation statements occur which may equally well advocate the Copernican and the Tychonian view.

59 Kierulius, Petrus Bartholinus, *Apologia pro observationibus, et hypothesibus... Tychonis Brahe... Contra... Martini Hortensii Delfensis criminationes et calumnias, quas in Præfationem commentationum Præceptoris sui Philippi Lansbergii Middelburgensis, de motu terræ diurno et Annuo etc. consarcinnavit* (Hafniæ, 1632).

60 Kopernikus, Nikolaus, *Gesamtausgabe*, vols. I—II (München, Berlin, 1944—49).

61 Kraft, Jens, *Betænkninger over Neutons og Cartesii systemata, tilligemed nye Anmærkninger over Lyset* ("Videnskabernes Selskabs Skrifter", Tredie Deel, 1747), pp. 213—296. — A Latin version is found in "Scripta a Societate Hafniensi", pars tertia, 1747, pp. 273—348.

62 —, *Cosmologie eller anden Deel af Metaphysik* (Kjøbenhavn, 1752).

63 Kragius, Andreas Ripensis, *Theses de definitione et svbjecto physices* (Hafniæ, 1593).

64 Krogius, Nicolaus, *Disputatio physica de origine motus diurni terræ* (Hafniæ, 1719).— Krog reproduces the Cartesian hypothesis.

65 Lang, Hans, *Cosmosistatia sive commentarius contra Aristotelem de fabrica mundi* (1606) — see *112*: 124.

66 Langus, Wilhelmus, *De Annis Christi libri duo. Primus varios variarum Gentium annos et tempora exponit: Secundus, in duas partes divisus, priori Epochas nobiliores, quæ Christi Domini passionem antecedunt; posteriori ipsum Dominicæ Nativitatis et Passionis tempus demonstrat* (Lugduni Batavorum, 1649).

67 —, *Exercitationes Mathematicæ VII. De annua emendatione et motu Apogei Solis...* (Hafniæ, 1653).

68 Lansbergen, Philippus, van: *P. Lansbergii Commentationes in Motum Terræ diurnum, et annuum, et in verum adspectabilis cæli typum... Ex Belgico sermone in Latinum verse, a M. Hortensio... una cum ipsius præfatione, in qua astronomiæ Braheanæ fundamenta examinantur; et cum Lansbergiana astronomiæ restitutione conferuntur* (Middelburgi, 1630).

69 Lesleus, Claudius Nicolai, *Geometria sive ars metiendi simul tractatus de Quadratura Circuli et Quæstiones nonnullæ discussu dignissimæ de Motu terræ, et Cometis. Item Unicus sacer sermo super Textum Luc: c. 2, v. 34* (Lubecæ, 1661).

70 Longomontanus, Christianus Severini, *Systema mathematicum, thesibus expositum* (Hafniæ, 1609).

71 —, *Disputatio prima astronomica, de præcognitis: In qua Definitio materiæ Coeli adeoque loci cuncta corpora mundana majora, suo gremio complectentis discutietur: Vna cum Natura, et forma ipsorum corporum, imprimis, qua motibus suis apta sunt* (Hafniæ, 1611).

72 —, *Disputatio philosophica, quæ secvnda astronomiæ est, de sphæræ coelestis legitima constitutione, officio, et multiplici utilitate* (Hafniæ, 1612).

73 —, *Disputatio tertia astronomica, de systematis mundani triplici hypothesi* (Hafniæ, 1614). — I have not succeeded in finding this no doubt interesting dissertation.

74 —, *Disputatio prima de astrologia seu prædictionibus ex astris, in qua Primum contra adversarios, certitudo... convincitur. Deinde generalis causarum indicatio... subnectitur...* (Hafniæ, 1621).

75 —, *Astronomia Danica... in duas partes tributa; Quarum prior Doctrinam de diurna apparente siderum revolutione... duobus libris explicat: posterior Theorias de motibus Planetarum ad observationes D. Tychonis Brahæ, et proprias, in triplici forma redintegratas, itidem duobus libris complectitur, Cum Appendice de... Stellis Novis et Cometis* (Amsterdami, 1640; other editions 1622 and 1663).

76 —, *Disputatio de matheseos, indole, methodo, et utilitatibus, velut incitamentis, quibus ingeniosæ juventuti in Scholis Gymnasiis, et Academiis discenda commendatur* (Hafniæ, 1636).

77 —, *Introductio in Theatrum Astronomicum. Quod in honorem Coelestium Opificis D. O. M. Nec non totius Orbis utilitatem, auspicio... Christiani IV. Havniæ... modo instauratur,... Cui accedit brevis Discursus de I.Æquatione Diei naturalis, II. Parallaxibus Siderum, III. Refractionibus siderum* (Havniæ, 1639).

78 Lous, Christianus Carolus, *Dissertatio historica sententias veterum varias de systemate mundi Exhibens* (Hafniæ, 1744). — The author argues that the idea of the movable Earth may be traced back to Pythagoras who had, perhaps, taken it over from a still elder Italian tradition (pp. 15–18).

79 Manfredii, Eustachius, *De novissimis circa fixorum siderum errores observationibus. Ad... Antonium Leprottum... epistola* (De Bononiensi scientiarum et artium instituto atque academia Commentarii, 1731), pp. 599–639.

80 Mercator, Nicolaus, *Institutionum astronomicarum libri duo, de motu astrorum... Secundum Hypotheses Veterum et Recentiorum præcipuas; deque Hypotheseon ex observatis constructione: cum Tabulis Tychonianis Solaribus, Lunaribus, Lunæ-Solaribus, et Rudolphinis Solis, Fixarum, Et Quinque Errantium; ... Quibus accedit appendix De iis, quæ Novissimis temporibus Coelitus innotuerunt* (Londini, 1676).

81 Morin, Jean Baptiste, *Coronis Astronomiæ jam a fundamentis ... restitutæ; qua respondetur ad introductionem in Theatrum Astronomicum C. Longomontani* (Parisiis, 1641).

82 —, *Defensio Astronomiæ a fundamentis... restitutæ, contra ... G. Frommii... Dissertationem Astronomicam* (Parisiis, 1644).

83 Mullerus, Nicolaus, *De indubio solis motu immotaque telluris quiete dissertatio* (Kiel, 1734).

84 Orthophilus, Philolaus, see *5*.

85 Orthunganus, Johannes Claudii, *Collegii physici Disputatio Septima De coelo sidereo et stellis in genere* (Hafniæ, 1651). — Sub præsidio Jani Jani Bircherodii.

86a Peper, Johan, *Tanker om Jordens Dreining eller... Bevis, at Jordkloden, ikke kan dreie sig...* (Kjøbenhavn, 1792). — This pamphlet gave rise to two anonymous rejoinders (*b, d*) both of which were in turn answered (*c, e*) by Johan Peper, a self-taught Copenhagener tailor striking pseudonymously a desperate blow for stopping the whirling Earth.

86b *Kort Kritik over og Medhold for Johan Pepers nyebagte astronomie* (Kjøbenhavn, 1792).

86c —, *Svar paa det imod min udgivne skrift sammenvævede Piat* (Kjøbenhavn, 1792).

86d *En liden betragtning over Johan Pepers piece: ... samt ... Beviis, at Buclerne paa forfatterens Paryk ... kan beholde deres Facon uagtet Jorden dreier sig...* (Kjøbenhavn, 1792).

86e —, *Betænkning over den saa kaldede liden Betragtning ... samt nye Anmærkninger over Jordens Dreining, og tillige en kort Anviisning... at Solen kan gaae* (Kjøbenhavn, 1793).

87 Ramus, Joachimus Friedericus, *Jupiter planeta cum quattuor satellitibus in forum mathematicum productus* (Hafniæ, 1713). — *Joachim Friderik Ramus* (1685 (?)—1769) was a pupil of Rømer's. His dissertation shows by a simple diagram the Copernican explanation of Jupiter's retrogradations.

88 Resenius, Johannes Pauli, *Lutherus triumphans Jubilæis sancte continuandis, per septentrionem, ipso die Omnium Sanctorum quotannis inde ab anno 1617* (Hafniæ, 1617).

89 Ripensis, Christianus Johannis, *Theses de sphæræ definitione et divisione* (Hafniæ, 1604).

90 —, *Theses de circulis sphæræ* (Hafniæ, 1605).

91 —, *Theses de Zonis* (Hafniæ, 1606).

92 —, *Theses de climatibus et parallelis* (Hafniæ, 1608).

93 Rosen, Edward, *Three Copernican Treatises* (New York, 1959).

94 Rossingius, Adamus Levinus, *Hypothesin Keplerianam paucis complectens Dissertatio astronomica* (Hafniæ, 1713).

95 —, *De ortu et progressu astronomiæ ellipticæ dissertatio* (Hafniæ, 1717).

96 Rømer, Ole, *Adversaria* (København, 1910). — Edited by Thyra Eibe and Kirstine Meyer.

97 Skive, Laurentius Th., *Theses Mathematicæ* (Hafniæ, 1703). — *Laurits Thomsen Skive* (1677—1711) was a pupil of Rømer's. Without further comment he asserts: "The Copernican philosophy does not, strictly speaking, maintain that the Sun rests, nor that the Earth moves. Vain are the arguments ordinarily brought forward abundantly against the perspicuity of the Copernican system".

98 Sperling, Johannes, *Institutiones physicæ* (Lubecæ, 1647).

99 Stenbuchius, Christianus, *Exercitatio physico-mathematica* (Hauniæ, 1648). — Sub præsidio Georgii Frommii.

100 von Steuben, Christian Ludewig, *Kurze Beschreibung des Welt-Gebäudes wie es Copernicus beschrieben ... Imgleichen Kurze Beschreibung, einer neu erfundenen Astronomischen Machine Darinnen sich alle himlische Cörper befinden ...* (Copenhagen, 1752).

101 Søeborg, Georgius, *Dissertatio de Phænomenis, ex motu terræ diurno pendentibus* (Hafniæ, 1779). — A rather trivial dissertation.

102 Thorlacius, Gislaus Islandus, *Collegii Physici Disputatio Octava De stellis fixis et errantibus* (Hafniæ, 1651). — Sub præsidio Jani Jani Bircherodii.

103 Tormius, Ericus Olai, *... Disquisitio Mechanica, in qua De Instrumentorum Necessitate ... sic ... disputatur ut I. Vera sententia ... adstruatur. II. Controversiæ, quæ ... inter Tychonicos et nostri seculi Clarissimos Mathematicos agitantur, examinentur ...* (Hafniæ, 1643).

104 —, *Disquisitionis Mechanicæ Continuatio, In qua de Instrumentorum, ... præcipue de Globo Tychonico sic ... disputatur ut ...* (Hafniæ, 1645).

105 Wilczeck, G. F. J., *Lehrsatz von der Lage der Erden* (Budissin, 1762). — A discussion concerning this book formed the starting point of the polemic between Gutfeld and Riese [5].

106 Windefontanus, Johannes Johannis, *Anatome Quæstionum Paradoxarum Claudii Nicolai Leslei, de Motu terræ, Anima hominis ... et Servatore Jesu die Jovis passo ...* (Hafniæ, 1662).

107 Wolfius, Christianus, *Elementa Matheseos Universæ, vols.* I—V (Halæ Magdeburgicæ, 1741—1756).

108 Worm, Olaus, *Cosmologica disceptatio secunda, De mundo et coelo ejusque partibus et Astris* (Hafniæ, 1620).

109 —, *Exercitationum physicarum Prima De naturalis philosophiæ constitutione* (Hafniæ, 1623).

110 —, *Liber Aureus philosophorum aquilæ Aristotelis de Mundi Fabrica ... nova analysi notis, com-*

mentariis doctiss. quibus controversiæ variæ ... enodantur, illustratus ab (O. W.) (Rostochii, 1625). – *Ole Worm* (1588–1654) was primarily an archaeologist, and within natural philosophy he follows his brother-in-law Caspar Bartholin.

111 Zytphen, W., *Solens Bevægelse i Verdensrummet* (Kjøbenhavn, 1863). – The author is a mad--man advocating the Tychonian world system.

Other works of reference

112 Nielsen, Niels, *Matematiken i Danmark 1528–1800* (København, Kristiania, 1912).

This bibliography of scientific literature in Denmark has formed the basis for the present treatise. I have consulted the following more general bibliographic works, too:

113 Nielsen, Lauritz, *Dansk Bibliografi 1551–1600* (København, 1931).

114 Bruun, Chr. V., *Bibliotheca Danica. Systematisk fortegnelse over den danske litteratur fra 1482 til 1830*, vols. I–V (København, 1961–63).

115 Houzeau, J. C. et Lancaster, A., *Bibliographie Générale de l'Astronomie jusqu'en 1880*, vols. I (in two parts) and II (London, 1964).

N. Nielsen (*112*) brings short biographical notices on the authors, and more detailed information as well as further references may in most cases be found in:

116 Bricka, C. F., *Dansk Biografisk Leksikon*, vols. I–XXVII (København, 1933–1944).

117 Meisen, V. (editor), *Prominent Danish Scientists through the Ages. With facsimiles from their works* (Copenhagen, 1932).

118 Petersen, Carl S., *Den danske Litteratur fra Folkevandringstiden indtil Holberg* (København, 1929).

In addition to which I shall call attention to a few writings on individual scientists:

119 Dreyer, J. L. E., *Tycho Brahe. A Picture of Scientific Life and Work in the Sixteenth Century* (New York, 1963).

120 Hofmann, Jos. E., *Nicolaus Mercator (Kauffman), sein Leben und Wirken, vorzugsweise als Mathematiker* (Akad. der Wissensch. und der Literatur, Abh. der Math.-Naturwissensch. Klasse, 1950, no 3), pp. 45–103.

121 Nielsen, Axel V., *Ole Rømer. En Skildring af hans Liv og Gerning* (Aarhus, 1944).

122 Pihl, Mogens, *Ole Rømers videnskabelige liv* (København, 1944).

123 Strömgren, Elis, *Ole Rømer som Astronom* (København, 1944).

One author has recently dealt with Copernicanism in Denmark during the first half of the 18th century, viz:

124 Spang-Hanssen, Ebbe, *Erasmus Montanus og naturvidenskaben* (København, 1965).

Finally I list a few writings to which I have made specific references in my text:

125 Carøe, K., *Den danske Lægestand*, vols. I–V (København, Kristiania, 1904–1922).

126 Dreyer, J. L. E., *A History of Astronomy from Thales to Kepler* (New York, 1953).

127 Hübertz J. R., *Aktstykker vedkommende Staden og Stiftet Aarhus*, vol. II (Kjøbenhavn, 1845).

128 Moesgaard, Kr. Peder, *The 1717 Egyptian years and the Copernican theory of precession* (Centaurus, 1968, *13*, pp. 120–138).

129 Neugebauer, O., *The Exact Sciences in Antiquity* (New York, 1962).

130 Peters, C. A. F., *Recherches sur la Parallaxe des Étoiles Fixes* (Mém. de l'Acad. de Sc. de Saint-Pétersbourg (VI), *5*, 1853, pp. 1–180).

131 Ravetz, J. R., *Astronomy and Cosmology in the Achievement of Nicolaus Copernicus* (Wrocław–Warszawa–Kraków, 1965).

132 Sandblad, Henrik, *Det Copernikanska Världssystemet i Sverige* („Lychnos", 1943, pp. 149–188, and 1944–45, pp. 79–131).

133 Zinner, Ernst, *Entstehung und Ausbreitung der Coppernicanischen Lehre* (Erlangen, 1943).

134 Brun, V., *Regnekunsten i det gamle Norge* (Oslo, Bergen, 1962).

SHIGERU NAKAYAMA
University of Tokyo

DIFFUSION OF COPERNICANISM IN JAPAN

Before we examine closely the diffusion of Copernicanism in Japan, it may be helpful (1) to define briefly what Copernicanism means by analyzing it into three conventional elements—astronomical, cosmological and physical — and (2) also to sketch the Far Eastern tradition with respect to each element. This juxtaposition will provide preparation for our main theme, the encounter of Copernicanism with a culture which did not share its historical roots. In so essentially foreign a milieu, it was perhaps inevitable that Copernicus's innovations played a radically different role from that which they played in the Western tradition.

WHAT IS COPERNICANISM?

Astronomical or observational. Copernicus is, without doubt, primarily a first-rate professional astronomer. His thorough mathematical treatment in *De revolutionibus* from Book II on must have challenged his astronomical successors, such as Erasmus Reinhold and Tycho Brahe. This aspect of Copernicanism may be called "astronomical Copernicanism" or "mathematical and quantitative Copernicanism" if not to say "observational Copernicanism". As regards practical astronomy, Copernicus was engaged in calendar reform, investigation of long-term variation of planetary positions, and eclipse predictions, which were the serious astronomical topics of his time[1].

Cosmological or geometrico-morphological. On the popular level, Copernicus is, of course best known as a cosmologist. The basic tenet of "cosmological Copernicanism" is obviously the heliostatic conception. His work on the order and relative distance of planetary orbits was only the outcome of this concept. Man's position in the universe, the taste for harmony, and even the idea of an infinite universe in the later Copernicanism are all closely associated with the cosmological element.

---

[1] Cf. Jérôme R. Ravetz; *The cosmology of Nicolaus Copernicus*, "Organon" No. 2, 1965, p. 54, and his *Origin of the Copernican revolution*, "Nature", March 11, 1961, pp. 859–860.

We must note that this aspect became weighty in popular thought only from the time of Galileo's recantation, although it must have been what drove Copernicus to the completion of the extensive *De revolutionibus*.

Physical or mechanical. Finally, "the moving earth" theory had important consequences for physics, as testified by those critics who denied it, and engaged in a controversy on how that motion, whether rotation or revolution, could possibly take place without causing strong winds on the surface of the earth. This physical argument constituted at the time a vulnerable point of Copernicanism.

THE FAR EASTERN BACKGROUND

Astronomy. Traditional Far Eastern astronomy, which remained centered in China, was oriented toward the special goal of providing the most accurate possible luni-solar calendar, and of predicting lunar and solar eclipses, since these predictions were an excellent means of checking the accuracy of any such calendar[2].

While calendrical astronomy was at the core of Far Eastern exact science, planetary astronomy, which was not highly developed, held a peripheral position, and played its main role in portent astrology.

It is hard to find in the development of Chinese astronomy any notable tendency towards conceptual schemes or mechanistic or geometrical models. The approach of the Chinese official astronomers was to represent numerically the course of the celestial bodies without depending upon a geometrical model. Their final aim was to reduce observations as accurately as possible to algebraic relations. Unlike Ptolemaic astronomy, Chinese astronomy showed no concern for the calculation of dimensions of the universe: for all mensurational purposes the sky was treated two-dimensionally.

Cosmology. During the Han and Six Dynasties period, there had been rivalry between cosmological systems, especially between one school which championed a spherical sky (*hun t'ien*)[a] and another which held that the sky was parallel to a flat or convex earth (*kai t'ien*)[b]. Later on, however, the controversy died out and astronomers lost their interest in it, occupying themselves solely with routine observations and calendrical calculations. In the T'ang period the Chinese official astronomers claimed that "Our business is exclusively calendrical calculations and observations in order to provide the people with the correct time. Whether flat or spherical cosmology is no concern of the astronomers!"[3]

---

[2] Shigeru Nakayama, *Characteristics of Chinese calendrical science*, "Japanese studies in the history of science", No. 4, 1965, pp. 124–131.

[3] *Li chih* (A) [Treatise on calendar-making] of the *Hsin T'ang shu* (B) [New standard history of the T'ang period], eds. Ou-yang Hsiu (C) and Sung Chi (D) (1060), quoted in Yabuuchi Kiyoshi (E), *Zuitō rekihō shi no kenkyū* [Researches in the history of calendrical science during the Sui and T'ang period; Tokyo, 1944], p. 6.

Although without effect on professional astronomical circles, there was a revival of cosmological interest among the Sung philosophers such as Chang Tsai[c] and Chu Hsi[d]. The Neo-Confucians favoured the *hun t'ien* theory. Chu Hsi, the chief figure of the school, argued that the *kai t'ien* theory could not explain how the sky remained in consonance with the earth. The sphericity of sky in the *hun t'ien* cosmology was, however, not paralleled by recognition of the sphericity of the earth. A flat earth, according to Chu Hsi's theory, was located in the middle of the universe, floating on water. Despite considerable developments of eclipse-predicting techniques, it is surprising that the sphericity of the earth was explicitly recognized neither by philosophers nor by astronomers, though the idea was hinted at in the writings of Shen Kua[e] and others.

In the absence of the idea of the sphericity of earth, sun-centeredness was simply inconceivable. The sun was accorded a most important role in cosmology, but was not allotted the central position in the geometrical cosmology.

Chu Hsi proposed a nine-layered stratification of the earth, sun, moon, planets and fixed stars, but did not work out in its astronomical details. A rigid and tightly designed universe like the medieval Western cosmos was not to the taste of the Chinese. Their cosmological outlook was not confined inside the shell of *hun t'ien* theory; implicitly or explicitly they admitted the plausibility of an infinite universe, as evidenced in outspoken arguments of the Buddhists.

Physics or mechanics. The main cosmological concern of the Chinese philosophers was not the shape of the world but its dynamic processes. Their discussions were almost always phrased in terms of the cosmic energetic fluid, *ch'i*[f], which permeated the universe; meteorological metaphors are more appropriate to it than that of mechanical clockwork. The most important Chinese principle of natural philosophy is the yin-yang[g] principle, which explained all phenomena in the universe in terms of alternating but complementary dynamic phases described by the rhythms of earth and heaven, rest and motion, female and male and so on.

Chinese geocentricism was not based on a sharply defined celestial-terrestrial dichotomy as in Aristotelian cosmology, nor was the cosmology complicated by theological requirements as in medieval Europe. For academic philosophers the earth was simply located at one end of a continuous spectrum running from most rapid (stellar background) to infinitesimally slow (earth) motion. Thus the earth was conceived as not absolutely and self-evidently at rest, but conditioned to be so by external circumstances. There remained a logical option to accept the motion of the earth once those circumstances were no longer accepted.

Contrary to philosophers' view, professional astronomers claimed that the star-bearing sky moves from east to west, and the sun, moon, and planets move from west to east against the stellar background. If to the central earth we extend this notion of eastward turning, the earth must rotate most rapidly against the westward-turning stellar background.

INTRODUCTION OF COPERNICANISM INTO JAPAN

Up to the early part of the eighteenth century, Japanese astronomy was still dominated by Chinese tradition. We can reasonably say that Copernicanism appeared in Japanese works only in the last quarter of the eighteenth century.

Compared with the date of the acceptance of Copernicanism in the West, its introduction into Japan was late. To explain this, the following three factors may be considered.

(1) Political action to limit free international communication,
(2) The language barrier to the translation of Western works,
(3) Ideological and technical difficulties in comprehending Copernicanism.

JESUIT INFLUENCE

The impact of the West finally began to make itself felt on the isolated islands of Japan in about 1543[4], the year in which *De revolutionibus* was published. In that year, shipwrecked Portugese introduced firearms into Japan and Jesuit evangelism followed.

Scholars who have studied the reports of such missionaries to China as Matteo Rici and Johann Adam Schall von Bell tend to project their picture of seventeenth-century Chinese science onto that of Japan in the corresponding period and often conjecture that Japanese science also was substantially affected by the early contributions of the Jesuits. However, circumstances in the two countries differed greatly.

While the Jesuits in China generally took a flexible, sometimes conciliatory, attitude toward the elite in Chinese bureaucracy and employed an indirect method to convert them to Christianity through the demonstration of the superiority of Western astronomy, the missionaries to Japan never attempted a systematic introduction of Western astronomy, but focussed their efforts on intensive direct exangelism. From the late sixteenth century onward, the Japanese government was suspicious of the Christians. Eventually it forbade any belief in Christianity and took steps to expel the Portuguese and Spanish missionaries from the country. Hence, the Jesuits' impact was relatively short-lived and the teachings of the missionaries in Japan were almost eradicated. We cannot say, therefore, that the Jesuits contributed to Japanese astronomy as much as Ricci and his successors did to Chinese astronomy.

AVAILABILITY OF WESTERN KNOWLEDGE UNDER SECLUSION POLICY

After 1638, the Chinese and Dutch were the only foreigners allowed to reside in Japan, and they were restricted to the city of Nagasaki for the pursuit of trade. This political action was paralleled by restrictions on the import of certain Chinese books, which included all works on Christianity and all works by Christian authors.

---

[4] A Portuguese source places the date at 1542. However, according to a Japanese source, the "Teppōki" (F), written between 1648 and 1651, the year is 1543. See C. R. Boxer, *The Christian century in Japan, 1549–1650* (Berkeley, 1951), p. 27.

It seems that Matteo Ricci was, in the eyes of the government censors, a most dangerous character. Any work by him, or associated with his name, was barred whether it concerned Christian tenets or not. The government had categorically forbidden the importation of all Sino-Jesuit (Chinese language) treatises.

The government, however, never took legal action against the importation of Western books. In early seventeenth-century Japan the fraction of the population that could read Western books was insignificant compared to the fraction that could read Chinese books. The prohibition of Christianity, the departure of foreign missionaries from the country, and the limitations on foreign trade left no opportunity for the ordinary intellectual to receive instruction in European languages.

The only exception was the group of official interpreters at Nagasaki. In view of their professional function, they were officially permitted to study European languages.

On the whole, the decree banning Christian writings was partly responsible for the predominance in Japan during the seventeenth and early eighteenth centuries of purely traditional astronomy in the Chinese pattern.

RELAXATION OF THE BAN – YOSHIMUNE

While the seventeenth century was mainly spent in catching up with traditional Chinese scholarship, the Japanese from the early eighteenth century on began to realize that the Chinese achievements did not suffice. The relations of Japanese astronomy with that of the West entered a new phase under the Shogun Yoshimune[h], ruler of Japan from 1716 to 1745, who relaxed in 1720 the ban on Sino-Jesuit treatises.

Intending to revise the contemporary Jōkyō[i] calendar immediately after his appointment, Yoshimune consulted astronomers and mathematicians. These men must have read some of the officially forbidden books which were preserved only in the shogunate library and found them superior to the traditional Chinese works, for they apparently persuaded Yoshimune to collect all Chinese translations and treatises on Western astronomy.

After the relaxation, Western knowledge on an advanced and professionally useful level was transmitted to Japan for the first time. Meanwhile, at the time of Yoshimune in Japan, scholars and officials had no source of information on Western astronomy other than Chinese works. While the Sino-Jesuit works were for decades relied upon by Japanese practical astronomers, modern Western astronomical ideas could not have been available until the difficulty of translating original Western treatises was overcome; the works in Chinese were practically useless in this respect. Thus, the late acceptance of Copernicanism in Japan was mainly due to the exclusive dependence until the early half of the eighteenth century on Chinese sources rather than to particular ideological obstacles.

BEGINNING OF THE "DUTCH LEARNING"

One consequence of this new receptivity was that in 1745 the interpreters were officially encouraged to learn to read Dutch books. Prior to this, their proficiency had been almost entirely restricted to translation of spoken Dutch. At this early stage, the task of introducing new ideas was left in the hands of only a few linguistic experts.

Circumstances of foreign relations confined Western-oriented Japanese to exclusive reliance on Dutch sources. After its golden age in the seventeenth century, Holland could no longer maintain its aureole; in the late eighteenth century its scientific efforts were at an ebb[5]. It was just than that the Japanese depended on Dutch translations of Western European works. In the 1770's, a generation after the reign of Yoshimune, a notable expansion of study of the Dutch langauge and science led to a move for translation of Dutch scientific works — or retranslation of Dutch translations of Western European works.

TWO SOURCES OF COPERNICANISM

Under the circumstances described above, Copernicanism entered Japan via two distinct routes:

(1) Chinese sources, i. e. the writings of Jesuit missionaries in China and their native sympathizers,

(2) Direct contact with Europeans in Japan, or direct translation from European sources.

Chinese sources. Since the introduction of heliocentricism into China has been exhaustively treated in another paper[6], it is necessary here only to note by way of summary that the Jesuit monopoly on scientific teaching in China greatly delayed it, and that by the time Copernicus's cosmological convictions were adequately described in Chinese by Michel Benoist in 1760, so many confusing statements had been made about the Polish astronomer's beliefs that the best Chinese astronomers of the time were justifiably little disposed to accept Benoist's description. Benoist's ideas did not enter Japan until they were referred to in 1846 by the Shogunal astronomer Shibukawa Kagesuke[j] (1787—1856) in his "Shinpō rekisho zokuhen"[7,k] (Sequel to calendrical treatise by the new method). Of course by this time adequate information on heliostatic cosmology had already entered Japan direct from the West.

Prior to that, no substantial Chinese influence in the matter of Copernicanism can be found. Shizuki Tadao[l] (1760—1806) in his *Rekishō shinsho*[m] (new treatise

[5] A. J. Barnouw and B. Lanaheer, ed., *The contribution of Holland to the sciences* (New York, 1943), p. 271.

[6] Nathan Sivin's paper, in print.

[7] Preserved in the Naikaku Bunko library.

on calendrical phenomena), Part 1 (drafted in 1798) states that "I have looked at a work entitled probably *Li-ch'i t'u-shuo*[n] (ritual implements, illustrated) and found many instruments made in England. Among them there were some which place the sun at the center and the planets and the earth outside. Hence, it may be that this (Copernican) theory is now adopted (in China)"[8].

Li-ch'i t'u-shuo is not found in any Chinese bibliography and "instruments made in England" does not sound genuinely Chinese, but I have located a *Li-ch'i t'u-shih*[o] (prefaced kin 1759 and 1766) which contains two illustrations of heliocentric planetaria kept in the Imperial palace. This must have been the work Tadao saw. Again, however, Tadao had been familiar with Copernican theory since his own translation of John Keill's work in the 1780's and hence this vague Chinese source merely strengthened his own conviction of Copernicanism and also his impression of its diffusion in other parts of the world.

Thus we may conclude that despite a tradition of heavy borrowing from Chinese calendrical astronomy until the Jesuits' time, the Japanese were not perceptibly influenced via China in the matter of Copernicanism. They found their own way towards Copernicanism through direct access to Western sources. This is one of the earliest instances of Japanese independence from her historically overwhelming intellectual dependence upon China[9].

Western sources. During the Jesuit century in Japan, some works of missionary origin conveyed an Aristotelian kind of cosmology. We have two such sources extant now. One is *Nigi ryakusetsu*[p] (Outline theory of celestial and terrestrial globes; basically a translation of *De sphaera* by the Spanish missionary Pedro Gomez) primarily for use in instructing Japanese students at a Jesuit *collegio*. The other is *Kenkon Bensetsu*[q] (Western cosmography with critical commentaries) consisting of an original text by the apostate Jesuit missionary Christovao Ferreira, and indented annotations by a Japanese Confucian commentator[10]. Both of them are merely popular accounts of Aristotelian cosmology. The latter has a passage refuting the concept of the rotation of the earth, saying that if it were so, everything on earth would be whirled out into space. This looks like a physical argument on Copernicanism but there was no mention of the heliocentric system or even of the name of Copernicus. Presumably, the author was not concerned with Copernicanism, since Ferreira left Europe before the Copernican issue came to be noteworthy;

---

[8] *Tentairon* (G) in part one of *Rekishō shinsho*, reprinted in *Nihon tetsugaku shisō zensho* (H) [Source book of Japanese philosophy], ed. Saigusa Hiroto (I) and Shimizu Ikutarō (J), (Tokyo, 1956), vol. 6, p. 142.

[9] Hirose Hideo (K), *Kyū Nagasaki tengakuha no gakutō seiritsu ni tsuite* (L), *Rangaku shiryō kenkyūkai kenkyū hōkoku* (M), No. 184 (1966). The manuscript is reprinted with commentaries in *Kinsei kagaku shisō* (N), No. 2, ed. by Hirose Hideo, Nakayama Shigeru (O) and Otsuka Yoshinori (P), (Tokyo, 1971) in *Nihon shisō taikei* (Q) [Source-books of Japanese thought], No. 63.

[10] In *Bunmei genryū sōsho* (R) [Series on the origins of civilization; Tokyo, 1914], vol. 2, pp. 1–100.

he simply repeated such traditional Aristotelian and Ptolemaic arguments as prevailed in the school curricula at the time[11].

The numerical value adopted in *Kenkon bensetsu* for the period of precession is 25,798 years, closero the Copernican value than to the Alphonsine-Clavius-Ricci 49,000 years. This is not a traditional Chinese value, either. It may be that some astronomical aspect of Copernicanism had by chance infiltrated.

From that time the ban on Christianity and Western knowledge set in. Even after Yoshimune's relaxation, he and his followers did not seem to have conversant with cosmological issues. Nishikawa Masayoshi[r] (1694–1756) of Nagasaki, who had access to the residences of Dutch merchants was, on account of a high reputation for knowledge of Western astronomy, appointed Shogunate astronomer, but his writings are still based on a sketchy knowledge of such quasi-Tychonic cosmology as appeared in the *T'ien-ching huo-wen*[s] (Queries on the classics of the heavens, ca. 1675) by Yu I.[t] For the advocacy of Copernicanism, we have to await the emergence of pioneers from a non-astronomical group.

MOTOKI RYŌEI'S TRANSLATION WORKS ON HELIOCENTRICISM–COSMOLOGICAL ASPECT

Translations of Dutch works by Motoki Ryōei[u] (1735–1794) are significant not only as the first Japanese sources on the Copernican heliocentric system, but also as a landmark in the advancement of the study of Western languages in Japan. Ryōei belonged to the third generation of a hereditary family of Nagasaki interpreters. Besides discharging routine duties, he was perhaps the first to receive official requests to translate Dutch works. During his lifetime Ryōei produced many translations, mainly in the fields of astronomy and geography.

He once stated that since there had never been a professional translation of a Dutch work in Japan, his predecessors were sceptical about the possibility of even literal translation and were reluctant to attempt it[12]. He recognized that if he were to fail in his initial translation, his own ability would be brought into question and his inherited position jeopardized[13]. His attitude indicates that the main

---

[11] Imai Itaru (S), *Kenkon bensetsu zakki* (T) [Miscellaneous notes on the Kenkon bensetsu], *Tenkansho* (U) [Private journal of Imai Itaru] *22*, 14–16 (mimeographed, 1957).

[12] Motoki Ryōei, *Seijutsu hongen taiyō kyūri ryōkai shinsei tenchi nikyū yōhō ki* (V) [The basis of astronomy, newly edited and illustrated, on the use of celestial and terrestrial globes according to the heliocentric system; 1792–1793], vol. 2, reprinted in *Tenmon butsuri gakka no shizenkan* (W) [Japanese astronomers' and physicists' views of nature], in *Nihon tetsugaku shisō zensho* (X) [Source book in Japanese philosophy], ed. Saigusa Hiroto (Tokyo, 1936), vol. 8, p. 342.

[13] It is recorded that in 1791 seven interpreters at Nagasaki were dismissed for having made inaccurate translations of Dutch documents. See Otsuki Nyoden (Y), *Shinsen yōgakis nenpyō* (Ā) [A newly edited chronology of Western learning in Japan; Tokyo, 1926], p. 76. For an English translation, see C. C. Krieger, *The infiltration of European civilization in Japan during the eighteenth century* (Leiden, 1940), pp. 94 ff.

difficulty in the introduction of heliocentricism lay in the linguistic barrier and the conservative intellectual atmosphere, in which an innovator hesitated to present anything extravagant. Official requests did, of course, encourage Ryōei, although not all his translations originated in this way. The choice of subject matter for translation must have reflected official concern for the material aspects and products of Western culture but the choice of original texts was his own. Ryōei was primarily a faithful translator, but his own favourite subject was heliocentricism.

The *Oranda chikyū zusetsu.* Ryōei's first translation referring to the heliocentric theory was drafted in 1772 under the title *Oranda chikyū setsu*[14,v] (A Dutchmen's view of the earth.) There is another copy under the title "Oranda chikyū zusetsu"[15,w] (A Dutchmen's view of the earth, illustrated.) These are referred to hereinafter as "first manuscript" and "second manuscript" respectively.

Their original was the Dutch translation *Atlas van Zeevaert en Koophandel door de geheele Weereldt* (Amsterdam, 1745, abb. "Dutch edition") of the *Atlas de la navigation et du commerce qui se fait dans toutes les parties du monde* (Amsterdam, 1715, abb. "French edition"), by Louis Renard. A comparison reveals some interesting aspects of abhorrence of "God" in the introduction of Copernicanism into Japan.

Both the French and Dutch editions are big-scale marine charts with a guide for seamen in the margin. While the original French edition provides only explanations for charts, the Dutch edition incorporates a number of revisions. It also has to begin with some additional elementary accounts of such matters as the earth, the heliocentric system, constellations, longitude and latitude and other astronomical themes, wind and geography, the history of astronomy and geography, the use of the compass, and so on. These additions are considered to have been contributed by a Dutch editor, Jan van den Bosch Melchiorsz.

It was Melchiorsz's additions in which the Japanese translator was chiefly interested. The first manuscript has a summary of the "Preface" (from which could be identified its Dutch original) and a translation of the first five pages of the Dutch edition. The second manuscript, while translating the first sixteen pages, neglected the "Preface". The latter translation, revision of the former, is in somewhat better Japanese.

A comparison with the Dutch edition shows that the translator has omitted certain of the original paragraphs. The details of omissions are as follows. Both translations omit the opening paragraph of the first page of the Dutch edition. This common omission is indicated by (A) in the diagram given below. The second manuscript omits material amounting to almost a whole page on pp. 8–9 in the Dutch edition.

---

[14] Preserved in the Tenri Library under the series title of "Tenmon hisho" (AA) [Secret books on astronomy], 22 sheets.

[15] Preserved in the Nagasaki City Museum, 90 sheets divided into three volumes.

This omission is indicated (B) in the diagram.

Dutch edition	"P" 1 5 8 9 16 ________________
First manuscript	____ ____
Second manuscript	(A)_____ (B)_______

Omission (A) is translated as follows:

The place which God has given people for habitation is commonly called the world, the Earth sphere, the inhabited earth; the term world implies the entire created world, the universe; or, to be more precise, everything that is within the circle of the planet Saturn, of which circle the Sun is the center, and the radius of which extends from the Sun to Saturn. The Earth sphere is composed of earth and ocean, of land and water. It contains many different kinds of materials which God has created for the benefit of man, the quantity of which He has decided upon so that nothing could be added to or taken away from either by the human mind nor by natural causes, for the materials by their nature are essentially distinct from one another, indestructible, incapable of increase and unchangeable...[16]

The omission is apparently deliberate, motivated by the translator's abhorrence of the account of God's Creation, and what is more, Creation seems to be associated with the sun-centered universe. This is clear in contrast to his full and faithful translation of the purely scientific account of the heliocentric hypothesis, which immediately follows the omitted paragraph in the Dutch edition. Of course, the Dutch account itself was only a short outline of the daily rotation and annual revolution and the difference between true (heliocentric) and apparent (geocentric) motions.

The Dutch edition gave a brief history of astronomy up to the time of Copernicus and then proceeded to present the discussion involving Biblical issues caused by Copernican theory. The second manuscript translated the historical account up to Copernicus. It was in this connection that the name of Copernicus and the information that Copernicus's heliocentricism was now generally accepted among Western scholars appeared for the first time in Japanese literature.

However, the translator left out all the theological argumentation. This omitted portion is (B) translated below.

However, the lower classes, or the excessively prejudiced (was this the case among lower classes only?), find this feeling strange, which seems to them to be contrary to what they see; apart from the fact that they have peculiar ideas about this rotation...
Even though it had pleased the Creator of all things to make the Earth and other planets turn in the period of several months around the Sun, and each of them in the period of several hours around their own axis, instead of making the Sun and stars and the immeasurable sky turn with an utterly incomprehensible speed around the Earth for the Earth's benefit, the latter being in comparison hardly a single tiny speck, we still should see everything in the same form as we see it now...

However, we shall not go into this any further in order not to digress from the main subject, considering that this suffices to show that nothing more simple and comprehensive can be understood

---

[16] A full translation of part (A) is available in Shigeru Nakayama, *Abhorrence of 'God' in the introduction of Copernicanism into Japan*, "Japanese studies in the history of science", No. 3, 1964, pp. 62–63.

than this explanation of Copernicus; and that only ignorant prejudice and certain passages of the Holy Scriptures have caused this explanation to be discarded as being contrary to common sense... So there is nothing in the Holy Scriptures which could serve as an argument against this explanation of Copernicus...[17]

The heliocentric account terminates at this point and then the text goes on to discuss the use of the compass. Hereafter, the translator resumed his faithful translation again.

Published in the liberal Netherlands, the Dutch edition had no prejudice against Copernicanism; it assumed the tone of enlightening ignorant seamen. After all, it was written in the mid-eighteenth century, when Copernicanism was already established. Yet, some obsolete theological discussion was added in order to meet popular interest. Most likely, the translator learned from it the intimate relationship between Copernicanism and the Christian God, and took special pains to introduce only heliocentricism[18].

The "Tenchi nikyū yōhō". The second work by the same translator was the *Tenchi nikyū yōhō*[x] (the use of celestial and terrestrial globes) dated 1774. Its Dutch original was *Tweevoudigh onderwiis van de hemelsche en aardsche globen* (Amsterdam, 1666[19], first edition, 1620).

The book was written and prefaced by Willem Janszoon Blaeu (or Blaaw, 1572–1638), and edited and published by his son Johan. Willem Janszoon Blaeu was a renowned Dutch cartographer and an intimate friend and disciple of Tycho Brahe. He was also one of the early proponents of Copernicanism[20].

We have no way of proving when Blaeu's 1666 edition was brought to Japan. It may be that it was imported into Nagasaki long before the Japanese translation in 1774 and then buried in obscurity. But there may be a hidden reason on the part of the translator why he picked on this particular work for translation.

At the time of the publication of Blaeu's book, the first edition of which appeared as early as 1620, the Copernican controversy was raging furiously; and hence it was naturally full of theological arguments. It is manifest that Blaeu wrote the book with the intention of propagating the Copernican hypothesis.

While Blaeu's preface stated that his intent was to illustrate the Ptolemaic system first, because it was more familiar and more easily comprehensible, and then

---

[17] A full translation of part (B) in Nakayama, "Abhorrence", pp. 63–66.

[18] Shigeru Nakayama, *Motoki ryōei yaku 'Oranda chikyū setsu' ni tsuite* (AB) [On the "Oranda chikyū setsu" translated by Motoki Ryōei], *Rangaku shiryō kenkyūkai kenkyū hōkoku*, No. 112 (1962) and No. 162 (1964).

[19] Because of the long interval between the date of the original and that of the Japanese translation, Itazawa Takeo (AC) conjectured that the original was published in 1766 instead of 1666 in his *Edo jidai ni okeru chikyū chidōsetsu no tenkai to sono handō* (AD) [The development of the earth's sphericity and motion theory and the reaction to it during the Tokugawa period], *Shigaku zasshi* (AE) 52(1), 12 (1941), but I found the 1666 edition in the Library of Congress.

[20] Pierre Henry Bandet, *Leven en Werken van Willem Janszoon Blaeu* (1871); and Edward Luther Stevenson, *Willem Janszoon Blaeu* (New York, 1914), pp. 11–13.

go on to the true theory of Copernicus, Motoki Ryōei's own preface merely mentions the name of Copernicus as follows:

> About one hundred years ago, there was a man called Nicolaus Copernicus. Being in intimate communications with Tycho Brahe, the biggest figure in astronomical observations, Copernicus investigated this [heliocentric] theory thoroughly and finally out of opaque darkness reached enlightenment[21].

This extract is apparently an exposition of Copernican heliocentricism, but we have found no further mention of it in this work. While the preface of Blaeu's original had eight paragraphs, Ryōei translated and put into his own preface only the first four paragrahps. Paragraphs 5 through 7 discussed Blaeu's plan of arranging the Copernican system as opposed to the Ptolemaic, but these are all omitted from Ryōei's preface. Furthermore, Ryōei confused the historical relationship between Copernicus and Tycho Brahe. This was the source of confusion among later popularizers, who took Ryōei's writings as a basis.

Blaeu, with the intention of propagating the new theory, divided his work into two volumes.

Volume I: Astronomical principles of celestial and terrestrial globes based on the inadequate hypothesis of Ptolemy.

Volume II: Astronomical principles of globes based on the true hypothesis of Copernicus.

Ryōei's translation terminated near the middle of Volume I (at page 120 of 163 pages) the part based on Ptolemaic geocentric theory. Apparently he had no intention of extending his translation further. It is highly probable that he deliberately omitted the section on heliocentricism from the main text as well as from his own preface.

It seems unlikely that the Copernican heliocentric theory was, despite its unfamiliarity, insurmountably difficult for Ryōei to comprehend. Blaeu's astronomical writing was not particularly advanced; Kepler's contributions did not appear in it. Hence it would be more plausible to interpret this deliberate omission as being caused by Ryōei's precaution against any subversive or unorthodox thoughts. The translator, perhaps influenced by his own earlier experience of translation, replaced Blaeu's long theology-filled preface with his own brief summary of a purely physical nature and left entirely untranslated the second volume on the Copernican system[22].

The "*Shinsei tenchi nikyū yōhō ki*". A more detailed and truly comprehensive account of the Copernican system was translated in 1792–1793. It was entitled *Seijutsu hongen taiyō kyūri ryōkai shinsei tenchi nikyū yōhō ki* (the basis of astron-

---

[21] The preface is more fully translated in Shigeru Nakayama, *A history of Japanese astronomy; Chinese background and Western Impact* (Cambridge, Mass., 1969), pp. 175–176.

[22] Shigeru Nakayama, *Motoki Ryōei no tenmonsho honyaku ni tsuite* (AF) [On Motoki Ryōei's translations on astronomy], *Rangaku shiryō kenkyūkai kenkyū hōkoku*. No. 66 (1960).

omy, newly edited and illustrated; on the use of celestial and terrestrial globes according to the heliocentric system), and consisted of seven volumes[23]. The Dutch original has been identified as *Gronden der sterrenkunde, gelegd in het zonnestelzel bevatlijk gemaakt; in eene beschrijving vant'n maaksel en gebruik der nieuwe hemel- en aard-globen* (Amsterdam, 1770), 470 pages[24].

Its author, George Adams the elder (died 1773), was a maker of mathematical instruments under King George III. He had a worldwide reputation as a maker of celestial and terrestrial globes. The ultimate original of Ryōei's translation, Adams' *Treatise describing and explaining the construction and use of new celestial and terrestrial globes* (London, 1766), passed through thirty editions in England and was also printed in America[25].

Ryōei's translation included the first 325 of the 360 paragraphs of the Dutch original; only a portion on the use of globes was left untranslated. The original began with a straightforward description and explanation of the solar system, in which the relationship between the apparent and true courses of the planets was expounded on the basis of the heliocentric scheme. It was already free of time-honoured religio-cosmological controversy; its purely scientific arrangement seemed to have appealed to the translator. It was translated in full and a truly comprehensive account of the Copernican system, for the first time, became available in Japan. He even ventured to put "heliocentric system" into the title of his translation.

The original was not an advanced treatise for professional astronomers, but a text-book for navigators. The arrangement of themes was, however, strikingly different from that of traditional treatises of calendrical astronomy. From the outset, the earth was treated as a member of the solar system. First, detailed instructions were given for the reduction from geocentric to heliocentric coordinates; then, the behavior of the planets and satellites was expounded. But, lack of accurate detail made this treatise of little use to Japanese practical astronomers.

Ryōei also gave brief summaries of *Philosophische onderwijzer* and *Beginselen der Natuurkunde*[26]. The original of the former is Benjamin Martin's *The philosophical grammar* (first edition 1738, Dutch edition 1744); that of the latter is the *Anfangsgründe der Physik*[27] by Johann Heinrich Winkler (Dutch edition 1768). These works elucidated the Newtonian laws of mechanics, but they were beyond Ryōei's concern

---

[23] Vol. I is reprinted in *Nihon tetsugaku zensho*, vol. 8 (1936).

[24] Itazawa, *Edo jidai*, p. 10. A first edition and an edition of 1771 were preserved in the shogunate government library. *Yōgaku kotohajime ten, rangaku no shokeifu to Edo bakufu kyūzōbon* (AG) [Catalogue of books, exhibited at the "Yōgaku kotohajime ten", on earlier phases of Western civilization in Japan], ed. Ōkubo Toshiaki [AH], pp. 17–18.

[25] *Dictionary of national biography*, s. v.

[26] See Johannes van Abkoude, *Naamregister van de Bekendset en meest in Gebruik Zynde Nederduitsche Boeken* (Amsterdam, 1787), p. 579.

[27] First edition published in Leipzig in 1753. An English translation was published in London in 1757.

or comprehension[28]. He could only briefly compare the Ptolemaic, Tychonic and Copernican systems.

The transformation from geocentricism to heliocentricism did not, in the absence of certain religious and philosophical presuppositions, raise difficult technical problems. Thus, unlike the physical aspect of Newtonianism, the introduction of the cosmological or geometrico-morphological aspect of Copernicanism never raised difficult questions for Ryōei.

SHIZUKI TADAO – THE PHYSICAL ASPECT

While Motoki Ryōei was thoroughly loyal to his official duty and interested in Copernicanism merely as an assignment, his pupil Shizuki Tadao (1760–1806), also born in a family of Nagasaki official interpreters, renounced his hereditary position at the age of eighteen and devoted the rest of his life to the subjects of his own interest—natural philosophy and cosmology[29].

Immediately after his retirement from office, he undertook the introduction of Newton's doctrines for the first time into Japan. The original version used by Tadao was a Dutch translation by Johan Lulofs, *Inleidinge tot de waare Natuur-en Sterrekunde* (Amsterdam, 1741), of John Keill's (1671–1721) *Introductiones ad veram Physicam et veram Astronomiam* (London, 1739).

Tadao spent more than twenty years on his translation. His preparatory notes were made in three drafts, *Tenmon kanki*[y] (Astronomical collection; 1782), *Dōgaku shinan*[z] (Guide to mechanics; n. d.), and *Kyūshinryoku ron*[aa] (on attraction; 1784)[30]. These were revised with substantial amendment into the final monograph, entitled *Rekishō shinsho* (new treatise on calendrical phenomena), which appeared in three volumes (completed 1798, 1800, and 1802).

Tadao's work was not a literal translations at all, but rather a collection of his notes with abundant commentaries of hs own.

Tadao was not a mere linguistic expert but the most profound philosophical mind of his day. Although a number of more systematically modernized text-books on natural philosophy, such as B. Martin's, were available to him, he picked up the then outdated work by Keill since Tadao, from his personal propensity for metaphysics, was interested in its polemical manner in early Newtonian days.

Keill's work had the typical post-Newtonian deistic tone; unrestrained praise of God's creation, harmony, order, symmetry, beauty, and so forth is lavishly distri-

---

[28] Kuwaki Ayao (AI), *Reimeiki no Nihon kagaku* (AJ) [Japanese science at the dawn; Tokyo, 1947], pp. 105–107.

[29] Watanabe Kurasuke (AK), *Oranda tsūji Shizukishi jiryaku* (AL) [Outline biographies of the Shizuki family, hereditary Dutch interpreters; Nagasaki, 1957], pp. 32–34.

[30] Ōsaki Shōji (AM), *Rekishō shinsho tenmei kyūyakubon no hakken*, (AN) [The discovery of manuscript translations preliminary to the Rekishō shinsho], *Kagakushi kenkyū* (AC), Nos. 4 und 5 (1943), p. 101.

buted throughout. These embellishments were eliminated from Tadao's translation, except in one place where he rendered "the idea of God's design" as "the agency of limitless *prajña* (a Buddhist term, literally "wisdom")[31].

To Tadao, the location of the sun, whether geocentric or heliocentric, was not a basis issue of Copernicanism; the relativity of location and motion was the most impressive and admirable feature of the heliocentric theory. Motion or rest depended solely on the point of reference, and the two were therefore intrinsically indistinguishable. There was neither absolute rest nor absolute motion. For geographical location, or even for family relations, there was no absolute measure. If the point of observation was the sun, the earth was in motion, and vice versa. Thus there was no reason to prefer either the ancient Chinese or the Western theory concerning the motion of the sun or earth.

From this relativistic notion the plurality of worlds followed easily, since the immediate world had no particular priority. To support this theory, Tadao quoted Buddhist writers and Lieh-tzu[ab], an ancient and possibly legendary Chinese Taoist thinker, whose grounds for this view were intuitive and metaphorical. Tadao cited a Chinese statement that "things [fluid] pure and light go up and from the heavens, while the turbid and heavy come down and make the earth[32]", and argued that every star and planet was composed of the same kind of turbid matter as the earth. Thus, he claimed that the plurality of worlds was known before the Westerners advanced it. (The original content of this statement, which is perhaps a Neo-Confucian saying, had nothing to do with the plurality problem as such.) While he saw that the Western theories of the earth's motion and the plurality of worlds were significant because they were based on properly related reasoning and observation, he was also anxious that the contributions of his own cultural tradition be recognized.

In the appendix to Volume I, Tadao presented his own account of the heliocentric system, entitled *Tentairon*[33,ac] (discussions on the heavenly bodies). Tadao did not evince any enthusiasm for the theory he was introducing. His intent was to reconcile modern Western theory with traditional Chinese views.

In the appendix to Volume II, entitled *Fusoku*[ad] (Immensurability), he felt obliged to justify his cosmologic views in terms of Confucian morality:

> However, in everything there always exists a governing center. For an individual, the heart; for a household, the father; for a province, the government; for the whole country, the imperial court; and for the whole universe, the sun. Therefore, to behave well, to practice filial piety toward one's father, to serve one's lord well, and to respond to the immensurable order of the haevens are the ways to tune one's heart to the heart of the sun. This is the way to admire the sov-ereign of the universe[34].

---

[31] In *Rekishō shinsho*, reprinted in *Nihon tetsugaku shisō zensho*, vol. 6, p. 117.

[32] *Nihon tetsugaku shisō zensho*, vol. 6, p. 141.

[33] *Nihon tetsugaku shisō zensho*, vol. 6, pp. 135–142.

[34] *Nihon tetsugaku shisō zensho*, vol. 6, p. 246.

When confronted with the incompatibility of the Copernican theory and the traditional notion, which identified the sky with yang and motion and the earth with yin and rest, Tadao sought to preserve the respectability of ancient Chinese concepts by quoting an ancient passage which, interpreted very freely, referred to the motion of the earth. And when the conservative Chinese attitude toward the new theory was attacked on the grounds that even though the name of Copernicus appeared in the *Li-hsiang k'ao-ch'eng*[ae], the Chinese still did not adopt the Copernican theory, Tadao again defended the Chinese. In a rather fair apologia, he pointed out that the Chinese were concerned only with observations and predictions of the apparent courses of the heavenly bodies, and not with theory; therefore they had no compelling reason to adopt heliocentricism.

At the end of Volume II, Tadao raised the question as to why all the planets rotate and revolve in the same direction, in planes not greatly inclined to the ecliptic. By way of an answer, he proposed at the very end of the treatise, in a section entitled *Kenkon bunpan zusetsu*[af] (the separation of opposites in the generation of the cosmos illustrated), a hypothesis concerning the formation of the planetary system. He claimed it as his own idea, saying: "It may be that this theory has already been formulated by some Western scholar, but we have never heard of it"[35].

Tadao's hypothesis immediately recalls the celebrated hypotheses of Kant and Laplace. Of these Laplace's is scientifically the most advanced. Tadao's argument was not Laplacian abstraction from a cautious synthesis of all relevant observational data, but was somewhat closer to the rationalistic inferences of Kant.

In view of the relative inaccessibility of Western treatises, it is unlikely that Tadao borrowed his idea from anyone else. His hypothesis, considering his background in Neo-Confucian ideas, was not a titanic leap. Many aspects of it were already present in the Neo-Confucian vortex cosmogony, which claims that beginning with primordial chaos the light fluid tends to float to the surface and heavy matter to precipitate at the center in the course of one-way revolution. Hence, a small portion of the ideas of attraction and centrifugal force provided Tadao with a more elaborate mechanical hypothesis, formulated in accordance with the heliocentric system.

DIFFUSION OF COPERNICANISM

At the time, no work by Motoki Ryōei and Shizuki Tadao was ever published in printed form. For the most part, their works were preserved in manuscript form and circulated as handwritten copies.

It was Shiba Kōkan[ag] (1747?–1818) who popularized the Copernican theory through three printed books[36]: *Chikyū zenzu ryakusetsu*[ah] (An outline world atlas;

---

[35] *Nihon tetsugaku shisō zensho*, vol. 6, p. 288.

[36] Kōkan's works are reprinted in Nakai Sōtarō (AP), *Shiba Kōkan* (Tokyo, 1942).

1793); *Oranda tensetsu*[ai] (Dutch astronomy: 1795); and *Kopperunyu tenmon zukai*[aj] (Copernican astronomy illustrated; 1805). A literary dilettante and gifted free-lance painter in the Western style, Kōkan enjoyed more freedom than did the official interpreters and astronomers. He frankly acknowledged Western achievements and superiority, but never apologized, as Tadao did, for Sino-Japanese tradition. Emancipating himself from the predominant notion that Western knowledge was valuable only from a utilitarian point of view, he incorporated Western astronomical concepts into his thought and used them as one basis of his unique materialistic cosmology, in which fire was the fundamental element[37]. Still the depth of his knowledge of Western astronomy did not exceed that of Motoki Ryōei, whose work was his main source.

Shizuki Tadao's physical ideas were taken over by the freethinker Yamagata Bantō[ak] (1748–1821), the head clerk of an Osaka business firm and an outspoken polemist with conventional beliefs. While Tadao remained introspective and critical to anything extravagant, open-minded Bantō extended Copernicanism freely into his "great universe" picture, in which plural worlds were arranged in hierarchical order, like Olbers' conception of the universe[38].

Many other popularizers such as Hoashi Banri[al] and Yoshio Hisasada[am] followed in expounding Copernican and Newtonian thought, but Shizuki Tadao's contribution was so outstanding and penetrating that his work remained unsurpassed until the middle of the nineteenth century.

THE ASADA SCHOOL – ASTRONOMICAL ASPECTS

Among the professional calendrical astronomers, Asada Gōryū[an] (1734–1799) and his school first fully recognized the technical superiority of Western astronomy over the traditional Chinese type, and gave effect to it in the Kansei[ao] calendar reform (1798). This recognition followed upon the importation of such impressively voluminous Sino-Jesuit works as the *Hsi-yang hsin-fa li-shu*[ap] and *Li-hsiang k'ao-ch'eng*.

Miura Baien[aq] in his *Kizan roku*[ar] (Notes during a trip to Nagasaki) quoted a letter from his close friend Gōryū in 1778 to the effect that: "later a man called Copernicus appeared. On the nine-layered spheres of the cosmos he superimposed other two heavens to account for the east-west and south-north precessions, building up a total of eleven celestial spheres"[39]. Thus at this time, Gōryū still held an erroneous picture of Copernicus derived from the *Hsi-yang hsin-fa li-shu*. On the other hand,

---

[37] Muraoka Tsunetsugu (AQ), *Zoku Nihon shisōshi kenkyū* (AR) [Studies in the history of Japanese thought, sequel; Tokyo, 1939], pp. 239 ff.

[38] Takamichi Arisaka (AS), *Chidōsetsu no denrai to shin uchūron no shutsugen* (AT) [An introduction to the theory of earth's motion and the emergence of a new cosmology], *Nihonshi no kenkyū* (AU) [Researches on Japanese history; Kyoto, 1970], pp. 281–301.

[39] Volume II. Reprinted in *Nihon tetsugaku zensho*, vol. 8, p. 185.

Gōryū must have learned from Baien about the heliocentric hypothesis, the gist of which had been transmitted to the latter by Mastumura Suigai[as], a collaborator of Motoki Ryōei, during his trip to Nagasaki in 1778[40].

It is, of course, quite possible that Gōryū was unable to identify the heliocentric Copernicus with the conventional technician mentioned in the Sino-Jesuit treatises. This would have been virtually impossible in view of the limited information provided by the missionaries in China.

Neverthless, Gōryū and his school committed themselves to Copernican heliocentricism. His pupils claimed for him the honor of having discovered independently, ca. 1796, although he did not publish it, the relationship between the distance of planets from the sun and the periods of their revolution (in other words, Kepler's third law)[41]. The independence of his discovery is quite doubtful[42], but in any case Kepler's third law could not have been reached without starting out from the Copernican heliocentric scheme. Hence, at least Gōryū and such advocates of his discovery as Takahashi Yoshitoki (1764–1804) and Hazama Shigetomi (1756–1816) must have been convinced of the cosmological and physical truth of the Copernican scheme. In fact, Takahashi Yoshitoki in the opening part of his *Zōshūshō chō hō*[43],[at] (Hsiao--ch'ang method, revised and augmented, 1798) advocated Copernicanism enthusiastically, although it was completely irrelevant to his *shōchō hō*[ah] concept, which deals exclusively with the matter of astronomical parameters. Around the same time, Shigetomi wrote *Tenchi nikyū yōhō hyōsetsu*[av] (A commentary on Motoki Ryōei's translation work on heliocentricism, 1798), attempting to correct the astronomical errors of Ryōei, who knew no professional-level astronomy.

Why was it that calendrical astronomers, who had maintained a professional indifference toward cosmology, developed a keen interest in Copernicanism? It is true that Copernicanism included some computational novelties which calendrical astronomers eagerly adopted, although in general they saw its cosmology as merely a matter of transformation of the geocentric coordinates of Tycho into a heliocentric frame of reference. At least as Ryōei understood and introduced it, it did not show any potential for new solutions to classical Japanese problems, and thus had little reference to the traditionally central issue of calendar reform.

Furthermore, by the late eighteenth century, when Copernicanism reached Japan, the observational and astronomical innovations of Tycho, Kepler and also Cassini,

---

[40] *Nihon tetsugaku zensho*, vol. 8, p. 174.

[41] Takahashi Yoshitoki (AV), *Shinshū goseihō zusetsu, furoku* (AW) [Appendix to the illustrated planetary theory, newly revised, ca. 1802, preserved in Inō Museum]; Hazama Jūshin (AX), *Senkō Taigyō sensei jiseki ryakki* (AY) [Outline of the work of my father Hazama Shigetomi, n. d.], reprinted in Watanabe Toshio (AZ), *Hazama Shigetomi to sono ikka* (BA) [Hazama Shigetomi and his family, 1943], p. 456.

[42] Shigeru Nakayama, *On the alleged independent discovery of Kepler's third law by Asada Gōryū*, "Japanese studies in the history of science", No. 7 (1968), pp. 55–59.

[43] Preserved in Tokyo Astronomical Observatory.

Delahire and others were already available through Sino-Jesuit works. There was no particular reason for the Japanese to pay particular regard to the astronomical aspects of Copernicus's own work.

The answer must involve less obvious factors than simple computational convenience. Perhaps the most important element is that Copernican heliocentricism functioned as the symbol for a novelty-oriented school of calendrical astronomers — that of Gōryū and his followers — who were advocating the superiority of modern Western astronomy over traditional one. The situation was similar to that in medicine around the same period, when men like Sugita Genpaku[aw] (1733–1817) and his associates enthusiastically advocated Western achievements in anatomy although they had no direct application as Japanese medicine was then conceived.

Furthermore, their willingness to diverge from tradition led to a new interest in planetary theory. In China, in the *Li-hsiang k'ao-ch'eng hou-pien*[ax] (1742), the Keplerian ellipse was applied only to the movements of the sun and moon; extending them to the planets would have required Copernican coordinates, which the missionary astronomers hesitated to adopt. Takahashi Yoskitoki, challenged by this lacuna, took upon himself the task of extending the Keplerian ellipse into planetary theory. He thus rejected Tychonic coordinates, which the *Hou-pien* still firmly maintained, in favor of the Copernican frame of reference; otherwise, with the planets turning around the sun, which in turn turns around the earth, the result would have been a complicated and ugly double elliptic scheme, and the aesthetic advantage of adopting the ellipse would have been largely lost. Yoskitoki's scholarly devotion to planetary theory was a remarkable departure from traditional calendrical concerns towards an astronomy based on the concept of the solar system.

Yoshitoki's son, Shibukawa Kagesuke (1787–1856), the last of the great calendrical astronomers, adopted planetary elliptic orbits in the last luni-solar calendar reform in 1843, but even in the final attempt at a traditional official treatise, *Shinpō rekisho*[44],[ay] (A calendrical treatise by the new method, 1846), the framework and construction remained traditional; the new cosmology could not be truly integrated in it. Only the last five sections of the "zokuhen" (Sequel) were devoted to the new Copernican and Newtonian approach, which were thus a mere appendix.

In the preface to the sequel, Kagesuke stated his intention to clarify from the view-point of a professional astronomer points of confusion caused by the introduction of heliocentricism. He emphasized that geostatic motion is apparent while heliostatic motion is physically true, but that this is merely a matter of coordinate transformation; the astronomical implication are all the same, and both theories have an equal *logical* claim to verisimilitude. This is the attitude of an astronomical technician towards Copernicanism, not that of a man free to cultivate cosmological or physical interests. Unlike such independent popularizes as Kōkan, or such pioneering enthusiasts as Yoshitoki, the head of the Shogunate's bureau of astronomy,

---

[44] Preserved in Naikaku Bunko Library.

Kagesuke, had to defend traditional bureaucratic responsibilities against modish Western learning, and never accepted any Western theoretical novelty without reservations. But among the professional astronomers of Kagesuke's generation, Copernicanism was accepted as practically self-evident.

IDEOLOGICAL REACTIONS TO WESTERN COSMOLOGICAL THEORIES

The reactions of Buddhists, Confucians and Shintoists to Western cosmologic theories varied widely, resistance to innovation being caused by a conflicting world-view and unyielding commitment to Eastern culture.

The reactions of the Buddhists. By the middle of the eighteenth century, the Aristotelian type of cosmology had been diffused through the printed editions of *T'ien-ching huo-wen*. Some Buddhists could not tolerate the diffusion of European cosmological ideas, which were incompatible with Buddhist beliefs. To refute these ideas, a treatise entitled *Hi tenkei wakumon*[az] (Contra *T'ien-ching huo-wen*), by the learned Buddhist monk Monnō[ba], was published in 1759. Another of his works, *Kusen hakkai tōron*[bb] (A discussion of the theory of the Nine Mountains and Eight Seas), expounded *Sumeru* cosmology, which sets at the center of the earth Mount Sumeru, round which the sun, the moon, and the stars revolved; this idea originated in Jaina cosmography and was taken over by the Buddhists[45].

Generally speaking, Buddhist cosmology and astronomy had a much wider scope than corresponding Western disciplines and enjoyed more freedom in the infinity and plurality of worlds than the rigidly constructed Aristotelian world, but this breadth and freedom was due to vagueness and lack of conviction concerning the existence of an underlying regularity in the phenomenal world, and the possibility of discovering one.

Some of Monnō's criticisms were interesting. For instance, according to him, the universe is a limitless void, no bounds being even conceivable. The infinity of the universe is a consequence of the emptiness of the phenomenal world. Attempts to measure the dimensions of the universe are therefore ridiculous. If there are nine spheres within the empyrean, he argued, why not an equal number of spheres outside it?

Monno's arguments were not very constructive, however. He denounced Aristotelian cosmology, but was unable to replace it with a more consistent system.

The creativity of Japanese Buddhism was in eclipse in the seventeenth century, and intellectual leaderhip shifted to Confucianism, the state orthodoxy. During the eighteenth century, as Western astronomy proved to Japanese intellectuals its superiority and as knowledge of the Copernican system was disseminated through the various popular editions of Shiba Kōkan, a feeling of anxiety and crisis was aroused among some sects of Buddhists. While the Confucians were not so strong-

---

[45] Cf. Sukumar Ranjan Das, *The Jaina school of astronomy*, "Indian historical quarterly" 8, 30–42. 565–570 (1933).

ly opposed to the achievements of Western astronomy, the Buddhists tried to restore their own intellectual aura by scoring a victory for Buddhist cosmology and by defending it against Kōkan's ridicule.

Among this group of Buddhists the most zealous and influential figure was Entsū[bc] (1754–1834). His thirty years of labor in defense of Buddhist astronomy culminated in his masterpiece, the *Bukkoku rekishō hen*[bd] (On the astronomy and calendrical theory of Buddha's country; five volumes, 1810)[46], which was followed by a number of other writings in the same vein.

From the outset, Entsū's single purpose of defending Buddhist doctrine from the invasion of Western scientific ideas was marked. His motivation was a fear that Christianity would undermine Buddhist teaching in Japan. In this respect, he might be following the precedents of Chinese Buddhists' experience under Jesuit influence. This basic attitude was not very different from that of his predecessors such as Monnō, but Entsū's investigations covered an extensive literature, not only in cosmology, but also in the technical aspects of calendar-making. His sources ranged from Buddhist sutras in Chinese translation and ancient and modern Chinese astronomical treatises to Sino-Jesuit works and works on Western astronomy. Although he was said to have had some background in Dutch learning, there is no evidence that he was able to read Dutch. He probably relied on books written by his Japanese contemporaries.

Entsū was not essentially opposed to Western influence in calendrical science, since this traditional activity did not bear on cosmology. He argued, however, that the only worthwhile aspects of calendrical science had originated in India. The Chinese had appropriated the advanced Indian calendar during, for instance, the T'ang period[47]. Citing a superficial Chinese account in the Ming history, he credited India with the origina of Islamic astronomy, and thus also of European astronomy, since the European ephemeris was practically the same as the Islamic. Thus distorting the course of historical development of science, he claimed eventually the Western appropriation of indigenous Indian science.

On cosmological questions, Entsū would yield nothing to the European system, his argument being almost identical to that of his predecessor Monnō, but more amplified and emphatic[48]. From beginning to end, he rigidly and literally supported Sumeru cosmology. Furthermore, he made use of the ancient Chinese flat-earth (*kai-t'ien*) theory, which also propounded a nonspherical model of the universe. The *kai-t'ien* theory differed from Sumeru cosmology in that it had an empirical basis and was virtually free from anthropomorphic and mythological elements, while Buddhist cosmology was replete with religious fantasy and lacked an observational

---

[46] A good summary of this work in English is Y o s h i o Micami's *A Japanese Buddist's view of the European astronomy*, "Nieuw Archief voor Wiskunde" 11, 1–11 (1912). In it the name "Entsū" appears as "Yentsū".

[47] *Bukkoku rekisho hen*, vol. 1.

[48] *Bukkoku rekisho hen*, vol. 2.

foundation. However, Entsū rationalized the difference by explaining that whereas the Chinese were skillful at rational investigation, the Indian sages had penetratins "spiritual eyes" which were given only to superior beings.

Entsū was not rabidly anti-scientific; his object was simply to defend the sacred Buddhist cosmology by all available means. Though he found many adherents among Buddhist monks, his biased views never achieved orthodoxy, but were frowned on by more modest people. He attempted to obtain a Buddhist imprimatur for the publication of the *Bukkoku rekishō hen*, but the aged commissioner Sen'yō[be] would not grant it, on the grounds that "since Entsū was too involved in astronomy, he confused the essentials of Buddhism with astronomical science. His opinions might cause incalculable damage to genuine Buddhism"[49].

Other monks also denounced Entsū. It would be dangerous for Buddhism to commit itself to Sumeru cosmology. Properly speaking, the Sumeru theory was not of Buddhist origin, but was derived from older Indian ideas.

Some, while in basic agreement with Entsū, considered his views too extreme[50], and others condemned his attempt to substitute for Sumeru the mathematically elaborate *kai-t'ien* theory[51]. To clarify this point, a number of commentaries on *kai-t'ien* theory appeared at that time[52]. Inō Tadataka[bf], a prominent surveyor in the Asada school, promptly came out with his *Bukkoku rekishō hen sekimō*[bg] (A refutation of the *Bukkoku rekishō hen*; 1816 or 1817), a bitter denunciation of Entsū's unscientific and misleading dogma. Most top-rank astronomers, however, merely disregarded or ridiculed Entsū's work[53].

Some of Entsū's writings were suppressed in the 1820's[54]; but other exponents of the same philosophy arose. Even during the Meiji period, when explosive westernization had almost eradicated traditional attitudes, a prolonged effort to defend Buddhist cosmology was made. The most notable apologist was Sada Kaiseki[bh]. His main work to explain all celestial phenomena in terms of a complicated mechanism, the *Shijitsu tōshōgi shōsetsu*[bi] (A detailed account of an instrument by which the apparent and real courses of the heavenly bodies are explained) appeared as late as 1880.

The sensation caused by these unusually sharp Buddhist reactions to Western

---

[49] Kōda Rohan (BB), *Kagyūan yatan* (BC), [A night tale of Kagyūan; Tokyo, 1907], pp. 67–76.

[50] For instance, Kanchū (BD), *Shiji idō ben* (BE) [The four seasons compared; 1843].

[51] An example is found in Kojima Tōzan (BF), *Bukkoku rekishō benmō* (BG) [On the absurdity of the *Bukkoku rekishō hen*; 1818].

[52] For instance, Ishii Kandō (BH), *Shūhi sankei seikaizu* (BI) [True illustration of the *Chou-pi suan-ching*; 1813] and Shinohara Yoshitomi (BJ), *Shūhi sankei kokujikai* (BK) [A Japanese annotated edition of the *Chou-pi suan-ching*; 1819].

[53] Mikami Yoshio (BL), *Nihon kagaku no tokushitsu*: *tenmon* (BM) [The Characteristics of Japanese science: astronomy], in *Tōyō shichō no tenkai* (BN) [The development of Oriental thought; Tokyo, 1936], pp. 60–63.

[54] Miyatake Gaikotsu (BO), *Hikkashi* (BP) [History of the suppression of literature in Japan, Tokyo, 1911], p. 69.

cosmology prompted the circulation of a great deal of controversial literature on astronomy. Thus, the Buddhists, by trying to block the advance of Western ideas in Japan, actually stimulated the propagandising of Copernicanism. The result was a heightened public interest in cosmology.

The reaction of the Confucians. By contrast with the pronounced, violent opposition of the Buddhists, we find relatively little reluctance on the part of the Confucians to accept the Copernican system. Some were outspokenly hostile to everything Western, and others claimed Western appropriation of ancient Chinese ideas. Neo-Confucianism (Chu Tzuism), the Japanese state orthodoxy, formed its cosmological background as an integral part of an unitary principle, and therefore, violations in Western cosmology of this unity between human and physical nature were often criticized; however, the Neo-Confucians did not, like the Buddhists or the medieval Church, maintain a detailed religious cosmos, and in general were not concerned with the appearance and morphological aspect (shape) of the universe. Therefore, there was no clear point of conflict with Western cosmology.

The Neo-Confucian idea of cosmological unity was challenged by the Ancient Learning (*kogaku*)[bj] school of Itō Jinsai[bk] and Ogiu Sorai[bl]. They accused Neo-Confucians of indulging in fruitless speculation concerning the heavens, when they should be studying only moral and social problems. Regarding astronomy as a technique unrelated to Confucian moral values, they excluded both physics and natural philosophy from their world-view.

Kan Sazan[bm] (1748–1828) commented that "there is no other use for astronomy than the determination of the correct time; other concerns (viz., cosmology and general astronomy) are merely useless argument and dull speculation"[55]. It would not be far from the truth to assume that his attitude was largely shared by other Confucian scholars, whose pragmatic interest in social and ethical problems excluded a disinterested concern with physical nature.

Influenced by the Ancient Learning school, Neo-Confucians themselves ceased to defend the idea of cosmic unity in the vigorous manner of earlier generations. Thus the Confucian framework of ideas became sufficiently flexible enough to tolerate the reception of Western cosmology without serious ideological difficulty. The acceptance of Western learning was facilitated by two assumptions: that it was historically of Chinese origin, and that as a mere technique it supplemented Eastern values without threatening them[56].

---

[55] Kan Sazan, *Fude no susabi* (BQ) [Writing for amusement's sake], reprinted in *Nihon zuihitsu taikei* (BR) [A comprehensive collection of Japanese informal essays, Tokyo, 1927], I, 80.

[56] Nakayama Shigeru, *Edo jidai ni okeru jusha no kagakukan* (BS) [Confucian views of science during the Tokugawa period], *Kagakushi kenkyū*, No. 72, 157–168 (1964). See also George H. C. Wong, *China's opposition to Western Science during the late Ming and early Ch'ing*, 'Isis' 54, 29–49 (1963) and his *China's opposition to Western religion and science during late Ming and early Ch'ing* (University Microfilms, Inc., Ann Arbor, Mich. L. C. Card no. Mic 58–7381; 1958); and N. Sivin, *On China's opposition to Western science during late Ming and early Ch'ing*, 'Isis' 56: 201–205, (1965).

The belief that current Western scientific theories were originally Chinese was propounded by contemporary Chinese intellectuals, who aimed at apologias for their own tradition and also at the revival of native science, especially mathematics and astronomy. Some of the Japanese Confucians were, as spokesmen for Chinese culture, also in a position to defend the Chinese scientific tradition and its abiding value. Some orthodox Neo-Confucians, Asaka Gonsai[57,bn] and Yasui Sokken[58,bo], both teachers at the Shōheikō[bp], the official shogunal school and a stronghold of conservatism, were typical in this respect. They merely imitated the Chinese way of a grandizing classical achievements by extensive investigation of ancient writings.

On the matter of Copernicanism, they welcomed Shizuki Tadao's interpretation. Deeply versed in Chinese classics, they often quoted an ambiguous passage, "the earth has four displacements", from the *Shang-shu wei k'ao-ling-yao*[bq] (An Investigation of the numinous luminaries; first century B. C.) to support their contention that the heliocentric theory is of Chinese origin.

This classical phrase received the attentions of Chinese philosophers and was interpreted in various ways by later commentators, but it was Japanese scholars who first connected it with Copernicanism, a little earlier than the Chinese. The Chinese priority in the idea of the earth's motion was claimed only on the premise that the truth of Copernicanism was fully recognized. Thus the alleged Western appropriation marks the time when Copernicanism was accepted among intellectuals in general.

This scholarly defense by Japanese Confucians of Chinese culture, though clearly directed against the aggression of Western science, was far less fanatic than the Buddhist reaction. It was also more objective than the contemporary Chinese attitude, which was limited by ethnocentricity. "It is the vice of the Chinese not to acknowledge the strong points of other countries and always to insist that everything worthwhile comes from China", wrote Ikai Keisho[br] (died 1845)[59].

Rather than merely apologize for Western science, certain private schools of Neo-Confucians acclaimed it outright. Notable was the Kaitokudō[bs] school in Osaka, which produced Yamagata Bantō and Hoashi Banri, famous exponents of Western science. Unencumbered by the responsibilities of public office, they could freely develop their interests and criticism.

The idea that Western learning was a mere technique and as such did not conflict with Confucian values was perhaps insisted upon more and earlier in Japan than in China. Japan thus felt free to choose between Chinese and Western techniques, without being inhibited by cultural ties.

---

[57] Asaka Gonsai, *Nanka yohen* (BT), vol. I (circa. 1837), reprinted in *Nihon jurin sōsho* (BU) [Source books of Japanese Confucianism], vol. 2 (1927).

[58] Yasui Sokken, *Suiyo manpitsu* (BV), in *Nihon jurin sōsho*, vol. 2.

[59] *Ikai Keisho sensei shokan shū* (BW). Undated correspondence; reprinted in *Nihon jurin sōsho* vol. 3.

The reaction of the Shintoists. During the eighteenth century, a group of Neo-Shintoists (*kokugaku*[bt], literally, "national learning") gradually became influential. Strongly opposed to the speculations of Buddhists and Confucians, these scholars maintained a more or less positivistic attitude toward scholarly problems and were generous and sympathetic to Western learning[60].

In 1790 Motoori Norinaga[bu] (1730–1801) wrote a critical essay denouncing Monnō and Sumeru cosmology in favor of the spherical earth theory. He singled out, one by one, the absurdities of Buddhists cosmology and concluded that the Buddhists were so envious of the Western theory that they book unfair advantage of the Chinese *kai-t'ien* theory to strengthen their position. Unlike the tradition-bound Confucians and Buddhists, he plainly acknowledged the advanced state of modern theory, saying: "The motive for which Western people study astronomy and geography is not merely to succeed in scholastic debate or calendrical work. Their science is of crucial importance for daily use in navigating the oceans; even a small error would result in a grave accident"[61].

But he was still far from comprehending the view point of Western science. In his *Tenmon zusetsu*[bv] (An illustrated description of astronomy; 1782), he wrote, "the treatment of the five planets has nothing to do with calendrical science. Astronomers should not regard it as an important concern of theirs".

By the next generation, general knowledge of Western science, including the Copernican and Newtonian theories, was more widely diffused. At the same time, because of increasing foreign threats, a nationalistic spirit and a desire for independent identity were prevalent. Hirata Atsutane[bw] (1776–1843) and his followers tried to establish a doctrinal basis for this nationalism out of the ancient native mythopoetic tradition, including elements of Christianity, Confucianism, Buddhism and whatever else was available to him. They emphasized native contributions to Japanese thought, which were uncontaminated by Chinese and Buddhist influences, and also attacked current Confucian and Buddhist ideas.

Unlike the Buddhists and Confucians, the Neo-Shintoists did not have a quasi-scientific tradition that required apology. The ambiguity of their mythology invited free interpretation. In the absence of historical domination and foreign authority, they could "create" their own tradition and include aspects of Western scientific thought.

Atsutane and his pupils, Satō Nobuhiro[bx] (1769–1850) and Tsurumine Shigenobu (1786–1859)[61],[by], were all acquainted with fruits of Dutch learning and had an unreserved appreciation for modern Western science. Earlier attempts had been made to amalgamate Neo-Confucian cosmology with the native creation myths, but Atsutane was bolder and quite cleverly utilized the most up-to-date Western

---

[60] Muraoka Tsunetsugu, *Nihon shisōshi kenkyū* (Tokyo, 1930), pp. 297 ff.

[61] *Shamon Monnō ga kusen hakkai tōron no ben* (BX) [A confutation of the monk Monno's argument of the nine mountains and eight seas], reprinted in *Zōho Motoori Norinaga zenshū* (BY) [Complete works of Motoori Norinaga, revised edition, Tokyo, 1926], vol. 10, pp. 131–136.

theories available. The result was a curious combination of primitive myth and modern science[62]. The process of world creation, ignored in modern science, was explained by traditional myths: the creators, a god and goddess, formed the universe from primordial chaos and gradually modeld the heliocentric system[63]. Thus Atsutane and his followers refuted Entsū's Sumeru argument and took full advantage of Western science in their attempt to systematize primitive mythology into a consistent cosmology.

THE IDENTIFICATION OF COPERNICUS

In China, the Jesuits' distorted account of Copernicus created a confused picture reflected in Juan Yuan's[bz] commentary on Copernicanism. This confusion was still worse in Japan, where the name of Copernicus came to be known not only through the Sino-Jesuits' works but additionally via quite another route, their own translations of Western works.

Motoki Ryōei, a professional interpreter, was not acquainted with the Chinese astronomical writings and hence in his translations he freely employed his own technical terms and transliterations without attempting to identify them with the traditional terminology of calendrical astronomy and standard Chinese transliterations for the names of such Western astronomers as Copernicus and Tycho.

As already indicated, Ryōei, in the preface of his second heliocentric translation, apparently made the error of considering Copernicus a contemporary of Tycho. To his third translation, he added an appendix consisting of summaries from Benjamin Martin's *Philosophical Grammar* and Johann Heinrich Winkler's *Anfangsgründe der Physik* (both in Dutch translations), in which three systems, the Ptolemaic, Copernican and Tychonic, were well illustrated. He did not commit any apparent historical error, though passages relating to relationships of the three figures are still somewhat clumsy. He well enough understood the major differences between the Copernican and Tychonic systems. He placed Tycho's scheme after that of Ptolemy, while Copernican was given full credit among "contemporary" astronomers.

The confusion was increased still further by thoughtless popularizers such as Shiba Kōkan and Honda Rimei[ca]. Kōkan is credited as the first to print an identification of Copernicus as mentioned in Ryōei's translation with the man identified by a Chinese transliteration of his name in the *Li-hsiang k'ao-ch'eng*[64]. But Kōkan

---

[62] Fujiwara Noboru (BZ), *Edo jidai ni okeru kagakuteki shizenkan no kenkyū* (CA) [A study of scientific view of nature in Edo period; Tokyo, 1966], p. 77 ff.

[63] Hirata Atsutane, *Tama no mahashira* (CB) (1812), reprinted in *Hirata Atsutannshūe ze* (CC) [Complete works of Hirata Atsutane], vol. 2 (Tokyo, 1911); Sato Nobuhiro, *Yōzō kaiku ron* (CD) [On the creation and formation of the world; ca. 1825], reprinted in *Shinchū kōgaku sōsho* (CE) [series in Nipponology, re-edited; Tokyo, 1927], vol. 10; Tsurumine Shigenobu, *Ame no mihashira* (CF) [The sacred heavenly pillar; 1821].

[64] Though I suspect that Tadao identified them still earlier in his unpublished manuscripts.

confused Copernicus with Kepler taking "Kopperunyu", the Japanese reading of the Chinese transliteration of "Kepler", as the equivalent of "Copernicus". He even included this erroneous name in the title of his book, *Kopperunyu tenmon zukai* (1805). Honda Rimei in his *Seiiki monogatari* (ca. 1798) wrote that "Copernicus is a pupil of Tycho Brahe"[65].

Shizuki Tadao took a much more cautious and scholarly attitude toward this identification problem in the first part of *Rekishō shinsho*[cc] (drafted in 1798), noting that "someone said that Kopeni (Japanese reading of Chinese Ko-po-ni) spoken of in the *Li-hsiang k'ao-ch'eng* must be identical with the Copernicus spoken of in Western sources"[66].

Finally the shrewd sense of a professional astronomer settled this identification problem. Takahashi Yoshitoki in a letter to Hazama Shigetomi in 1800 compared the relative distance of the solar perigee (or perihelion) given by various astronomers in the Jesuit treatises with those given in Ryōei's third translation (based on Adams), concluding that the Sino-Jesuits' Copernicus was identical with the Copernicus who appeared in the Dutch sources[67], despite the fact that the latter writings portrayed his cosmology as heliocentric and the former as geocentric.

"MOVING-EARTH" THEORY

It is interesting to note that how and why the present-day Japanese term "*chid setsu*"[cd] and its equivalents in modern Chinese and Korean came to represent Copernicanism or heliocentricism. Literally, *chidō setsu* is "theory of the earth's motion", but no such set phrase existed in any of the Dutch sources consulted by Motoki Ryōei and Shizuki Tadao. Western sources employed such words as "Copernican theory", "heliocentric hypothesis", or "sun-centered".

Generally, whenever the word "*chidō*" ("*ti-tung*" in Chinese) is found in the Chinese classics, its interpretation is usually "earthquake". Some writers define *ti-tung* as a major earthquake, as distinguished from one of ordinary magnitude *ti-chên*[ce] (earth tremor)[68].

Even if we interpret *ti-tung* liberally as equivalent to *ti-chuan*[cf] (earth-turning), occasionally found in early Chinese cosmological writings, it is not at all clear whether this referred to the rotation or the revolution of the earth. Even in its present usage, the word *chidō* is too vague to be employed as a scientific term. As a matter of fact, a Japanese apologist for Buddhist cosmology, Entsū, attacked the word *chidō* as confusing and susceptible of various divergent interpretations.

---

[65] Part I, reprinted in *Nihon shisō taikei 44* (Tokyo, 1970), p. 95.

[66] *Nihon tetsugaku shisō zensho 6*, p. 141.

[67] *Seigaku shukan* (CG) [Notes and correspondence on astronomy; ed. by Shibukawa Kagesuke] in *Nihon shisō taikei 63*, p. 213.

[68] I am indebted for this remark to Professor Jeon Sang-woon (CH), a historian of Korean science.

Then why was this misleading term adopted to denote such an important concept as the Copernican world picture? Presumably the man who coined the phrase *chidō no setsu* (the theory of the earth's motion) was Shizuki Tadao, who in part I of his *Rekishō shinsho* (drafted in 1798) employed this phrase in his own commentary on the appendix "tentairon" (on the heavenly bodies)[69]; in the preceding straightforward translation of John Keill's work, he simply transliterated the word "Copernican" to describe the theory.

From the context of "tentairon", we can deduce two reasons why Tadao coined this phrase:

1) In order to show that in ancient China, "there was a theory of the earth's motion, preceding its Western counterpart", and thereby to defend the Asian scientific tradition by demonstrating its priority.

2) In order to place Copernicanism in the framework of traditional natural philosophy — nemely, the system based on the polar conception of yin and yang. Arguments concerned only with local and morphological relationships, whether heliocentric or geocentric, appeared rather superficial to traditional Asian thinkers. The cosmic polar concepts of motion (to correspond to yang) and rest (to correspond to yin) must have appealed to Shizuki Tadao as more profound and fundamental in their implications. In other words, Copernicanism was comprehensible to him primarily in terms of the physico-dynamical principle of "motion-rest". In the traditional list of various dichotomies associated with "yang-yin", undoubtedly a very important place is occupied by "motion-rest"; while "heaven-earth" or "sun-moon" does form a dichotomy, "sun-earth" never does.

Tadao's attitude towards Copernicanism was welcomed and shared by Neo-Confucian orthodox philosophers such as Asaka Gonsai and also by the official atronomer, Shibukawa Kagesuke[70].

We shall now turn to popular treatises in order to estimate the degree of dissemination of the term "theory of the earth's motion". In earlier works of the most illustrious Copernican advocate, Shiba Kōkan, such as *Chikyū zenzu ryakusetsu* (1793) and *Oranda tensetsu* (1796), the term "earth's motion" does not appear explicitly, but *Oranda tsūhaku*[cg] (Dutch oversea activities, 1805), which appeared after the completion of *Rekishō shinsho*, identified the Copernican view by saying that "this is called the theory of the earth's motion!"[71]. In still later works, however, he did not use "earth's motion" again. It may be that Kōkan, thoroughly devoted to things Western, might have had a distaste for the chauvinism of Shizuki Tadao.

Perhaps the book most instrumental in disseminating the term "the theory of the earth's motion" was Yoshio Hisatada's *Ensei kanshō zusetsu*[ch] (An illustrated treatise on Western astronomy, first edition 1823). At the end was appended "chidō

---

[69] *Nihon tetsugaku shisō zensho 6*, p. 136 ff.

[70] *Shinpō rekisho zokuhen.*

[71] Volume I, in Nakai, *Shiba Kōkan*, p. 116.

wakumon"[ci] (Queries on the erath's motion), in which he tried to expound further what Tadao had said in *Rekishō shinsho*. This work was reprinted again and again, becoming the most standard textbook of astronomy in mid-nineteenth-century Japan. Thus the term "earth's motion" entered the popular vocabulary. In the succeeding Meiji period, when the modern educational system was established, the phrase was adopted for mass enlightenment at the elementary text-book level and has remained current since.

On the Chinese scene, "ti-tung" (earth's motion) appeared in Juan Yuan's *Ch'ou-jen chuan*[cj] (Biographies of Chinese mathematicians and astronomers, 1799) in reference to Michel Benoist's exposition of heliocentricism, though Juan himself did not accept it. In the sequel of that book (prefaced in 1840), Juan tried to identify the origin of heliocentricism in Chang Heng's[ck] *ti-tung*[cl] (earth's motion instrument, second century A. D., presumably a kind of seismometer) and thereby defend his own tradition on the ground that Copernicanism was accepted. The set phrase *ti-tung shuo* (theory of the earth's motion) appears only as late as 1859 in *T'an t'ien*[cm] (Discussion on the heavens), Li Shan-lan's[cn] translation of John Herschel's book. It is impossible to conceive of any influence of Tadao on these Chinese works, but they shared the same traditional mentality with its dichotomy of yin-yang and motion-rest.

In sum, among the various implications of Copernicanism, the physico-dynamic aspect was most attractive, much more so than the geometrical or astronomical aspects, because of the Chinese inclination to use the yin-yang dichotomy and its various equivalents at the most fundamental level of explication of natural phenomena. In this context, "theory of the earth's motion" received public recognition as a popular as well as professional term.

CONCLUSION

To sum up, the Copernican system did not evoke bitter ideological opposition in Japan except in Buddhist circles. Even the latter could not exert such a commanding reactionary influence as was wielded by the Renaissance Church in Europe. The main cause of the delay in the introduction of heliocentric theory into Japan was the seclusion policy of the government, rigidly maintained until the early part of the eighteenth century and, secondly the linguistic barrier, which remained formidable until the last quarter of that century.

Interest in Western cosmology was initiated not by the camp of traditional astronomers but by linguistic experts, who began with the introduction of the cosmological aspect of Copernicanism. The principal concern with Copernicanism was shown as regards its physico-dynamical aspect, which was interpreted in terms of Eastern *Naturphilosophie*.

It seems that the question of the reception of Copernicanism was more or less used in the more general problem of the superiority of Western learning. Copernicanism never played a pivotal role in overthrowing traditional ideologies or recognizing Western superiority. In this respect, recognition of Western superiority in the astronomical domain had been well established earlier through Sino-Jesuits works, and on the basis of this recognition Copernicanism, in spite of some ideological incompatibility and conflict, had a rather smooth reception in the course of the nineteenth century.

59

a 渾天
b 蓋天
c 張載
d 朱熹
e 沈括
f 氣
g 陰陽
h 吉宗
i 貞享
j 澁川景佑
k 新法暦書續編
l 志筑忠雄
m 暦象新書
n 禮器圖說
o 禮器圖式
p 二儀略說
q 乾坤辯說
r 西川正休
s 天經或問
t 游藝
u 本木良永
v 和蘭地球說
w 阿蘭陀地球圖說
x 天地二球用法
y 天文管闚
z 動學指南

aa 求心力論
ab 列子
ac 天體論
ad 不測
ae 暦象考成
af 乾坤分判圖說
ag 司馬江漢
ah 地球全圖略說
ai 和蘭天說
aj 刻白爾天文圖解
ak 山片蟠桃
al 帆足萬里
am 吉雄尙貞
an 麻田剛立
ao 寛政
ap 西洋新法暦書
aq 三浦梅園
ar 歸山錄
as 松村翠崖
at 增修消長法
au 消長法
av 天地二球用法評說
aw 杉田玄白
ax 暦象考成後編
ay 新法暦書
az 非天經或問

60

ba 文雄
bb 九山八海嘲論
bc 圓通
bd 佛國暦象編
be 仙薬
bf 伊能忠敬
bg 佛國暦象編斥妄
bh 佐田介石
bi 現實等象儀詳説
bj 古學
bk 伊藤仁齋
bl 荻生徂徠
bm 菅茶山
bn 安積艮齋
bo 安井息軒
bp 昌平黌
bq 尚書緯考靈曜
br 猪飼敬所
bs 懐徳堂
bt 國學
bu 本居宣長
bv 天文圖説
bw 平田篤胤
bx 佐藤信淵
by 鶴峯戊申
bz 阮元
ca 本田利明
cb 西域物語
cc 歌白尼
cd 地動説
ce 地震
cf 地轉
cg 和蘭通舶
ch 遠西觀象圖説
ci 地動或問
cj 疇人傳
ck 張衡
cl 地動儀
cm 談天
cn 李善蘭

61

A 曆志
B 新唐書
C 歐陽修
D 宋祁
E 藪内清
F 鐵炮記
G 天體論
H 日本哲學思想全書
I 三枝博音
J 清水幾太郎
K 廣瀬秀雄
L 旧長崎天學派の學統成立について
M 蘭學資料研究會研究報告
N 近世科學思想
O 中山茂
P 大塚敬節
Q 日本思想大系
R 文明源流叢書
S 今井溱
T 乾坤辯說雜記
U 天官書
V 星術本源太陽窮理了解新制天地二球用法記
W 天文物理學家の自然觀
X 日本哲學思想全書
Y 大槻如電
Z 新撰洋學年表

62

AA 天文秘書

AB 本木良永譯和蘭地球説について

AC 板澤武雄

AD 江戸時代における地球地動説の展開とその反動

AE 史學雜誌

AF 本木良永の天文書翻譯について

AG 洋學事はじめ展, 蘭學の諸系譜と江戸幕府旧藏本

AH 大久保利謙

AI 桑木彧雄

AJ 黎明期の日本科學

AK 渡邊庫輔

AL 阿蘭陀通詞志筑氏事略

AM 大崎正次

AN 歴象新書天明旧譯本の發見

AO 科學史研究

AP 中井宗太郎

AQ 村岡典嗣

AR 續日本思想史研究

AS 有坂隆道

AT 地動説の傳來と新宇宙論の出現

AU 日本史の研究

AV 高橋至時

AW 新修五星法圖説附錄

AX 間重新

AY 先考大業先生事跡略記

AZ 渡邊敏夫

63

BA 間重富とその一家

BB 幸田露伴

BC 蝸牛庵夜譚

BD 環仲

BE 四時異同辯

BF 小島濤山

BG 佛國暦象辯妄

BH 石井寛道

BI 周髀算經正解圖

BJ 篠原善富

BK 周髀算經圖字解

BL 三上義夫

BM 日本科學の特質 天文

BN 東洋思潮の展開

BO 宮武外骨

BP 筆禍史

BQ 筆のすさび

BR 日本随筆大系

BS 江戸時代における儒者の科學觀

BT 南柯餘編

BU 日本儒林叢書

BV 睡餘漫筆

BW 鵜飼敬所先生書簡集

BX 沙門文雄が九山八海嘲論の辯

BY 增補本居宣長全集

BZ 藤原暹

64

CA	江戸時代における科學的自然觀の研究
CB	靈能眞柱
CC	平田篤胤全集
CD	鎔造化育論
CE	新註皇學叢書
CF	天の御はしら
CG	星學手簡
CH	全相星

JOHN L. RUSSELL
Heythrop College, London

THE COPERNICAN SYSTEM IN GREAT BRITAIN

I. THE COPERNICAN SYSTEM IN ENGLAND

During the sixteenth and early seventeenth centuries England was, from the scientific point of view, relatively isolated from the Continent. There had been a modest but promising revival of learning in the early sixteenth century, in which John Colet, John Fisher, William Grocyn, Thomas Linacre and Thomas More had played a prominent part. They were helped by such international scholars as Desiderius Erasmus and Luis Vives, who visited England and were on the friendliest terms with the English scholars. This circle, apart from its theological interests, was mainly concerned with humanistic studies and with the propagation of Greek studies in the Universities and elsewhere. They were less concerned with philosophical and scientific problems. The revival was brought to an end by the increasingly autocratic behaviour of King Henry VIII, culminating in the execution of Fisher and More in 1535 on religious grounds. For the next hundred years or more, English scientists, with the notable exceptions of William Gilbert and William Harvey, contributed little to the European cultural community. This is not to say that there was no scientific activity during this period. There was a lively interest in practical applications, especially in the fields of surveying and navigation. But the published works were mainly textbooks or popular expositions, some of which reached a high standard in their own field but which contributed little to the advancement of science. In any case, they were normally written in English and were scarcely known on the Continent.

The Sixteenth Century

The first English writer to refer to the Copernican theory was Robert Recorde (c. 1510–1558) in his book *The Castle of Knowledge* (1556)[1]. This was an elementary treatise on astronomy written for students and set in the form of a dialogue between

---

[1] London, Reginalde Wolfe, 1556.

a scholar and his teacher. The treatment was on traditional Ptolemaic lines, as one would naturally expect of an introductory textbook, but towards the end there was a reference to Copernicus. Recorde has been giving the traditional arguments that the earth is in the centre of the universe and then passes to the question whether it is in motion or at rest:

But as for the quietness of the earth, I need not to spend any time in proving of it, since that opinion is so firmly fixed in most men's heads, that they accompt it mere madness to bring the question in doubt. And therefore it is as much folly to travail to prove that which no man denieth, as it were with great study to dissuade that thing which no man doth covet, neither any man alloweth: or to blame that which no man praiseth, neither any man liketh.

Scholar. Yet sometimes it chanceth, that the opinion most generally received, is not most true.

Master. And so do some men judge of this matter, for not only Eraclides Ponticus, a great Philosopher, and two great clerks of Pythagoras school, Philolaus and Ecphantus, were of the contrary opinion, but also Nicias [Hicetas] Syracusius, and Aristarchus Samius, seem with strong arguments to approve it: but the reasons are too difficult for this first Introduction, and therefore I will omit them till another time. And so will I do the reasons that Ptolemy, Theon and others do allege, to prove the earth to be without motion: and the rather, because those reasons do not proceed so demonstrably, but they may be answered fully, of him that holdeth the contrary. I mean, concerning circular motion: marry, direct motion out of the centre of the world seemeth more easy to be confuted, and that by the same reasons, which were before alleged for proving the earth to be in the middle and centre of the world.

Scholar. I perceive it well: for as if the earth were always out of the centre of the world, those former absurdities would at all times appear: so if at any time the earth should move out of his place, those inconveniences would then appear.

Master. That is truly to be gathered: how be it, Copernicus, a man of great learning, of much experience, and of wonderful diligence in observation, hath renewed the opinion of Aristarchus Samius, and affirmeth that the earth not only moveth circularly about his own centre, but also may be, yea and is, continually out of the precise centre of the world 38 hundred thousand miles: but because the understanding of that controversy dependeth of profounder knowledge than in this Introduction may be uttered conveniently, I will let it pass till some other time.

Scholar. Nay sir in good faith, I desire not to hear such vain fantasies, so far against common reason, and repugnant to the consent of all the learned multitude of Writers, and therefore let it pass for ever, and a day longer.

Master. You are too young to be a good judge in so great a matter: it passeth far your learning, and theirs also that are much better learned than you, to improve [i. e. disprove] his supposition by good arguments, and therefore you were best to condemn nothing that you do not well understand but another time, as I said, I will so declare his supposition, that you shall not only wonder to hear it, but also peradventure be as earnest then to credit it, as you are now to condemn it[2].

It will be seen from this that Recorde was quite prepared to accept a diurna rotation. The arguments of Ptolemy, Theon etc against it can be fully answered He was at least sympathetic to the full Copernican system but his description o it was very incomplete. The reader is told that Copernicus accepted a diurnal rotation and put the earth "continually out of the precise centre of the world" by nearly 4 million miles, but not that he put the earth in orbit round the sun. Evidently he

---

[2] *Castle*, pp. 164–65. I have modernised the spelling.

thought that a full statement of the theory would unduly confuse students at this stage in their studies. *The Castle of Knowledge* enjoyed a steady popularity until at least the end of the century. A second edition was published in 1596. No doubt it stimulated the interest of its readers in the work of Copernicus even if it did little to satisfy their curiosity. It is unfortunate that Recorde did not live to write the more advanced treatise referred to in the text.

Recorde, as a young man, had graduated at Oxford University, after which he transferred to Cambridge where he took the degree of Doctor of Medicine in 1545. The two most prominent Cambridge scholars at this time were Sir John Cheke (1514–1557) and Sir Thomas Smith (1513–1577)[3]. Both were primarily classical scholars but Smith in particular had an interest in astronomy and accumulated a good scientific library which included a first edition of *De Revolutionibus*. It is quite possible that Recorde came to know of the Copernican system through him, but there is no definite evidence on this point.

Shortly after the publication of Recorde's book, the work of Copernicus was referred to by two other mathematicians: John Dee (1527–1608) and John Feild (1520–1587). Dee, in a preface to Feild's *Ephemeris* for 1557[4], deplored the inadequacy of the older astronomical tables:

> A multis, et illis quidem clarissimis Mathematicis, non solum satis fuisse decantatum, Tabellas veteres, & eorundem canones, haud amplius cum Phaenomenis conuenire: Verum etiam ab excellentissimis artificibus, insigni veritatis demonstrandae modo, longissime, eius generis errores, fere omnes, profligatos arbitrabar. Sperabam etiam alios, illos praesertim qui in Astronomicis tum multa, tum magna tractant, & moliuntur, de COPERNICI, aut Rhetici & Reinhaldti scriptis, vel eorum saltem nominibus, auditione tandem aliquid accepisse: praeclaramque horum famam, istorum hominum aures iam circumsonasse diutius. Illius quidem, ob labores plus quam Herculeos, in coelesti disciplina restauranda, eademque firmissimis rationum momentis corroboranda, ab eodem exantlatos: (cuius de hypothesibus nunc non est disserendi locus). Horum vero, propter eam quam ostenderunt strenuam in illius insistendo vestigiis diligentiam.

Dee therefore praised Copernicus "for the more than Herculean labours which he endured in giving a new impetus to the study of the heavens and confirming it most strongly by his calculations", but he added that "this is not the place to discuss his hypotheses". He was clearly concerned more with the accuracy of the Copernican tables than with their theoretical basis, on which he expressed no opinion. In spite of his respect for Copernicus he never, in fact, committed himself to his system and two years later, in his *Propaideumata aphoristica*[5], he described the universe in pre-Copernican terms. In aphorism 58, for instance, he referred to the diurnal motion of the heavens from east to west as the swiftest of all motions. Aphorism 66 spoke of the sun as moving through the ecliptic "per proprium suum motum".

---

[3] See Francis R. Johnson, *Astronomical Thought in Renaissance England*, Baltimore, Johns Hopkins Press, 1937, pp. 87–90.

[4] *Ephemeris anni 1557 currentis iuxta Copernici et Reinholdi canones... supputata*, London, Thomas Marshe, 1556, Sig. A2^r–A2^v.

[5] London, 1558.

Feild, on the other hand, apparently accepted the views of Copernicus without qualification. In his own foreword to the *Ephemeris* he wrote:

"Quapropter hanc tibi peruulgavi Ephemeridem Anni 1557, in ea authores sequutus N. Copernicum et Erasmum Reinholdum, quorum scripta stabilita sunt et fundata veris, certis, et sinceris demonstrationibus"[6]. It does not seem that Feild is praising merely the greater accuracy of the Copernican and Reinholdian tables. It is more reasonable to suppose that the "writings" he refers to, which have been "established and based upon true, certain and genuine demonstrations" include also the theoretical basis of the tables.

Dee was the most prominent English mathematician of his day. He accumulated a first class scientific library which contained two copies of *De Revolutionibus* as well as works of Nicholas of Cusa, Ptolemy and many others of astronomical interest. His house at Mortlake attracted many visitors and became the main centre of scientific activity in England during the third quarter of the 16th century. It is probable that the Copernican theory was freely discussed within this circle and that Dee himself helped to spread some knowledge of it[7].

Neither Recorde nor Dee nor Feild had given any precise indication of what Copernicus had actually taught, nor had any of them asserted unambiguously that they accepted his theory. The first English writer to publish a clear exposition of the system, and the first who publicly accepted it as true, was Thomas Digges (c. 1546–1595). Thomas was the son of a distinguished mathematician, Leonard Digges. After his father's death in 1559 he became a pupil of John Dee and thereafter remained in close touch with him. In 1573 he published a short treatise[8] on the new star which had appeared in Cassiopeia during the previous year. In it he gave data for the position of the star which were commended for their accuracy by no less an authority than Tycho Brahe. Most of the book was devoted to a mathematical treatment of diurnal parallax and its measurement but Digges also discussed briefly the possibility of using the new star to test the Copernican theory, to which he was obviously favourably inclined. He suggested two lines of investigation:

Praepostere etiam Antiquos progredi perspexi ex Theoricis scilicet fictis Parallaxeis et distantias venari veras, cum inuerso ordine procedere potius debuissent, et ex Parallaxibus obseruatis et cognitis, Theoricas examinare: et hac ratione haud difficile esset si diu perseuerauerit Phoenomenon istud mirabile, exacto iudicio discernere an Terra immobilis in Mundi centro quiescat, et ingens illa Orbium erraticarum et fixarum moles rapidissimo cursu 24 horarum spacio in gyrum rotetur, seu potius fixarum illa immensa sphaera vere fixa maneat, et apparens ille motus tantummodo ex Terrae circulari super Polis suis rotatione contingat... Fuit igitur causa praecipua cur Copernicus vir admirandi ingenii, & industria singulari, aliis hypothesibus uti, et nouam Coelestis Machinae Anatomiam eruere conatus fiet: at prolixa nimis oratione et huic loco parum conuenienti opus esset, vt dilucide collatis vtriusque generis hypothesibus veritas elucesceret, hoc saltem admonere statui ansam oblatam esse, et occasionem maxime opportunam experiendi an Terrae motus in Co-

---

[6] *Op. cit.*, Sig. A3^{r}.

[7] See Johnson, *op. cit.*, pp. 137–39.

[8] *Alae seu Scalae Mathematicae*, London, Thomas Marshe, 1573.

pernici Theoricis suppositus, sola causa fiet cur haec stella magnitudine apparente minuatur, nam si ita fuerit in Aequinoctio verno semper decrescens minima sua magnitudine conspiceretur. Post vero si durauerit paulatim crescens in Iunio sequente eiusdem fere fulgoris erit quemadmodum in prima sua apparitione, at in Aequinoctio Autumnali insolitae magnitudinis necnon splendoris videbitur: eiusmodi autem quantitatum apparentium diuersitas nulla poterit alia assignari causa' quam ipsius a terra elongationes, quoniam augeri aut minui stellam in Coelo, non solum Physicis prorsum fundamentis contrarium esset, sed manifestis etiam mensuris hac arte adhibitis aliter esse deprehendetur[9].

Digges's argument is confused and suggests that he had not, at this stage, fully understood the implications of the Copernican theory. His first suggestion is that astronomers should try to discover a diurnal parallax of the star in order to discover whether it is the earth or the heavens which have a daily rotation. He does not explain himself further and one is left with the suspicion that he has failed to realise that the diurnal parallax will be exactly the same whether the earth is rotating and the heavens are at rest or vice versa. The second suggestion is that the diminution in brightness of the star is due to the earth's recession from it in its orbit round the sun. If this is the case, then the brightness should vary according to an annual cycle, reaching a minimum at the spring equinox and then increasing to a maximum in the autumn. Digges was, of course, writing shortly after the star had appeared, without waiting to see whether the prediction would be fulfilled.

Digges's book was subjected to a lengthy criticism by Tycho Brahe in his *Astronomiae Instauratae Progymnasmata*[10]. Tycho pointed out that Digges had completely overlooked the fact that on the Copernican theory there should be not only a diurnal but also an annual parallax. If the nova was near enough to shew a diurnal effect (and this would put it, on Tycho's reckoning, no further away than the planet Mars), then it must have a very large annual parallax which could not possibly have been overlooked. The absence of any such effect implied that, if Copernicus was right, the star must be far beyond the orbit of Saturn, in the region of the fixed stars. This being so, it must be too far away for its apparent brightness to be affected by the earth's position in its orbit round the sun. With some justification, therefore, Tycho dismissed Digges's proposed tests as irrelevant.

Digges returned to the Copernican system much more fruitfully in 1576. In this year he prepared for publication a new edition of a popular work on astronomy written by his father and first published in 1553[11]. Leonard Digges had made no mention of the Copernican theory and his son now decided to rectify the omission. He therefore added, as an appendix to the book, a translation of part of the *De Revolutionibus* (Book I, chs. 10, 7, 8), in which Copernicus had summarised the main features of his system and had answered the traditional objections against the earth's

---

[9] *Alae*, Sig. A2ᵛ—A3ᵛ.

[10] Prague, 1602, pp. 653—90. Reprinted in Tychonis Brahe Dani *Opera Omnia*, III, 167—203, (Ed. J. L. E. Dreyer).

[11] *A Prognostication euerlastinge of righte good effecte... Published by Leonard Digges Gentleman... Lately corrected and augmented by Thomas Digges his sonne*. London, Thomas Marsh, 1576.

motion. (He omitted ch. 9, however, which dealt with the movements of precession and 'trepidation'.) The translation was free but substantially accurate and was interspersed with some material added by Digges himself. It was preceded by a preface highly commending the system and insisting that Copernicus had propounded it as true — not simply as a convenient mathematical device. It was also accompanied by a diagram of the universe which would shew the reader at a glance what the system involved. Digges's diagram has a special interest in that it shewed the fixed stars at varying distances from the centre of the universe (the sun), and that the description of the outermost sphere, inserted into the diagram, ran:

> This orbe of starres fixed infinitely up extendeth hit self in altitude sphericallye, and therefore immovable the pallace of foelicitye garnished with perpetuall shininge glorious lightes innumerable, farr excellinge our sonne both in quantitye and qualitye the very court of coelestiall angelles devoyd of greefe and replenished with perfite endlesse joye the habitacle for the elect.

Digges, therefore, definitely held that the universe is infinite in extent. This view was not, of course, new. It had been accepted by Anaximander and Democritus among the Greeks. Nicholas of Cusa had taught that the universe is unlimited though he would not apply the term 'infinite' to it. Nevertheless, Digges was the first to propound this view within the context of the new astronomy: Copernicus himself having left the question open. Eight years later it was to receive more vigorous support from Giordano Bruno in his *De l'infinito universo e mondi*, published in London, 1584. In a sense, both Nicholas and Bruno were more 'modern' than Digges since both denied that the universe has a centre and both rejected the theory that celestial matter is intrinsically immutable whereas Digges represented the heavenly regions as an infinite sphere centred upon the sun and he apparently still held that the celestial bodies are incorruptible.

The revised *Prognostication* of 1576 was the first actual description of the Copernican system in English or by an Englishman. As such it was, for many years to come, the principal means by which English readers came to know of the system. It must have reached a very wide circle since no less than seven editions were published between 1576 and 1605, all with the Copernican appendix.

The Copernican theory was mentioned briefly by John Blagrave of Reading in his work: *The Mathematical Iewel* (1585)[12], in which he described a new type of astrolabe he had invented. Referring to the complexity of the Ptolemaic planetary system he remarked:

> Insomuch that of late yeares that singuler man Copernicus affirmeth that the sunne is the fixed centre of the world, about whom the earth moueth (not the sunne about the earth) and that all the rest of the planets moue regularly about the centre of the sunne sauing the moone which like an epicicle moueth about the earth in the speere of the earth 13 times in his yearely motion. But omitting the inuentions of Copernicus, and a number of the rest, I will only heere shew a figure of those which haue always bene before his time...[13]

---

[12] London, Walter Venge, 1585.

[13] *Ibid.*, p. 11.

There was no reason, in this purely practical work, why he should commit himself for or against the Copernican theory and, in fact, he gave no indication of his attitude. Eleven years later, in his *Astrolabium Uranicum Generale* (1596)[14] he more definitely accepted at least a semi-Copernican theory. In this work he described a new astrolabe designed on the supposition that the earth rotates on its axis and the fixed stars are at rest. The heavens were engraved on a fixed spherical plate instead of on the traditional moving *rete* and the earth, or rather, the horizon of the terrestrial observer, was represented by a moving ruler and scale. He pointed out that his instrument did not differ from those described by John Stoeffler and Geoffrey Chaucer except for the fact that:

> they according to the auncient Astronomers, appointed the Starry Heauens to mooue rightwards from East towards West, vppon the earth or fixed Horizon of the place. An I according to *Copernicus* cause the earth or Horizon to moue leftwards from West towards East, vppon the Starry Firmament fixed: In so much, that if in this my Astrolabe you hold still that perticular moouer with one hand, and with your vnder hand turne about the Celestiall, then is it iumpe *Stophler* againe. In which motion (a pretty thing to note) one that standeth by shall hardly perceiue any other but that the Reete mooueth, although in deede you turne about the Mater, strongly confirming *Copernicus* argument, who sayth, that the weakenesse of our senses do imagine the Heauens to mooue about euery 24 houres from East to West by a *Primum mobile*, where as in deede they haue been alwayes fixed, and it is the earth that whirleth about euery 24 houres from West to East, of his owne propper nature allotted vnto him, as most fit for the receptacle of all transitory things, being appointed in a place where nothing is to stay him from his continuall moouing[15].

Blagrave thus explained how his astrolabe could be used by those who regarded the earth as fixed, but he himself clearly preferred the 'Copernican hypothesis' of a rotating earth. There was no mention in this book of the heliocentric system: it would not, of course, have been relevant since it could not have been represented in an instrument of this type. Johnson is therefore exaggerating when he says that "with the publication of Blagrave's *Astrolabium Uranicum Generale*, sound information about the mechanical details of the new heliocentric astronomy became readily available to all his countrymen"[16]. The instrument did, however, graphically illustrate the fact that the two rival theories of a rotating heaven and a rotating earth were observationally equivalent.

In chapter 2, Blagrave explained how the precession of the equinoxes could be allowed for on his instrument. He did not explicitly state that this phenomenon was due to a movement of the earth's axis but it would be clear to the user that it could be explained on this supposition.

Thomas Blundeville, a friend and contemporary of Blagrave, took a less favourable view of the Copernican theory. He wrote a popular treatise on arithmetic and cosmography entitled *M. Blundevile His Exercises* (1594)[17] in which he rejected the

---

[14] Printed by Thomas Purfoot for William Matts, 1596.

[15] Sig. Fl^{v}.

[16] Johnson, *op. cit.*, p. 210.

[17] London, John Windet, 1594.

theory on philosophical and theological grounds, while celarly recognising its mathematical advantages. He said of it:

Some also deny that the earth is in the middest of the world, and some affirme that it is moueable, as also *Copernicus* by way of supposition, and not for that he thought so in deede: who affirmeth that the earth turneth about, and that the sunne standeth still in the midst of the heauens, by helpe of which false supposition he hath made truer demonstrations of the motions and reuolutions of the celestiall Spheares, then euer were made before, as plainely appeareth in his booke *de Reuolutionibus* dedicated to *Paulus Tertius* the Pope, in the yeare of our Lord 1536. But *Ptolomie, Aristotle*, and all other olde writers affirme the earth to be in the middest and to remaine immooueable and to be the very Center of the world, proouing the same with many most strong reasons not needefull here to be rehearsed, because I thinke fewe or none do doubt thereof, and specially the holy Scripture affirming the foundations of the earth to be layd so sure, that it neuer should mooue at any time: Againe you shall finde in the selfe same Psalme these words, Hee appointed the Moone for certaine seasons, and the Sunne knoweth his going downe, whereby it appeareth that the Sunne moooeth and not the earth[18].

Blundeville was a populariser rather than a professional mathematician and his work was evidently widely read since it went through a number of editions, the 7th and last appearing in 1638. Some additional matter was added to the 2nd edition but the judgment on Copernicus remained unchanged to the end.

Blundeville's assertion that "few or none" of his contemporaries had doubts about the geostatic universe underestimated the extent to which Copernicus, or at least a semi-Copernican theory was accepted by the mathematicians of his day but it was no doubt substantially true of the 'intelligent layman' for whom he was writing.

In 1602 Blundeville published a further work on astronomy: *The Theoriques of the Seven Planets*[19]: the first English work which gave methods for computing planetary orbits. In the preface he says that it had been "collected, partly out of *Ptolomey*, and partly out of *Purbachius*, and of his Commentator *Reinholdus*, also out of *Copernicus*, but most out of *Mestelyn*, whom I haue cheefely followed, because his method and order of writing greatly contenteth my humor"[20]. It was a purely practical treatise and did not discuss the theoretical basis of planetary motion.

The Copernican theory was briefly referred to in a treatise on iatrochemistry published by Richard Bostocke in 1585[21]. It supported the Paracelsian system of chemistry and medicine but, at the same time, said of Paracelsus:

He was not the author and inuentour of this arte as the followers of the Ethnickes phisicke doe imagine, as by the former writers may appeare, no more then *Wicklife, Luther, Oecolampadius, Swinglius, Caluin, &c.* were the Author and inuentors of the Gospell and religion in Christes Church, when they restored it to his puritie, according to Gods word, [...] And no more then *Nicholaus Copernicus*, which liued at the time of this *Paracelsus*, and restored to vs the place of the starres accord-

[18] *Exercises*, fol. 181r.

[19] London, A. Islip, 1602.

[20] Sig. A3r.

[21] *The difference between the auncient Phisicke... and the latter Phisicke.* London, R. Walley, 1585.

ing to the trueth, as experience & true obseruation doth teach is to be called the author and inuentor of the motions of the starres, which long before were taught by *Ptolomeus* Rules Astronomicall and Tables for Motions and Places of the starres...[22]

Johnson's remark, on the strenght of this passage, that "Bostocke was obviously a Copernican"[23] is hardly justified. He may well have been asserting no more than that Copernicus had rectified errors which had crept into the medieval astronomical tables through faulty observation. On the other hand, it may be an example of the tendency, noted by Thorndike[24] among some writers of this period, to regard Copernicus as having restored and developed the Ptolemaic system as against the medieval Arabic and Alphonsine corruptions. It was, for instance, generally regarded as a blemish on the late medieval system that it required two or even three spheres outside the eighth sphere of the fixed stars in order to account for the (partly spurious) phenomena of precession and trepidation. The fact that Copernicus could eliminate these unwanted spheres by transferring their movements to the earth's axis helped to restore the conceptual simplicity of the original Greek model. It must be remembered also that many astronomers accepted only the semi-Copernican theory that the earth rotates on its own axis at the centre of the universe. This could be regarded as a comparatively minor modification of the Ptolemaic system, and one which eliminated more than one blemish which it had either always possessed or had acquired in the course of the Middle Ages.

Another example of the tendency to merge the Ptolemaic and Copernican theories is to be found in Richard Forster's *Ephemerides Meteorographicae* (1575):

Languet apud nos in ipso pene exortu Mathematum disciplina, quae apud Anglos primum renasci coepit, e tenebris in lucem emersa, per solertissimum Mathematicum nostratem Ioannem Dee, nouarum hypothesium, & Ptolemaicae doctrinae acerrimum vindicem. Et nisi vir ille ingenue Atlanti humeros supposuerit, breui tandem fiet, vt tota cum Copernici & Rheinoldi coelo corruat, tanta est apud nos in artem grassatio imperitorum, & impunitas, vti hanc disciplinam Vraniae sacram, temerare nihili aestimatur[25].

It is not quite clear what Forster means when he describes Dee as "the keen champion of new hypotheses and Ptolemaic teaching". His further remark that if it were not for his support the whole science of mathematics (in England) would collapse "together with the heavens of Copernicus and Reinhold" suggests that he regarded Dee as upholding both Ptolemy and Copernicus. It is possible, however, that the distinction between Ptolemaic *doctrina* and Copernican *hypothesis* is meant to imply that Ptolemy taught the true theory and that Copernicus had simply proposed a mathematical device for simplifying the calculation of planetary paths.

---

[22] Sigs. H7^v–H8^r.

[23] *Op. cit.*, p. 183.

[24] Lynn Thorndike, *A History of Magic and Experimental Science*, Vol. V, New York, Columbia University Press, 1941, p. 422.

[25] *Ephemerides meteorographicae ad annum 1575*, London, J. Kingston, [1575], last page.

Apart from Blundeville, already mentioned, the only other 16th century astronomical writer of any significance who explicitly rejected the Copernican system was Thomas Hill, who died about 1575. In his treatise on astronomy: *The Schoole of Skil*, published postumously in 1599 he wrote:

> *Aristarchus Samius*, which was 261 yeares, before the byrth of Christ, tooke the earth from the middle of the world, and placed it in a peculiar Orbe, included within *Marses* and *Venus* Sphere, and to bee drawne aboute by peculiar motions, about the Sunne, which hee fayned to stande in the myddle of the worlde as vnmoueable, after the manner of the fixed stars. The like argument doth that learned *Copernicus*, apply vnto his demonstrations. But ouerpassing such reasons, least by the newnesse of the arguments they may offend or trouble young students in the Art: wee therefore (by true knowledge of the wise) doe attribute the middle seate of the world to the earth, and appoynte it the Center of the whole, by which the risings, & settinges of the stars, the Equinoctials, the times of the increasing and decreasing of the dayes, the shadowes, and Eclipses are declared[26].

Two works of some importance from the earlier part of the period under review failed to make any mention of Copernicus, perhaps because their authors had not heard of his theory. The first was by Leonard Digges: *A Prognostication of Right Good Effect*[27], of which the first extant edition was dated 1555 although the title page and text indicate that this was a revised and enlarged edition of a work first published in 1553. A third edition, still further enlarged, appeared in 1556 when its title was changed to *A Prognostication Euerlasting* and this was reprinted unchanged in 1564 and 1567. In 1576, as we have already seen, it was revised by his son, Thomas Digges, who added an appendix expounding the Copernican system. The *Prognostication* was a perpetual almanac containing, in accordance with the custom of the time, a variety of astronomical, astrological and meteorological information. It gave a short account of the Ptolemaic system whose validity was taken for granted without discussion.

The second was William Cuningham's *The Cosmographical Glasse* (1559)[28]: a popular treatise on mathematics, astronomy and geography; the first two books of which covered much the same ground as Recorde's *The Castle of Knowledge*. It was a less scholarly work than Recorde's and evidently fell short of it in popular estimation since no further editions of it were called for.

The number of scientific books written and published by native English writers during this period was relatively small but they were supplemented by translations of various works by Continental authors which we shall now briefly survey. An early example was a short treatise on navigation by Martin Cortes, first published in 1551; translated by Richard Eden and published in London, 1561, under the title: *The Arte of Nauigation*[29]. The first part consisted of an elementary introduction to astro-

---

[26] *The Schoole of Skil*, London, T. Judson for W. Jaggard, 1599, sigs. A3^r—A4^r, pp. 42—43

[27] London, T. Geminus, 1555.

[28] London, John Day, 1559.

[29] London, Richard Jugge, 1561. The translation was popular and went through at least 8 editions up to 1615.

nomy on traditional lines. In chapter 6 the author mentioned that "The *Pithagorians* & other auncient naturall Philosophers (as saith Aristotle) were of opinion that the earth dyd moue". This view was then briefly discussed and refuted by the usual Aristotelian arguments reinforced by a quotation from Scripture. There was no mention of Copernicus and no indication that any modern astronomers accepted the theory.

A more important work, from the point of view of cosmology, was a long Latin poem by Marcellus Palingenius entitled *Zodiacus Vitae*, first published in Venice about 1531. It was a typical renaissance work aiming to give a summary of all learning. Astronomy figured quite largely in it, especially in the eleventh book *Aquarius*. It was popular in England both in the original Latin, of which at least 9 editions were published in this country between 1572 and 1620, and in an English translation by Barnaby Googe, of which the first three books appeared in 1560, the first six in 1561 and a complete translation in 1565. Further editions of the complete work were published in 1576 and 1588. The Latin poem was used as a prescribed text in many grammar schools during this period. It was therefore widely read and must have exerted considerable influence in this country.

Since no attempt was made to bring it up to date there was, of course, no mention of Copernicus. The astronomy was generally traditional, with the earth motionless in the centre of the universe. Palingenius accepted the Aristotelian distinction between the celestial spheres which are immutable, and the sublunary world of earth, water, air and fire which are subject to death and decay. But there were some significant departures from Aristotle. He accepted the neo-Platonic view that beyond the outermost rotating sphere — the primum mobile — there is an infinitely extended space, empty of matter but filled with light, the abode of the highest and most perfect immortal spirits. He also suggested that the stars are inhabited, though by a superior type of being to ourselves. More important, perhaps, he vigorously maintained that such questions must be judged by reason, not by the authority of Aristotle or anyone else.

Zodiacus Vitae, therefore, did nothing to spread a knowledge of the Copernican theory but it gave some encouragement to the reader to free himself from traditional categories and to think for himself[30].

Another foreign work which had some influence in England at this time was *La Semaine, ou Création du Monde*, by Guillaume Du Bartas, first published in its entirety at Paris in 1578. It was a popular work in this country; various portions of it were translated into English in the late 16th century and a complete translation by Joshua Sylvester was published in London in 1605[31]. There were many subsequent editions throughout the 17th century. Du Bartas referred briefly to the Co-

---

[30] For further details see Foster Watson, *The Zodiacus Vitae of Marcellus Palingenius Stellatus*, London, 1908.

[31] *Bartas*: *His Deuine Weekes and workes Translated*... by Iosuah Sylvester, London, Humfrey Lownes, 1605.

pernican system and decisively rejected it, giving a few of the traditional arguments against it. This work would hardly have impressed mathematicians but many less instructed readers may have been influenced by it.

Finally, mention must be made of two foreign scholars who exercised an influence on cosmological thought in England although none of their relevant writings were translated into English. The first and more important was Giordano Bruno who visited this country in 1583–85 and in 1584 published, in London, *Cena de la Ceneri*[32] — a vigorous defence of the Copernican system in which he taught definitely that the universe is infinite in extent. This was followed, shortly afterwards, by *De l'Infinito Universo e Mondi*[33] in which the infinity of the universe was again asserted. Bruno's ideas were known and discussed within a fairly limited circle in England, particularly, perhaps, by Thomas Hariot and his friends.

The second was Pierre de la Ramée, or Petrus Ramus, who discussed some cosmological questions in his *Prooemium Mathematicum* (1567)[34]. He had an evident respect for Copernicus as a mathematician and astronomer but did not accept his or any other current theory of the heavens. He wished, indeed, to eliminate all hypotheses from astronomy and to return to the purely numerical methods of computing planetary orbits which had been practised by the Chaldeans and Egyptians.

Ramus' works were studied by small but influential groups at Cambridge during the late 16th and early 17th centuries and no doubt helped to loosen the grip of Aristotelianism at that university[35].

The English Universities. The Universities of Oxford and Cambridge shewed little interest in science during the 16th century. In Oxford the normal teaching in the Arts School was based almost entirely on Aristotle. Hence any astronomy that was taught would have closely followed his system of the heavens with, perhaps, some medieval accretions. There was, however, a certain discretion allowed to the young regent masters who were bound to give courses of 'ordinary lectures' for two years after their inception as Masters of Arts. These courses would normally have been thoroughly conventional but there were some exceptions. Henry Savile, for instance (who was later to found the Savilian Professorships in Astronomy and Geometry) broke with tradition in 1570 by giving advanced lectures on Ptolemy's *Almagest*[36]. It is quite possible that other inceptors who had acquired competence in mathematics followed his example but if so, the fact has not been recorded. Some elementary astronomical lectures were given in the Faculty of Medicine on account of the importance of astrology in the medical practice of the time but these would

---

[32] [London; J. Charlewood], 1584.

[33] [London, J. Charlewood], 1584. Bruno's visit to England and his attitude to the Copernican system are discussed by Frances A. Yates, *Giordano Bruno and the Hermetic Tradition*, London, Routledge & Kegan Paul, 1964, chs. 12 and 13.

[34] Paris, Andreas Wechelus, 1567, pp. 299–301.

[35] See Johnson, *op. cit.*, pp. 186–96.

[36] See Anthony a Wood, *Athenae Oxonienses*, (Ed. P. Bliss), vol. II, London, 1815, p. 310.

not have been attended by the ordinary Arts students. There is no reason to suppose that such lectures would have dealt with the more modern theories.

Cosmological questions were sometimes debated in the public disputations which candidates for the degree of Master of Arts had to undertake. Among these were: *An terra quiescat in medio mundi*? (1576); *An materia sit in coelo*? (1581); *An sint plures mundi*? (1588)[37]. Johnson draws the conclusion that such questions as the Copernican theory, the contemporary observations on comets and so on were being discussed and that the fundamental principles of Aristotle were being questioned[38]. This is possible but, in my view, unlikely. All the above questions were carefully discussed by Aristotle himself in *De Coelo*, and were debated at considerable length by medieval scholastics such as John Buridan and Nicholas Oresme. They would probably have been making occasional appearances in the Oxford disputations ever since the 14th century. There is no good evidence that Copernican ideas had made any serious impact on the university at this time.

At Cambridge, mathematics and astronomy were, if anything, even more neglected than at Oxford. In the Statutes of Edward VI (1549) undergraduate students were required to attend an elementary course on these subjects but this provision was changed in the Elizabethan Statutes of 1572. Thereafter, until the end of the century, no mathematical or scientific instruction was given for the B. A. degree. The three extra years of study required for the M. A. degree did, however, include some elementary lectures on mathematics. Though nominally compulsory, they seem to have been poorly attended in practice[39].

However, the opportunities for becoming acquainted with the newer learning may have been slightly greater at Cambridge than at Oxford since a few prominent scholars such as Everard Digby, William Temple and Gabriel Harvey publicly supported the teaching of Petrus Ramus from 1570 onwards and, like him, stressed the importance of mathematics in education. Gabriel Harvey refered to Copernicus with respect in his *Marginalia* but there is no evidence that he or any of the other Cambridge Ramists accepted his theory. They were, in any case, unable to influence the official teaching of the university though they must have encouraged some of the students to take an interest in contemporary scientific developments. Most of the leading scientists of the early 17th century were, in fact, educated at Cambridge.

London. There was, of course, no university in London until the 19th century. However, in 1588 the City of London embarked on a small venture in adult education. They instituted a lectureship in mathematics, primarily in order to foster the arts of surveying and navigation but also to give general instruction, at an elementary level, in astronomy and mathematics. The first (and only) holder of the

---

[37] Andrew Clark (Ed.), *Register of the University of Oxford*, vol. 2, part 1, Oxford, 1887.

[38] *Op. cit.*, p. 181.

[39] See James B. Mullinger, *The University of Cambridge from the Royal Injunction of 1535 to the Accesion of Charles the First* (*History of University of Cambridge*, vol. 2), Cambridge 1884, pp. 109–111, 401–404

post was Thomas Hood, who published several good elementary textbooks to cover the various parts of his course[40]. He did not discuss the theory of celestial motion and took for granted the geocentric system, as was to be expected in a book of this type. He did, however, draw attention to one modern discovery when he described the nova of 1572 and asserted that it was situated above the sphere of the moon. Nevertheless, he apparently still accepted that comets are normally produced in the upper regions of the air.

Hood's lecture courses were continued for at least four years but were terminated some time before 1596. In 1597 a more permanent centre for adult education was founded in the City of London. This was Gresham College: named after its founder, Sir Thomas Gresham. Its teaching staff consisted of six Professors who taught, respectively, Divinity, Astronomy, Geometry, Music, Physic [i. e. Medicine] and Rhetoric. Judging by their published works, most of the Professors of Astronomy and Geometry were more interested in the practical aspects of their subjects – navigation, dialling and surveying – than in cosmological theory. It is unlikely that any of them, before the mid 17th century, paid much attention to the Copernican system in their lectures. The first who was clearly committed to the new cosmology was Seth Ward (*Geometry*, 1643) though Henry Briggs (*Geometry*, 1597–1619), Henry Gellibrand (*Astronomy*, 1626–36) and Samuel Foster (*Astronomy*, 1636–37, 1641–52) evidently favoured it. Their views will be considered in the next section.

In addition to its formal lecture courses, Gresham College provided a valuable meeting place for scholars, where new scientific ideas could be and were discussed. It was particularly active in this respect from the 1640's onwards.

The Sixteenth Century: Summary. The impact of the Copernican system, as reflected in published work, was rather small before 1600. Thomas Digges (1576) had accepted it wholeheartedly, Robert Recorde (1556) was sympathetic to it; John Blagrave (1596) certainly accepted the diurnal rotation of the earth and perhaps also its orbital motion; John Feild (1556) accepted it but John Dee, though respectful, was apparently unconvinced. Richard Forster (1575) and Richard Bostocke (1585) both admired the achievement of Copernicus and may have accepted his system. There were others, such as William Gilbert, Edward Wright and Thomas Hariot, who were either full or semi-Copernicans during this period but who did not publish anything on the subject before 1600. These will be discussed in the next section.

Among non-mathematicians the new theory had, up to this time, been generally ignored. Apart from Gabriel Harvey, who has already been mentioned, the only reference that I am aware of is in a poem by John Davies: *Orchestra* (1596)[41] in which he referred briefly to the possibility that the earth moves but expressed no opinion on it. There seems to have been little consciousness of, or interest in the new cosmology among the general public.

[40] See Johnson, *op. cit.*, pp. 200–205.

[41] London, J. Roberts for N. Ling, 1596.

The Seventeenth Century I: 1600–1640

William Gilbert. England produced only one scientist of major importance whose work was done almost wholly within the 16th century. This was William Gilbert (1540–1603). He was born in Colchester and graduated at St John's College, Cambridge where he was elected fellow of the college in 1561. In 1569 he left Cambridge and practised as a physician in London until his death in 1603. The only work of his which was published in his lifetime was *De Magnete*[42]. After his death, some of his manuscript papers on cosmology and meteorology were edited by his half-brother, also called William Gilbert, and published in Amsterdam under the title *De Mundo* in 1651[43].

One of the main purposes of *De Magnete* was to prove that the earth itself is a great magnet which, in its magnetic properties, does not differ essentially from a spherical loadstone. In the sixth and last book of the treatise he considered the wider cosmic aspects of his theory and discussed, more specifically, the movement of the earth. Gilbert's treatment of this question was much more detailed and carefully reasoned than that of any of his predecessors. It will therefore be necessary to examine his views in detail since they exerted a strong influence on his contemporaries and immediate successors.

Gilbert regarded the diurnal rotation of the earth as having been established beyond any reasonable doubt. The reasons he gave were mainly negative and were based upon certain unsatisfactory or rather, in his opinion, patently absurd features of the traditional geostatic theory. His main objections to the latter were: (a) the lack of any reasonable proportion between the so-called first and second motions of the heavenly bodies. The second, or orbital, motions shewed a clear general relation between distance from the earth and period of the orbit such that the nearer a body was to the earth the shorter was its period: ranging from 27 days for the moon to 30 years for Saturn and 36000 or (according to Copernicus), 25816 years for the fixed stars. It should be noted that here Gilbert was somewhat inconsistently arguing on the basis of the Ptolemaic theory; he himself followed Copernicus in transferring the precessional motion from the fixed stars to the earth. All these bodies, moreover, rotated in the same direction from West to East. But then, inexplicably, beyond the fixed stars was the outermost sphere or primum mobile which revolved in the opposite direction, from East to West, in the incredibly short time of 24 hours. (b) His second objection was a mechanical one: that no physical structure could stand the strain imposed upon the outermost sphere by its enormous

---

[42] Guilielmi Gilberti Colcestrensis, Medici Londinensis, *De Magnete, Magneticisque Corporibus, et de Magno magnete tellure, Physiologia noua*, London, Peter Short, 1600. Later editions: Stettin, 1628 & 1633; Berlin, 1892. English translations by P. Fleury Mottelay, New York, 1893; reprinted by Dover Publications, New York, 1958); and by the Gilbert Club, London, 1900.

[43] Guilielmi Gilberti Colcestrensis, Medici Regii, *De Mundo nostro sublunari, Philosophia Nova*, Amsterdam, L. Elzevir, 1651.

rotational speed. Even on the old Ptolemaic system the speed was implausibly high, owing to the great distance of the primum mobile. But Gilbert, influenced no doubt by Thomas Digges, maintained that the universe must be much larger than previously supposed. He argued that the stars are not all fixed in the inner surface of a sphere but are at different distances from the earth: the fainter being, in general, more distant than the brighter. There must be countless numbers which are so far away as to be invisible. This will make the required velocities still more incredible.

Gilbert's own view was that the universe is infinite, in which case the idea that it rotates is absurd: "At infinitatis atque infiniti corporis motus esse non potest, neque idcirco vastissimi illius primi mobilis diurnus"[44]. But even if a finite world is granted, the speeds are too high to be acceptable.

Has the rotation of the earth been proved with certainty? In the heading of book 6, chapter 3 Gilbert said: "De terrestris globi diurna reuolutione magnetica, aduersus primi mobilis inueteratam opinionem, probabilis assertio"; but towards the end of the chapter he made the stronger claim: "Ex his igitur rationibus, non probabilis modo, sed manifesta videtur terrae diurna circumuolutio, cum natura semper agit per pauciora magis, quam plura; atque rationi magis consentaneam vnum exiguum corpus telluris diurna volutationem efficere potius, quam mundum totum circumferri". In *De Mundo* he again took the more cautious line: "Terram circumvolvi diurno motu, verisimile videtur"[45]. This probably represents his more considered opinion but the degree of probability which he attached to the theory was obviously very high.

Gilbert's cosmology was closely linked with his magnetic theories but the precise nature of the connexion is not always clear. One point of contact, however, was insisted on strongly by him. He was very conscious of the traditional objection to a rotating earth; that it required the earth to have two or more diverse natural motions, thus contradicting the basic Aristotelian principle that each type of matter could have only one natural motion. Earth has undoubtedly a natural tendency to move vertically downwards; how then could it also have a natural tendency to rotate? Gilbert countered this by pointing out that a compass needle, or his small spherical loadstone, tended to rotate naturally in order to take up its correct orientation with respect to the earth's magnetic poles. Hence magnetic bodies clearly do have a natural movement of rotation, and this should apply also to the earth[46].

The analogy may seem a little weak to the modern mind. A magnet only rotates until it reaches its correct alignment and then stops; the earth continues to do so indefinitely. Gilbert, however, approached the problem from an angle very different from our own. Whereas we think of magnetism as a purely mechanical interaction, he took a much more teleological view. The earth and the heavenly bodies were for

---

[44] *De Magnete*, Lib. VI, c. 3.
[45] *De Mundo*, p. 135.
[46] *De Magnete*, VI. 4.

him, as for very many of his contemporaries, animated beings which moved for a purpose: either for their own good or for that of the whole system to which they belonged. Magnetic force was an expression of this natural tendency to seek the good. An ordinary compass needle achieved a better state by aligning itself with the earth's magnetic field; it therefore ceased to rotate when it had achieved this end. But the good of the earth requires (a) that it should rotate continually so that the warmth and life-giving influence of the sun is received by every part of its surface; (b) that its axis should be inclined to the plane of the ecliptic in order to produce the regular succession of the seasons; and (c) that the axis should precess in order that the stars, by slowly changing their declination, should distribute their influences more evenly over the earth's surface. Hence the same magnetic force which gives a limited power of rotation to the compass needle can give a permanent and more complex power to the earth[47].

Gilbert extended his concept of magnetism to explain the orderliness of the planetary system, which he thought of as a society of animated beings 'inciting' each other to the production of a harmonious and mutually beneficial order. The sun exerted the main controlling influence but all the other planets and the earth made their contribution as well. The stars, on the other hand, were beyond the sphere of the solar magnetic influence and were, to that extent, immobile, though he did not exclude the possibility that they might themselves be magnetic systems with intrinsic movements which we cannot detect owing to their great distances[48]. Thus when Gilbert ascribed the constant orientation of the earth's axis in space to magnetic force he did not mean, as we might suppose, that there were magnetic poles situated in the distant heavens, relative to which the earth aligned itself. Indeed he expressly rejected this supposition. The constancy arose rather from the mutual ordering of the earth and the planets[49].

Gilbert's attitude to the heliocentric theory is very difficult to assess. He had no doubt about the diurnal rotation of the earth but he never clearly committed himself either for or against an orbital motion around the sun. In *De Magnete* he did, in one place refer to the sun as if it were one of the planets with its own orbital motion but the sentence is — no doubt deliberately — ambiguous[50]. And there is one passage where he seems to come out in favour of Copernicus:

Tellurem circulariter moueri super suum centrum posuimus, diem conficientem integra reuolutione ad solem. Luna menstruo cursu circa tellurem voluitur, & solis conjunctionem a priore synodo repetens, mensem constituit siue diem Lunarem. Medium orbis concentrici Lunae Copernici & recentiorum obseruationibus plurimis, inuenitur distare a centro telluris 29 diametris telluris & quasi 5/6. Reuolutio Lunae ad solem, fit 29 diebus, ½, & horae minutis 44. Motum obseruamus ad solem, non periodicum, quemadmodum dies est reuolutio integra telluris ad solem, non perio-

[47] *Ibid.*, VI. 4
[48] *Ibid.*, VI. 6
[49] *Ibid.*, VI. 4
[50] *Ibid.*, VI. 3

dica; quia sol causa motus est, tam terrestris, quam Lunaris: etiam quia (iuxta recentiorum hypotheses) mensis synodicus sit vere periodicus, propter telluris motum in orbe magno[51].

In this passage Gilbert is discussing an apparent relation between the diameter of the earth and its period of rotation on the one hand, and the radius of the moon's orbit and its periodic time on the other. Taking the diameter of the earth and the day as the units of length and time respectively, he points out that the mean distance of the moon from the earth is about 29 5/6ths diameters and the period of the lunar cycle is about 29½ days. The relationship, such as it is, required him to take the moon's synodic period (new moon to new moon) rather than the sidereal period (the time required for it to return to the same position relative to the fixed stars). He gives three arguments to justify this: (a) because we measure the day synodically; i. e. we use a solar rather than a sidereal measure for it: hence, by analogy, we should do the same for the moon; (b) because the sun is the cause of motion of both the earth and the moon; (c) finally, he gives his third argument: because "according to the hypotheses of recent astronomers, the synodic month is really periodic on account of the earth's motion in its great orbit". The argument is not altogether clear but the point seems to be that if, as on the traditional theory, the earth is at the centre of the universe and the sun moves around it, then the only intrinsic period of the moon's motion (i. e. the only one arising wholly from its own motion), is the sidereal period of 27 days. The additional two days required to make up the synodic month would be due to the sun's motion, not to the moon's. On the other hand, if the sun is stationary, then it is the lunar motion alone which produces the synodic cycle; in this case it will be a true periodic time.

This passage is evidence that Gilbert was prepared to be open-minded about the Copernican theory but it hardly amounts to a positive acceptance. It is only one of three arguments adduced and is qualified by the phrase: "secundum recentiorum hypotheses". He gives it for what it is worth but does not depend on it. Some historians of science have searched *De Magnete* for clues as to Gilbert's views on the earth's orbital motion but this is, in my view, an unprofitable exercise. It seems clear that he did not wish to come to any decision on the subject. A similar reticence is to be found in his *De Mundo*. Although this work dealt *ex professo* with the structure of the universe and although it defended the axial rotation of the earth, the question of orbital motion was deliberately left undecided: "Terram circumvolvi diurno motu, verisimile videtur: an vero circulari aliquo motu annuo cieatur, non hujus est loci inquirere". In chapter 20 he returned to the problem:

Copernici vero ratio magis incredibilis, licet minus in motuum convenientiis absurda; quod terram triplici oportebat motu agitari, tum vel maxime, quod nimis vastam capacitatem inter orbem Saturni & octavam sphaeram esse oportet, quae prorsus sideribus vacua relinquitur[52].

The explanation of Copernicus is less absurd in the way that it relates the planetary motions to

---

[51] *Ibid.*, VI. 6.

[52] *De Mundo*, p. 193.

each other but more incredible in so far as it gives a threefold motion to the earth and, more especially, because it requires such a vast empty space between Saturn and the eighth sphere (*De Mundo,* p. 135).

But there is no indication whether the 'absurdity' of the one theory or the 'incredibility' of the other is to be preferred. Similarly, in the diagram of the planetary system which he gave on p. 202, it is left undetermined whether it is the sun which is in orbit round the earth or vice versa.

Gilbert's Influence. *De Magnete* was highly esteemed by his contemporaries and immediate successors. Galileo praised it and discussed it at some length in his *Two World Systems*[53]. Kepler referred to it in terms of warm appreciation in his *Astronomia Nova*[54], using it as a basis for his own cosmological theories. However, Kepler's application of magnetic principles differed significantly from Gilbert's. Gilbert had stressed the spontaneity with which the earth and the planets move: each seeking its own good by virtue of an inner striving. They are 'incited' thereto by the sun and other heavenly bodies; the sun, indeed, is called the cause of the earth's rotation but the causality envisaged is more closely analogous to a psychological influence than to a mechanism. The idea of mechanical force is not entirely absent but it plays a very subordinate role; the universe as a whole is more like a community of souls stimulating each other to activity than to a purely mechanical system. Kepler, on the other hand, while still prepared to endow the earth with a soul, shifted the balance much further towards the mechanical side. His explanation of planetary motion was primarily in terms of mechanical quasi-magnetic fibrils which surrounded the sun and entrained the planets, thus carrying them around in their orbits. The important difference between the two approaches was that Gilbert's was equally compatible with a geocentric and a heliocentric theory, so that he never felt called upon to decide between the two, whereas Kepler's necessarily involved the heliocentric. If the heavenly bodies move spontaneously, through an intrinsic desire, there is no reason why the sun should not do so, in any way that is for its own good and for that of the world as a whole. Whereas if motion is due to a mechanical and, in principle, quantifiable force, then it is obviously more plausible to regard the small earth as being moved around the much larger and more massive sun, rather than vice versa.

Kepler's planetary theory attracted little attention until the 1630's[55] whereas Gilbert's magnetic theories exerted considerable influence from the time of their publication. This may partly account for the prevalence of semi-Copernican theories and the neglect of heliocentrism during this period. Nevertheless, one must not see Gilbert and Kepler as radically opposed to each other. There are the beginnings

---

[53] *Dialogo... sopra i due Massimi Sistemi Del Mondo...* 3rd day, *Opere* (Ed. A.Favaro), VII, pp. 426–440.

[54] C. 34 and c. 57. *Gesammelte Werke* (Ed. M. Caspar), III, p. 246; 350–355. See also Kepler's Correspondence, G. W. XV, p. 232 and elsewhere.

[55] See J. L. Russell, "Kepler's Laws of Planetary Motion", *Brit. J. Hist. Sci.* II, 1964, pp. 1–24.

of a genuinely dynamic theory of planetary motion in Gilbert's work. It is not unreasonable to regard Kepler's dynamics of the solar system as a fruitful development of something already there in germ.

Gilbert's work exerted an immediate influence on many of his English contemporaries. One of his first supporters was Edward Wright (1558?–1615). Wright graduated at Caius College, Cambridge, where he was a Fellow until 1596. He is known mainly for his contributions to the mathematical theory of navigation. He contributed an enthusiastic preface to *De Magnete* in which he endorsed Gilbert's views and, in particular, proclaimed his own acceptance of the earth's rotation. Unlike Gilbert, however, he definitely rejected the earth's orbital motion: *In suo namque eodemque loco terra semper manere poterit, ut non vaga aliqua latione dimoueatur, aut extra sedem suam (in qua a diuino opifice posita primum fuit) transferatur. Nos itaque* [...] *experimentis & rationibus philosophicis non paucis inducti, satis probabile esse existimamus, terram, quanquam super centro suo, tanquam basi & fundamento immobili innixam, circulariter tamen circumferri*[56].

A year later, Gilbert's theory of a rotating magnetic earth was taken up by Nicholas Hill (1570?–1611) in his *Philosophia Epicurea*[57], in which he defended a form of Democritean atomism, propounded succinctly in a series of 509 propositions with a minimum of supporting argument. Among these were two which maintained the Copernican system. In prop. 20 he asserted: *Omnis apparentia Coelestis, diurna, menstrua, annua, secularis & periodica conuenientius, facilius, aptius per motum terrae suppositum, quam solis saluatur, & soluitur.* Clearly, therefore, he accepted not only the earth's diurnal rotation but also its orbital and precessional motions. In prop. 434 Hill briefly enumerated 19 arguments in favour of the system:

1. Magneticus confluxus grauissimorum. 2. Polorum magneticorum constantia & axeos. 3. Astrea terrae natura probabilis. 4. Φαινομένων explicatio optata. 5. Eccentricitatum mille absurditas sublata. 6. Continua generatio torporiferam quietem non admittens. 7. Aquarum terrae pene aequiponderantium motus manifestus. 8. Terrae in soluto & libero aethere suspensio absque basi ad quietem necessaria. 9. Terrae animalitas, & primordialitas non sine motu locali intelligibilis, qui generalis rerum vita est. 10. Terrae figura admissa, demonstrata. 11. Terrae cohaerentia & firmitas. 12. Grauitatis nullitas praesertim rerum in proprijs locis existentium. 13. Improbabilis centri, & medij puncti infinitas. 14. Medij in mundo infinito nullitas. 15. Terrae tam exiguae impotentia continui solis ferendi. 16. Ignei solis dissipabilis substantia. 17. Naturae compendiosa perficiendi ratio. 18. Necessaria homogeneitas primarum & notabilium mundi partium. 19. Rotatio partium separatarum.

Many of these arguments shew the influence of Gilbert: e. g. nos. 1, 2, 9, 19 and probably others. 15, however, suggests that his approach was more mechanistic. Like Gilbert, he accepted an infinite universe (n. 14).

Hill had been a student at St John's College, Oxford. According to Anthony Wood he had a reputation for eccentricity and for maintaining "fantastical notions

---

[56] *De Magnete*, Preface, sig. 5ʳ.

[57] *Philosophia Epicurea, Democritiana, Theophrastica, proposita simpliciter, non edocta*, Paris, 1601. Another edition was published at Geneva, 1619.

in his philosophy"[58] so we can hardly regard his book as evidence for any widespread interest in Copernicus at Oxford at this time. As we shall see, the available evidence does not support such a view.

Gilbert's magnetic theory of rotation received further support from Mark Ridley (1560–1624), a distinguished London physician who had been educated at Clare Hall, Cambridge. In 1613 he published: *A Short Treatise of Magneticall Bodies and Motions*[59] in which he further developed Gilbert's ideas, using for this purpose the telescopic discoveries of Galileo, Kepler and Fabricius. He used these as arguments in favour of the earth's rotation:

> ... it is lately obserued vnto our sences by helpe of the truncke-spectakle [telescope], both by *Galileus* and *Kepler*, famous Mathematitians, that the great body of the globe of Iupiter, being twelue times greater then the Earth, doth turne about in lesse time then a day vpon his *axis* and poles, who also haue obserued foure Moones, attendant on *Iupiter*, which moue round about him, the slowest in 14 dayes, the next in seuen dayes, and the rest in shorter time. So likewise *Iohn Fabricius* hath obserued, that the great globe of the Sunne, hauing three great spots, like continents in him, and being sixty times greater then the Earth, to moue about his *axis* and poles neere the time of ten dayes, or thereabouts [...] therefore it being certaine by obseruation, that the globe of *Iupiter* and the Sunne do turne about their *axis* and poles, whose materials we know not, we need not doubt that the Earth should haue a circular motion for her great good[60].

This passage shows that Ridley had up-to-date information about recent astronomical discoveries but he used his sources carelessly. Kepler, for instance, in his *Dioptrice* (1611)[61] deduced from his magnetic theory that Jupiter must have an axial rotation with a period of less than one day, but neither he nor Galileo claimed to have observed it. Its actual period of about 10 hours was first established by J. D. Cassini in 1665[62]. Fabricius did not assert that the sun has a period of rotation of 10 days; he said that 10 days elapsed between the disappearance of the third of his three spots behind one limb of the sun and the reappearance of the first of them at the other limb – an observation which is quite consistent with the true synodic period of 27 days[63]. In spite of these errors, however, Ridley was justified in claiming that the recent discoveries had greatly increased the plausibility of the earth's motion.

Ridley was silent about the heliocentric theory. He asserted positively that Mercury and Venus go round the sun but he did not say whether the other planets or the earth did so, nor does the diagram of the heavenly bodies on the title page of his book give us any clue. In a subsequent work: *Magneticall Animadversions* (1617) he briefly referred to "*Copernicus* and *Tichobrahe*, most perfect and exact Astronomers, who make the Sunne to be *centrum vniuersi & planetarum*"[64] without pass-

[58] *Athenae Oxonienses* (Ed. P. Bliss), II, London, 1815, pp. 86–87.
[59] London, Nicholas Okes, 1613.
[60] *Short Treatise*, pp. 14–15.
[61] Augsburg, 1611, p. 14; *Ges. Werke*, IV, p. 343.
[62] *Journal des Sçavans*, Paris, 1665, p. 69.
[63] *De Maculis in Sole Observatis*, Wittenberg, 1611.
[64] London, N. Okes, 1617, p. 9.

ing any judgment on this view. He evidently considered it was impossible to determine whether the earth or the sun is 'centrum universi'.

Ridley's main interest was in magnetism rather than astronomy and the same is true of William Barlowe (d. 1625), with whom he was soon to find himself in controversy. Barlowe was educated at Balliol College, Oxford, took Holy Orders and in 1615 was appointed Archdeacon of Salisbury. He had been experimenting in magnetism for many years when in 1616 he published a book entitled *Magneticall Aduertisements*[65] in which his own investigations, and Gilbert's, were summarised. He had a high respect for Gilbert, with whom he had been on friendly terms, but vigorously rejected the view that the earth rotates. He regarded this as contrary to Scripture. It was in reply to this work that Ridley wrote his *Magneticall Animadversions* mentioned above, in which he both defended himself against a charge of plagiarism which Barlowe had levelled against him and reasserted his cosmological principles. Barlowe returned to the fight with a rejoinder: *A Briefe Discovery of the Idle Animaduersions of Marke Ridley...*[66] in which he ridiculed the Copernican system and rejected it completely.

Gilbert's ideas on cosmic magnetism found another and perhaps more influential supporter in Nathanael Carpenter (1589–1628?), of Exeter College, Oxford. In his book: *Geography Delineated* (1625)[67] he defended the proposition: "It is probable that the terrestrial Globe hath a circular motion"[68] – i. e. a diurnal rotation. His main argument in favour, like Gilbert's, was the improbably high velocity which the *primum mobile* must have on the traditional view. The cause both of the earth's rotation and of the constant direction of its axis was its magnetism. Carpenter did not accept either the annual or the precessional motions because he considered it more reasonable that the earth, the sun and the system of fixed stars should each have one proper motion rather than that the earth should have three and the sun and stars should be stationary[69]. He also thought that the arguments from Scripture against an orbital motion of the earth were more powerful than those against a diurnal rotation. Even the latter was only proposed as a probable theory.

Apart from giving the earth an axial rotation, Carpenter's theory was Tychonic: the sun went round the earth and all the other planets round the sun. Throughout the discussion he gave the impression that the geocentric, and indeed the geostatic view was the one currently accepted in his day. For instance, after expounding (and generally approving) Ptolemy's arguments that the earth is in the centre of the universe he concluded: "But these demonstrations of *Ptolomy*, as I haue set them downe enlarged and explained by our later writers, may seeme sufficient, especially

[65] London, E. Griffin for T. Barlow, 1616.

[66] London, E. Griffin for T. Barlow, 1618.

[67] *Geography Delineated Forth in Two Bookes*, Oxford, J. Lichfield and W. Turner for H. Cripps, 1625. (Second edition, Oxford, 1635.)

[68] *Ibid.*, p. 76.

[69] *Ibid.*, p. 98.

in a matter of few called in question". And again, when he expresses his hesitation in proposing a new planetary theory: "I might seeme perhaps presumptuous beyond my knowledge, to reiect and passe by the draughts and delineations of *Ptolomy*, *Alphonsus* and their followers, which are commonly defended and in use"[70].

Carpenter had already discussed astronomical problems more briefly in his *Philosophia Libera* (Frankfurt, 1621), especially in the enlarged 2nd edition (Oxford, 1622)[71] and had reached similar conclusions. The 2nd edition is significant as the first published work by an Englishman in which Kepler's theory of planetary ellipses was discussed. Carpenter treated the theory with respect but rejected it on the ground that circular paths seemed to be more consonant with the order and perfection of nature than non-circular[72].

Another writer who was directly influenced by Gilbert was Francis Godwin who graduated at Christ Church, Oxford in 1581 and subsequently became Anglican Bishop of Llandaff (1601) and Hereford (1617). His book: *The Man in the Moone*[73] was published postumously in 1638. The date of its composition is uncertain. Johnson[74] regarded it as having been written soon after the publication of *De Magnete* – certainly before 1610 – whereas H. W. Lawton[75] and G. McColley[76] both put it later than 1625. It is a work of science fiction in which a Spaniard, Domingo Gonsales, having made a voyage to the moon and back, describes the various appearances of earth, moon, etc, as they were observed by him in the course of his journey. The phenomena encountered were all based directly on Gilbert's theories. The earth and the moon were both found by the traveller to be great magnets and he was able, from his position in space, to observe the rotation of the earth on its axis. Gonsales, who was clearly a spokesman for Godwin himself, accepted the axial rotation as proved, but rejected both the orbital and precessional movements of the earth, the latter being ascribed to the fixed stars.

Godwin's book was very popular. There were at least 8 English editions between 1638 and 1768, mostly somewhat altered and abbreviated. It was translated into French (1648), Dutch (1651) and German (1659). At least 12 Continental editions had appeared by 1718.

Most of the leading English astronomers between 1600 and 1640 were strongly influenced by Gilbert. There were two, however, who stood somewhat apart from the main stream: Thomas Hariot and Jeremiah Horrox. Both were men of outstanding ability and both supported the Copernican system but each failed, for different reasons, to exercise the influence that his talents deserved. Thomas Hariot (or Har-

---

[70] *Ibid.*, pp. 110–111.

[71] Subsequent editions were published at Oxford 1636 and 1675.

[72] Pp. 384–6.

[73] London, J. Norton for J. Kirton and T. Warren, 1638.

[74] *Op. cit.*, p. 233.

[75] "Bishop Godwin's «Man in the Moone»", *Review of English Studies* VII (1931), pp. 35–37.

[76] "The Man in the Moone", *Smith College Studies in Modern Languages* XIX, No. 1 (1937).

riot) (1560–1621) was a mathematician of great distinction whose *Artis Analyticae Praxis* was published posthumously in 1631. He was interested in astronomy but published nothing on the subject. He had, however, a devoted circle of friends and pupils whom he helped and instructed. He read and evidently approved Kepler's *Astronomia Nova* soon after it appeared in 1609. A letter to him from his pupil William Lower, written in 1610, makes it clear that at that time both he and Lower accepted the Copernican system and the principle of elliptical orbits. There is evidence also that he and his pupils were making telescopic observations on the planets independently of Galileo prior to the publication of *Sidereus Nuncius*. But this work was never published and was known only to his own limited circle[77].

The heliocentric system was accepted without reservation by Jeremiah Horrox (1617?–1641), an astronomer of great ability whose achievement was frustrated by his early death. In 1633 he began to compile a set of astronomical tables using, initially, those of Philip Lansberg as the basis of his own. He soon found that these were too unreliable so, on the advice of his friend William Crabtree, he changed over to Kepler's Rudolphine Tables which were greatly superior. Thereafter, he was a convinced adherent of Kepler's elliptical orbits and of his physical theories of planetary motion. During his short life he made substantial contributions to the theory of lunar motion and discovered the 'great inequality' of the mean motions of Jupiter and Saturn. He was the only astronomer to predict, and then to observe, the transit of Venus in 1639. At the time of his death he was engaged in writing a book in support of Kepler which was published posthumously by John Wallis in 1673[78]. In this he strongly upheld the Copernican theory and insisted that it should be regarded as true, not merely as a convenient hypothesis for the purposes of calculation. Horrox's library contained a copy of Copernicus's *De Revolutionibus* as well as the more important works of Kepler, Lansberg and other 17th century astronomers[79].

Popular Astronomy. The Almanacs. The ordinary educated man would derive such knowledge of astronomy as he possessed, mainly from the popular almanacs which competed with each other for the public favour. These all followed substantially the same pattern: a calendar for the year with data on the rising and setting of the sun, the phases of the mon, astrological predictions and advice, together with miscellaneous items of astronomical information which varied considerably in extent and usefulness from one writer to another. Few of them expressed any opinion either for or against Copernicus. Johnson[80], in his extensive researches up

[77] For further details on Hariot's astronomical work, see S. P. Rigaud (1833) and Henry Stevens (1900) in the General Bibliography.

[78] *Opera Posthuma, viz Astronomia Kepleriana, defensa et promota...*, London, W. Godbid for J. Martyn, 1673.

[79] Horrox's work is discussed in more detail by A. B. Whatton, *A Memoir of the life and labours of the Rev. Jeremiah Horrox... to which is appended a translation of his celebrated discourse upon the Transit of Venus across the Sun*, London, 1859.

[80] *Op. cit.*, pp. 248–58.

to about 1640, found only two who definitely supported the heliocentric theory. These were Edward Gresham, whose almanacs for 1604, 1606 and 1607 are extant, and Thomas Bretnor for whom we have a whole series from 1605 to 1618. Gresham, in the preface to his almanac for 1607, remarked that he was regarded by some as a heretic because of his cosmological views: "And some (I heare) who (for that I am *paradoxall* in many things, but especially in the frame and *systeme* of the world, differing from all Phylosophers and Diuines in that poynt, as they thinke) absolutely condemne me of *Atheisme* and *Haeresie.* To these I reply, that *Apostasie* from *Errour* to *Truth,* is no good *Argument* of *Atheisme*..."[81] This can only refer to the heliocentric theory which he asserted explicitly in the body of his work.

Bretnor's Copernicanism was propounded more aggressively. In his almanac for 1615 he wrote: "This Brumal season, commonly called *Winter*, ... tooke its beginning the 11 of December last: for then (according to old dotage) did the Sun enter the first scruple of the cold and melancholick signe *Capricorne*, or rather according to verity this earthly planet entring the first minute of *Cancer*, and furthest deflected from the Sunnes perpendicular raies, did then receiue least portion of Sunshine, and greatest quantitie of shadow"[82].

Bretnor and Gresham both had a high reputation for learning; Bretnor in particular was regarded by his countrymen as the leading almanac maker of his time. Their support must have been an important influence in favour of Copernicus. Some other respected writers were, however, definitely anti-Copernican. Among these were Arthur Hopton in his almanac for 1613 and, more explicitly, in his larger work: *A Concordancy of Years*[83], and also Richard Allestree[84]. Most writers of this class, however, either gave no indication of their views or used geostatic language in a way that may have been purely conventional.

Anti-Copernicans: *1600–1640.* A balanced picture of the situation in England must, of course, take account of those who opposed as well as of those who supported the Copernican system. After 1600 the number of writers with any technical competence who rejected all movement of the earth was, in fact, very small. Two of them, Hopton and Allestree, have already been mentioned. Another was Nathanael Torporley (1564–1632) of Christ Church, Oxford, in his *Diclides Coelometricae*[85] (1602). Torporley rejected all the current systems — Ptolemaic, Tychonic and Copernican — and proposed instead a theory based upon the *Homocentricon* of Fracastorius (1538) in which a motionless earth was surrounded by a complex system of homocentric spheres. He referred with respect to both Copernicus and Tycho;

---

[81] *Gresham 1607. A new Almanacke and Prognostication for the yere of our Lord God 1607,* London, for the Company of Stationers, sig. B2^v.

[82] *Bretnor 1615. A Newe Almanacke and Prognostication for the yeare of our Lord God 1615,* [London], sig. C2^v.

[83] London, for the Company of Stationers, 1612. Later editions 1615, 1616, 1635.

[84] *Allestree 1620. A New Almanack... for this yeare of our Lord, 1620.* [London], sig. C3^r–C3^u.

[85] London, Felix Kingston, 1602.

he looked forward to the publication of Tycho's observational data which, he hoped, would shew whether his own theory was preferable to that of Copernicus or not.

Thomas Lydiat (1572–1646), of New College, Oxford, likewise rejected all current systems[86]. He proposed a modified Ptolemaic system in which the universe was filled with a fluid aether whose density decreased regularly with height. The planets floated in this medium at varying distances from the earth, which was at the centre and motionless. The world was surrounded by a solid firmament whose rotation produced a corresponding circular movement of aether and planets. Lydiat recognised that the distance of any given planet from the earth was variable; he explained this by supposing that the sun's rays acted upon the aether to produce local variations in density.

The different world systems were briefly discussed by Samuel Purchas (1575?–1626) of St John's College, Cambridge, in the 2nd edition of *Purchas his Pilgrimage* (1614), in which he also gave a short account of Galileo's telescopic discoveries. He expressed a preference for Tycho's cosmology but his final conclusion was sceptical: "A learned ignorance shall better content me, and for these varieties of motions, I will with *Lactantius* ascribe them to God the Architect of Nature and co-worker therewith by wayes Natu[r]all, but best knowne to himselfe"[87]. Tycho's system was also accepted by John Swan in his *Speculum Mundi* (1635)[88]. This was an encyclopaedic survey of contemporary scientific knowledge which achieved considerable popularity. The author rejected all movement of the earth on both philosophical and Scriptural grounds, though he followed Tycho in regarding the heavenly bodies as changeable. A more conservative writer was George Hakewill (1578–1649) of Exeter College, Oxford. His *Apologie or Declaration*... (1627)[89] was written with the purpose of praising the modern times and their achievement. Nevertheless he not only rejected the earth's movement but also accepted the Aristotelian doctrine of the celestial spheres and the incorruptibility of the heavenly bodies. As will be mentioned later, in the 2nd and later editions he reproduced a letter from Henry Briggs which commended the Copernican system but this did not lead him to modify his own views in any way.

Purchas, Swan and Hakewill were not professional astronomers, nor were they particularly well qualified to judge on cosmological questions. But they must have been representative of a large number of educated people who were sympathetic to the new learning but unwilling to abandon completely the principles on which the traditional cosmology had been built.

---

[86] *Praelectio Astronomica de Natura Coeli & conditionibus Elementorum: tum autem de Causis Praecipuorum Motuum Coeli & Stellarum*, London, J. Bill, 1605.

[87] London, W. Stansby for H. Fetherstone, 1614, p. 10.

[88] Cambridge, by the Printers to the University, 1635. Later editions: 1643, 1665, 1670.

[89] *An Apologie or Declaration of the Power and Prouidence of God in the Gouernment of the World*, Oxford, J. Lichfield and W. Turner, 1627. Later editions: Oxford, 1630; London, 1635.

The most prominent English philosopher-scientist who rejected the Copernican theory was Francis Bacon (1561–1626). His lack of sympathy with it is surprising since he was generally receptive to new scientific ideas; he believed, for instance, that the heavenly bodies were made of the same sort of matter as the earth and rejected the Aristotelian solid spheres. He also insisted that we must approach all such questions with an open mind, not allowing our judgment to be swayed by undue reverence for antiquity.

Bacon referred briefly to the Copernican system in *De Augmentis Scientiarum*[90], published in 1623 but mainly written in 1605. After subjecting the Ptolemaic theory to a lengthy criticism he concluded: "And it is the absurdity of these [Ptolemaic] opinions that has driven men to the diurnal motion of the earth; which I am convinced is most false". He repudiated it also in the *Novum Organon*, when criticising Galileo's theory of the tides:

> And it was upon this inequality of motions in point of velocity that Galileo built his theory of the flux and reflux of the sea; supposing that the earth revolved faster than the water could follow; and that the water therefore first gathered in a heap and then fell down, as we see it do in a basin of water moved quickly. But this he devised upon an assumption which cannot be allowed, viz. that the earth moves; and also without being well informed as to the sexhorary motion of the tide[91].

He elaborated his objections to Copernicus in another work: *Descriptio Globi Intellectualis*[92], written about 1612 but not published until 1653. In this he weighed the arguments on both sides and eventually decided against him but less decisively than in the *De Augmentis*. After giving some arguments in favour of his system he concluded:

> Nevertheless, in the system of Copernicus there are found many and great inconveniences; for both the loading of the earth with a triple motion is very incommodious, and the separation of the sun from the company of the planets, with which it has so many passions in common, is likewise a difficulty, and the introduction of so much immobility into nature, by representing the sun and stars as immoveable, especially being of all bodies the highest and most radiant, and making the moon revolve about the earth in an epicycle, and some other assumptions of his, are the speculations of one who cares not what fictions he introduces into nature, provided his calculations answer. But if it be granted that the earth moves, it would seem more natural to suppose that there is no system at all, but scattered globes... than to constitute a system of which the sun is the centre... But if the earth moves, the stars may either be stationary, as Copernicus thought or, as is far more probable, and has been suggested by Gilbert, they may revolve each round its own centre, in its own

[90] *De Dignitate et Augmentis Scientiarum Libri IX*, London, J. Haviland, 1623, with many later editions. Reprinted in *The Works of Francis Bacon*, (Ed. Spedding, Ellis and Heath), new edition, London, Longmans, vol. 1 1870. English translation: vol. 4, p. 348.

[91] Francisci de Verulamio, *Instauratio Magna*, London, John Bill, 1620. Reprinted in *Works*, vol. 1 (1870). English translation, vol. 4, p. 212.

[92] First published in: Francisci Baconi *de Verulamio Scripta in Naturali et Universali Philosophia*, Amsterdam, L. Elzevir, 1653. Later editions 1685, 1699. Reprinted in *Collected Works*, vol. 3; English translation, vol. 5.

place, without any motion of its centre, as the earth itself does; if only you separate that diurnal motion of the earth from those two supposititious motions which Copernicus superadded[93].

Bacon's main weakness as a philosopher of science was his failure to appreciate the role of mathematics in science. He regarded all attempts to obtain exact mathematical descriptions of planetary paths as a waste of time unless these were based on sound physical principles backed, where possible, by experiments. These principles, he thought, favoured a geostatic system though he did not exclude the possibility of a diurnal rotation.

The Universities. The writers whom I have so far considered, from 1600 onwards, were all or nearly all educated at Oxford or Cambridge but did not hold teaching posts in astronomy or mathematics. It is now time to ask what sort of instruction was in fact given in these subjects. Did the universities have any official attitude, favourable or unfavourable, towards the Copernican system? As a matter of convenience I shall carry this survey up to the end of the 17th century, although this will involve some overlapping with the next section. I shall also say something about Gresham College which did not have university status but was, nevertheless, concerned with higher education.

Gresham College. There is no evidence to shew whether the Gresham professors, up to 1640, ever discussed the Copernican system in their lectures. There would have been no absolute necessity for them to do so, since they were mainly concerned with the practical arts of navigation, surveying etc. The references to Copernicus in their published writings are, in fact, very few and indefinite. Henry Briggs (1561–1630), professor of geometry from 1597 to 1619 and subsequently Savilian Professor at Oxford until 1630, published nothing on the subject himself. He did, however, express his opinion in a letter written to George Hakewill and printed by the latter in the 2nd edition of his *Apologie* (1630), to which reference has already been made. Briggs was supplying Hakewill with ammunition for his defence of modern learning against that of the ancients; for this purpose he enumerated some scientific achievements of the present time. The first item on his list of *Mathematica ab antiquis minus cognita* was:

Astronomia Copernicana quae docet Terram esse centrum orbis Lunaris, Solem vero esse centrum reliquorum omnium planetarum... Docet etiam per motum Telluris diurnum, Ortus & Occasus omnium Syderum: Et per motum ejusdem annuum in Orbe suo magno, omnium Planetarum motus & distantias, eorumque in caelo progressus stationes & regressus, multo facilius & accuratius investigare, quam per Ptolomaei aut antiqui cujusquam Epicyclos aut alias Hypotheses[94].

It is reasonable to conclude that Briggs accepted the Copernican system which, as he says, explains the phenomena 'much more easily and accurately than any other'.

---

[93] *Descriptio*, [in:] *Coll. Works*, vol. 5, pp. 516–7.
[94] *Apologie*, 2nd ed., p. 263.

The attitude of Henry Gellibrand (1597–1636), professor of astronomy from 1621 to 1636, was more ambiguous. He referred to the Copernican theory in his work on terrestrial magnetism in the following terms:

> I will not here enter into a dispute concerning the cause of this sensible diminution [of the magnetic variation], whether it may be imputed to the Magnet, or the Earth, or both. It is not unknowne to the world, how the Greatest Masters of Astronomie, which this age hath afforded, for the more easy salving of the apparent anomalar motions of the fixed and erratique caelestiall lights, and avoyding that supervacaneous furniture of the Ancients, do with all alacrity embrace that admirable *Copernicean Hypothesis* of the diurnal, Annual, & Secular motions of the earth, in so much as conferring with that Great Astronomer *D. Phil. Lansberg* in *Zealand* about Astronomicall matters, did most seriously affirme unto me, he should never be disswaded from that Truth. This which he was pleased to stile a truth, I should readily receive as an *Hypothesis*, and so be easily led on to the consideration of the imbecillity of Mans apprehension, as not able rightly to conceive of this admirable opifice of God or frame of the world, without falling foule on so great an absurdity. Yet sure I am, it is a probable inducement to shake a wavering understanding[95].

The meaning of the passage is obscure but the general argument seems to be that it would be unprofitable to try to find a cause of the earth's magnetic variation since such explanations are beyond the reach of man's intellect. This is shewn by the fact that the Copernican hypothesis is the most plausible theory of planetary motion and yet is absurd. God's workmanship is beyond our comprehension. Such speculations are an "inducement to shake a wavering understanding" (and therefore, presumably, are best avoided). Gellibrand may be implying that he regarded the hypothesis as possibly true in spite of its (apparent) absurdity, or he may mean that it is an attractive hypothesis which is actually false. But on any interpretation he clearly did not wish to commit himself definitely to it.

Finally, towards, the middle of the century, Samuel Foster, professor of astronomy from 1636–37 and 1641–52, mentioned and approved the theory in an essay *Of the Planetary Instruments*[96] in which he wrote: "The way that I go is (in general) agreeable to Copernicus his frame of the world; and in particular to that which Kepler useth in his Rudolphine Tables. Only this difference there is: Kepler makes the Orbits of the Planets to be Ellipses, which is the better way; and I here do make them perfect circles which is the easier way". Evidently, then, he was teaching the Copernican system at some time before his death in 1652.

Oxford University. At the beginning of the 17th century Gresham College was the only institution in England where regular scientific instruction was given. In 1619, however, Sir Henry Savile founded professorships of geometry and astronomy at Oxford, both of which attracted distinguished occupants throughout the century. From this time onwards, regular university lectures were given in both

---

[95] *A Discourse Mathematical on the Variation of the Magneticall Needle*, London, William Jones, 1635, p. 20.

[96] Published posthumously in: *Miscellanies: or, Mathematical Lucubrations of Mr Samuel Foster...*, London, R. & W. Leybourn, 1659.

subjects. All students in the Faculty of Arts were required to attend a course in geometry. The professor of astronomy, however, lectured only to those who had already obtained the degree of B. A. and were studying for the M. A. degree. Concurrently with their astronomy lectures, these latter had also to attend a course on natural philosophy in which the cosmological section was based on Aristotle's *De Coelo.* The duties of the professor of astronomy were laid down by Savile and incorporated into the University Statutes as follows: *Astronomiae Professor ad suum Munus sciat necessario pertinere interpretationem totius Mathematicae Constructionis Ptolemai (Almagestum vocant), adhibitis, suo loco, Copernici, Gebri et aliorum Recentiorum inventis... Ita tamen, ut Sphaeram Procli, vel Hypotheses Planetarum Ptolemai, possit (introductionis ad interiorem Artem gratia) Auditoribus proponere.* He was also forbidden to practise divinatory astrology and was urged *ad imitationem Ptolemaei et Copernici* to make astronomical observations himself, this being the only way in which the ancient astronomy could be either established or, if necessary, emended[97].

Savile clearly envisaged that the teaching would be basically Ptolemaic but would take account of the discoveries of 'Copernicus, Geber and other more recent astronomers'. The coupling of Copernicus's name with that of the medieval astronomer Geber[98] suggests that he was thinking of the more accurate observations and techniques of 'recent' astronomers, rather than of radically new theories. His primary duty was to lecture on practical and mathematical astronomy. The theory of the heavens would have been dealt with, officially, by his colleague in the school of natural philosophy where the teaching throughout was based upon the works of Aristotle.

The first Savilian Professor of astronomy was John Bainbridge (1582–1643), who held the post from 1621 to 1643. In 1618 he had published a book on the comet of that year, in which he gave a guarded approval to the heliocentric theory:

> These considerations [concerning the orbit of the comet] bee only fit for those who haue beene rapt vp aboue the elementary regions of vulgar Schooles: and slept not in *Parnassus,* but *Olympus,* vnder the spangled canopy of *Vrania*; I can hardly keepe within the sphaere of this little Treatise, and scarsely refraine from the Samian philosophy of *Aristarchus* in the earths motion were it not I feared another *Aristarchus* his broach: and that I must reserve these mysteries for a more learned language[99].

He was obviously attracted by 'the Samian philosophy' but was afraid to expound it in a popular treatise. And in spite of his 22 years as a professor of astronomy, he never published anything more on it, even in Latin.

Bainbridge was succeeded in the chair of astronomy by John Greaves (1643–1648) whose published writings were entirely concerned with the history of ancient

---

[97] See: *Statutes of the University of Oxford codified in the Year 1636...* Edited by John Griffiths, Oxford, Clarendon Press, 1888, p. 245.

[98] I. e. Gabir ben Aflah (d. 1145), not to be confused with the alchemist Gabir ben Haijan.

[99] *An Astronomical Description of the late Comet*, London, E. Griffin for H. Fetherstone, 1618, p. 5. (*Broach* = 'lance' or 'spear')

astronomy and related topics. The views of Henry Briggs, the first professor of geometry (1619–1630) have already been discussed. He evidently approved the Copernican theory but published nothing on it himself. His successor, Peter Turner (1631–1648), gave no public indication of his view, so far as I know.

In view of the widespread interest in, and sympathy with the new cosmology in England during this period, the reticence of the Savilian professors before 1649 is at first sight surprising. They, more than anyone, should have felt themselves professionally concerned with it. They were, however, in a difficult position. By the Statutes they were bound to base their teaching substantially on Ptolemy's Almagest. By way of introduction they could explain the author's hypotheses but it was not their business to lecture, *ex professo*, on the theory of the heavens. No doubt if they had sincerely accepted the Ptolemaic theory they could have taken part in the current debates with a clear conscience. But to come out against it would have been, to some extent at least, in conflict with their statutory duties; they would also have been encroaching on the domain of their philosophical colleagues. They may well have decided that a diplomatic silence was preferable[100].

After 1648 the situation at Oxford improved. In 1649 Seth Ward and John Wallis were appointed Savilian professors of astronomy and geometry respectively. Together with John Wilkins, Christopher Wren and – a few years later – Robert Boyle and Robert Hooke, they formed the nucleus of a small scientific society which met regularly to discuss questions of interest and to perform experiments. All were thoroughly in sympathy with the new cosmology. Apart from Boyle who apparently regarded the systems of Copernicus and Tycho as equally probable, they were all Copernicans. From this time onwards, whatever may have been the official teaching given to students in the Arts Schools, the more modern ideas were freely examined and discussed, at least within a limited circle.

Nevertheless, the University remained officially committed to Aristotle's philosophy including, it would seem, his cosmology, until near the end of the 17th century. Until 1677 or later, successive Vice-Chancellors repeatedly promulgated ordinances to the effect that all who defended theses in the public disputations must support the doctrines of Aristotle in accordance with the University Statutes[101]. A study of the actual topics proposed for debate indicates that this was generally observed, at least on paper, though occasional exceptions were apparently allowed. In 1634, for instance, one candidate defended the (anti-Aristotelian) thesis that comets may be composed of celestial matter. But even in 1671 the question "An Terra sit mobilis?" was still being answered in the negative. One must not attach too great weight to these documents since there is good evidence that the disputations were not taken very seriously by the students, in spite of repeated exhortations by the authorities

[100] *Of the High Veneration Man's Intellect owes to God*, Oxford, M. F. for Richard Davis, 1685. Reprinted in *Robert Boyle's Collected Works*, London, 1772, vol. 5, pp. 130–57.

[101] For information concerning these disputations I have used mainly the Wood Collection in the Bodleian Library, Oxford, especially those in the volume *Wood 276a*, nn. 419 onwards.

that they should be conducted in a seemly manner. It must, however, be concluded that they represented the official teaching given in the School of Natural Philosophy. One piece of indirect evidence to this effect was the publication at Oxford in 1671 of a small work by Robert Sanderson: *Physicae Scientiae Compendium*[102] which gave, in an elementary form suited to students, a summary of physics and cosmology based almost entirely on Aristotle untouched by any modern discoveries. In it the heavens were said to consist of ten solid spheres rotating round a motionless earth and comets were situated in the upper atmosphere. It was not stated that the work was intended for use by university students but it is difficult to see who else would have had much use for it. It must have had a reasonably wide clientele since a second edition, substantially unchanged, was issued in 1690. Robert Sanderson, who died as Bishop of Lincoln in 1663, probably wrote the book before 1619 so he can hardly be blamed for its complete neglect of later developments.

Another indication of the traditionalism of Oxford philosophy at this period was the popularity of a small textbook by Christopher Scheibler: *Philosophia Compendiosa*. Six editions of this, all published at Oxford, are extant, the first being dated 1628 and the latest (the 10th) 1685. The book included a section on astronomy which was almost entirely Aristotelian and Ptolemaic; the one exception being that Copernicus's value for the precessional period of the fixed stars, 25,816 years, was given in preference to Ptolemy's 36,000 years. Whether or not this was an official textbook at the university, it must have been widely used.

Cambridge University. At Cambridge, there was no university course in mathematics or science for Arts students at the undergraduate level before the middle of the 17th century, but some of the colleges provided instruction at a very elementary level. The Elizabethan Statutes, still in force, prescribed an elementary course for M. A. candidates but even this was largely neglected[103]. An interesting light on the state of these studies in the university is given by an anonymous and unidentified writer, T. B., in a short work entitled *Nuncius Propheticus*[104]. At the end of it the appended the text of a lecture which he claimed to have given at Cambridge 'in Scholis Mathematicis' in May 1639. In this lecture he protested against the neglect of mathematics at the university which he described as an indulgent mother to all other arts but an unjust stepmother to mathematics. He deplored the fact that many of the students completed their studies successfully without having studied any astronomy at all. He praised Copernicus warmly and stated that his theory of the diurnal rotation of the earth was defended by "Astronomorum pars major et melior"[105].

---

[102] Oxford, H. Hall for R. Davis, 1671.

[103] J. B. Mullinger, *op. cit.*, pp. 401–3.

[104] *Nuncius Propheticus: sive, Syllabus Selectiss. Vaticiniorum Theologico-Mathematicorum...* London [No publisher], 1642. The title page is anonymous but the signature T. B. appears on p. 44.

[105] P. 52. See also p. 68.

The almost complete neglect of mathematics and science in the official teaching at Cambridge before 1650 is well established by the evidence. It is all the more surprising that so many of the leading scientists and mathematicians of the time came from this university. The list includes such names as Dee, Gilbert, Wright, Ridley, Briggs, Bainbridge, Horrox, Seth Ward and John Wallis. All these, except perhaps Dee, were either Copernicans or semi-Copernicans. Some, such as Gilbert, Ridley, Briggs and Bainbridge may have received some instruction in astronomy in the medical school, but the rest must have been largely self taught. The eventual revival of scientific studies probably owed much to the philosopher Henry More (1614—1687) who entered Christ's College in 1631 and remained there all his life, first as a student and then as a fellow. He defended the Copernican system in a poem *Psychathanasia* (1642)[106] and in other later writings. He accepted the system unreservedly and was scornful of those who opposed it. For many years he was a follower of Descartes but eventually came to reject much of his system. It may have been due to his influence that Descartes' cosmology was taught at Cambridge during the latter part of the 17th century, being eventually displaced by Newton's in the early 18th century, largely through the influence of Samuel Clarke.

Advanced mathematical teaching was officially introduced for the first time at Cambridge in 1663 when Isaac Barrow became the first Lucasian Professor of that subject. Barrow was a good mathematician in his own right and was the teacher of Isaac Newton who succeeded him in the professorship in 1669. It can safely be assumed, then, that the Copernican system was being taught at Cambridge by about 1670 at the latest, and probably for some years previously[107].

The Seventeenth Century II: After 1640

The early 1640's marked the beginning of a new scientific awakening in England. Among the astronomers, the Copernican works of Lansberg, Galileo, Kepler and (a little later) Boulliau were being more and more widely read and approved. In consequence, both the geostatic system and the semi-Copernican compromise of Gilbert and his followers fell out of favour. After 1640 no new scientific works of any importance were published in England in which the heliocentric system was definitely rejected. The more competent astronomers, without exception, accepted it.

---

[106] Published in *Psychodia Platonica: or A Platonicall Poem of the Immortality of the Soul...* By H. M... Cambridge, Roger Daniel, Printer to the University, 1642. The whole of Book III, canto 3 (pp. 80—99), of *Psychathanasia* was devoted to a defence of Copernicanism.

[107] The official textbook for many years was Jacques Rohault's *Physica* which was heliocentric and thoroughly Cartesian. First published (in French), Paris 1671; Latin translations were published in London 1682 and 1697 with at least three later editions. It was translated into English by John Clarke in 1723 (3rd ed. 1735). Most of these editions had notes by Samuel Clarke in which the Newtonian system was expounded. Further information about Cartesianism at Cambridge is given by Marjorie Nicolson, "The early stages of Cartesianism in England", *Studies in Philology*, vol. 26 (1929), 356—74.

Between 1640 and 1642 three writers expounded and defended the heliocentric system. The first, and most influential, was John Wilkins (1614-1672). Educated at Oxford, he later lived in London where he was one of the leaders of the scientific group which met at Gresham College from 1645. In 1648 he returned to Oxford and became the centre of a similar group, to which reference has already been made. In 1638 he published a small book: *The Discovery of a World in the Moone*[108], in which he referred briefly to Copernican system. He did not then express any opinion on it except to remark that "how horrid soever this may seeme at the first, yet is it likely enough to be true". The book achieved immediate popularity. When the third edition appeared in 1640 Wilkins added another treatise to it: *A Discourse concerning a New Planet: Tending to prove, That tis probable our Earth is one of the Planets*[109]. In this he defended the Copernican system in detail, dealing more particularly with the objections which had been broght against it by Alexander Ross in his violently anti-Copernican *Commentum*[110]. Wilkins's book was popular in style; it did not have much to say on the technical astronomical reasons for preferring the system but it did deal effectively with the traditional objections. In view of his influence in scientific circles, his support must have played an important part in making it acceptable to his countrymen.

The second writer of this group has already been mentioned in the section on Cambridge University. He was Henry More, whose *Psychathanasia* (1642) contained a detailed and vigorous defence of the heliocentric system.

In the same year, further support was given by Thomas White (1593—1676). White, who wrote under a variety of pseudonyms — Anglus, Albius, Blacloe and Blacklow — was a Roman Catholic priest who at one time lectured on philosophy and theology in the ecclesiastical seminary at Douai and later came to England to exercise his priestly ministry. He wrote a number of controversial theological works, some of which provoked strong opposition among his fellow Catholics. He also published three scientific works: *De Mundo* (1642), *Institutionum Peripateticarum* (1646) and *Euclides Physicus* (1657). The Copernican system was discussed at considerable lenght in the first of these and more briefly in the second. *De Mundo*[111] was clearly modelled upon Galileo's *Dialogo*, and was in dialogue form, so that White was not called upon to state his own view explicitly. It is quite clear, however, that he favoured the system. Of the three participants in the dialogue — Andabata, Ereunius and Asphalius — only Andabata, who corresponds to Simpli-

---

[108] *The Discovery of a World in the Moone: or, a discourse, tending to prove, that 'tis probable there may be another habitable world in that planet.* London, E. G. for M. Sparke and E. Forrest, 1638. [Anon].

[109] The two works were issued with a common title page: *A Discourse concerning a New World and Another Planet, in 2 Bookes*, London, J. Maynard, 1640 [Anon].

[110] See p. 230.

[111] *De Mundo Dialogi tres,... Authore Thoma Anglo*, Paris, 1642. The Copernican system was discussed in the 2nd dialogue.

cius in Galileo's work, was against it. Ereunius definitely accepted it and Asphalius was favourable. The Greek derivations of the names are a sufficient indication that White's sympathies lay with Asphalius (steadfast) and Ereunius (the searcher). An *andabata*, in ancient Rome, was a gladiator who fought in the arena blindfolded. In the dialogue Andabata confesses that he "prefers to remain on the ground in secure ignorance rather than seek dangerous knowledge by flying on the wings of the wind" (*Malo in secura ignorantia humi degere, quam in scientia periculosa super pennas ventorum volitare*)[112]. This was certainly not Thomas White's ideal.

The movement of the earth was more positively asserted by White in his *Institutionum*[113], published four years later. Here he was writing in his own name and made his position quite clear. After describing the rotational and orbital motions of the earth he continued:

Astronomers prove these motions of the Earth: because, otherwise, greater motions of greater bodies must be suppos'd; and these, neither themselves constant, nor proportion'd to the bodies, and besides, more entangled, both in the Stars and in the Sun it self, as is apparent by its *Spots*: Which if you say make not up a perfect Astronomicall Demonstration, that Maxime must be renounc'd upon which all Astronomy depends, *viz.* that *the Phenomena* (or *appearances*) are to be solv'd the best way we can.

Again; because there follows a variety in the fixed stars, from the diversity of the Earth's position in its *Orbis Magnus*; when there is once found out a *Telescope*, of like perfection as to be able to distinguish that variety, we may expect a Geometricall Demonstration: and because for the same reason, there must needs be a variety of reflection from *Mars* and *Iupiter*; when the laws & rules of light shall be better known, there will not want a Physical Demonstration.

White was strongly influenced by Galileo but he rejected the latter's explanation of the tides. He propounded a theory of his own, according to which the sun and, to a lesser extent, the moon act upon the earth's atmosphere to produce systematic prevailing winds in different regions. These set up ocean currents which, by pressing on the shore, not only cause the tides but also produce both the rotational and orbital motions of the earth. The movements of the other planets were similarly explained by the action of the sun on them. Hence the sun, ultimately, is the source of all movement in the solar system. White was, as I have said, involved in various theological controversies but none of these was concerned with his views on the Copernican system which, so far as I know, were never condemned by the ecclesiastical authorities.

The battle for Copernicus had been effectively won by 1650. The victory was finally sealed when English astronomers discovered, rather belatedly, the planetary theory of Kepler, together with Ismael Boulliau's modified version of it. The first English writer to accept and recommend Kelper's elliptical orbits was Vincent

[112] P. 133.

[113] *Institutionum Peripateticarum ad mentem summi viri, clarissimique Philosophi Kenelmi Equitis Digbaei..., Authore Thoma Anglo*, Lyons, 1646. A 2nd edition was published in London, 1647, and an English translation London, 1656, under the title: *Peripateticall Institutions*. The extract which follows is taken from this translation, p. 176.

Wing who, from 1641, was probably the best known and most competent compiler of almanacs. In his almanac for 1647[114] he was anti-Copernican but within the next four years he changed his views completely. In his *Harmonicon Coeleste* (1651)[115] he strongly supported Kepler and Boulliau, asserting that the movement of the earth was "clearly proved". Elliptical orbits were accepted also by Jeremy Shakerley (1653)[116], Seth Ward (1654[117], 1656[118]), John Newton (1657)[119], and Thomas Streete (1661)[120]. By this time the heliocentric system was so thoroughly taken for granted that few writers thought it necessary to discuss the question at all.

A few other astronomical works may be briefly mentioned. In 1661 Thomas Salusbury published a translation of Galileo's *Dialogo* in his Mathematical Collections and Translations[121]. A few years earlier, Joseph Moxon had translated William Blaeu's Copernican treatise: *Institutio Astronomica* under the title: *A Tutor to Astronomy and Geography* (1654)[122]. Moxon also published two astronomical works under his own name: (a) *A Tutor to Astronomie and Geographie*: *Or an Easie and Speedy Way to know the Use of both the Globes, Celestial and Terrestrial...* (1659)[123] and (b) *A Tutor to Astronomy and Geography*, *or*, *The Use of the Copernican Spheres* (1665)[124]. In spite of the similarity of title these two works were entirely distinct from each other. Both were textbooks, (a) being based on Tycho's system (b) on that of Copernicus. The fact that the former went through five editions before 1700 whereas the latter was never reprinted may indicate that the elementary teaching of astronomy remained predominantly geocentric until the end of the century.

The Copernican System in English Literature. As the 17th century moved forward, more and more people must have acquired at least some knowledge of the Copernican system. References to it in the literature of the period are not, however, particularly frequent. The specific question at issue — whether the earth is moving or at rest — does not seem to have aroused much interest. It was, however, one factor among many which played its part in the struggle between supporters of the ancient and the modern learning. To the more progressive minds it was a proof that knowledge really is increasing and that we can add substantially to the inheritance of

---

114 *An Almanack and Prognostication for... 1647*, London, J. Legatt for the Company of Stationers, 1647, sig. C6ᵛ–C7ʳ.

115 London, R. Leybourn for the Company of Stationers, 1651.

116 *Tabulae Britannicae. The British Tables...*, London, R. & W. Leybourn, 1653.

117 *Idea Trigonometriae... item Praelectio de Cometis et Inquisitio in Bullialdi Astronomiae Philolaicae Fundamenta*, Oxford, L. Lichfield, 1654–53.

118 *Astronomia Geometrica*, London, J. Flesher, 1656.

119 *Astronomia Britannica*, London, R. & W. Leybourn, 1657.

120 *Astronomia Carolina: a new Theorie of the Coelestial Motions*, London, for L. Lloyd, 1661.

121 London, W. Leybourne, 1661.

122 London, J. Moxon, 1654.

123 London, J. Moxon, 1659.

124 London, J. Moxon, 1665.

wisdom which has been handed down from the past. To the conservatives it was a sad sign of the break up of the traditional world system and its replacement by a general atmosphere of scepticism.

Robert Burton's attitude to the new cosmology in his well known and widely read work *The Anatomy of Melancholy*[125] is typical of some of the more moderate conservatives. Burton (1577–1639) was a student of Christ Church, Oxford where he remained for most of his life. He read widely, if rather superficially, in the science of his day, from which he quoted extensively in his book. His own personal library included a copy of the first edition of *De Revolutionibus*, Thomas Digges's *Alae seu Scalae...* (1573) and other important astronomical works. He discussed the new cosmology sympathetically but non-committally; he definitely rejected the Aristotelian solid spheres[126] but could not make up his mind on other points. He had no sympathy with the more frevid anti-Copernicans such as Alexander Ross[127]; on the other hand he found the heliocentric theory difficult: partly because of the enormous distances at which the fixed stars must be situated and partly on account of the implication that if the earth is a planet, then the other planets must also be inhabited — a conclusion which was hard to reconcile with the principle that all things were made for man[128]. He therefore regarded as more probable the semi-Copernican theory that the earth has a diurnal rotation only. But his final conclusion was sceptical; we cannot hope to discover the truth:

> In the meantime, the world is tossed in a blanket amongst them [the astronomers], they hoist the earth up and down like a ball, make it stand and go at their pleasures: one saith the sun stands, another he moves; a third comes in, taking them all at a rebound, and, lest there should any paradox be wanting, he finds certain spots and clouds in the sun... and so, whilst these men contend about the sun and moon, like the philosophers in Lucian, it is to be feared the sun and moon will hide themselves, and be as much offended as she was with those, and send another messenger to Jupiter, by some new-fangled Icaromenippus, to make an end of all those curious controversies, and scatter them abroad[129].

A similar attitude is to be found in a poem written by John Donne (1573–1631): *An Anatomie of the World*; *The First Anniversary* (1611)[130]:

> And new Philosophy calls all in doubt,
> The Element of fire is quite put out;
> The Sun is lost, and th'earth, and no mans wit
> Can well direct him where to looke for it.

---

[125] Oxford, J. Lichfield & J. Short for H. Cripps, 1621. It went though many subsequent editions, the earlier ones having been extensively revised by Burton himself. Page references are to the Everyman edition, London, J. M. Dent, 1948, 3 vols.

[126] Vol. II, p. 50.

[127] II, p. 57.

[128] II, pp. 53–55.

[129] II, pp. 57–58.

[130] London, S. Macham, 1611 (Anon.]. Reprinted in *John Donne's Complete Poetry and Selected Prose*, Edited by John Hayward, London, Nonesuch Press, 1929.

And freely men confesse that this world's spent,
When in the Planets, and the Firmament
They seeke so many new; then see that this
Is crumbled out againe to his Atomies.
'Tis all in peeces, all cohaerence gone;
All just supply, and all Relation[131].

It was the loss of the old certainties which distressed the poet, rather than any specific feature of the 'new philosophy'. He had, apparently, no serious objection to the Copernican system as such. In his satirical work *Ignatius his Conclave* (1611) he remarked that the opinion of Copernicus "may very well be true"[132]. These words are, admittedly, put into the mouth of the arch-villain Ignatius but the context suggests that they represented the view of Donne himself. In his other writings, however, when he had occasion to refer to cosmological theories, it was the traditional geostatic system which he seemed to prefer.

John Milton (1608–1674), in his epic poem *Paradise Lost* (1677) shewed himself to be familiar with the main outlines of the Copernican system as well as with Galileo's telescopic discoveries. In Book VII, for instance, the angel Raphael puts the question to Adam:

What if the sun
Be centre of the World, and other stars,
By his attractive virtue and their own
Incited, dance about him various rounds?
[...] and what if, seventh to these
The planet Earth, so steadfast though she seem,
Insensibly three different motions move
Which else to several spheres thought must ascribe?[133]

But once again, the final conclusion tended to scepticism:

But whether thus these things or whether not,
Whether the sun predominant in heav'n
Rise on the earth or earth rise on the sun [...]
Solicit not thy thoughts with matters hid:
Leave them to God above, him serve and fear...
[...] heav'n is for thee too high
To know what passes there; be lowly wise:
Think only what concerns thee and thy being;
Dream not of other worlds, what creatures there
Live, in what state, condition or degree
Contented that thus far hath been revealed
Not of earth only but of highest heav'n[134].

---

[131] Nonesuch ed., p. 202.
[132] London, N. Okes for R. More, 1611, [Anon.], Nonesuch ed., p. 366.
[133] VII, lines 122–31.
[134] VIII, lines 159–78.

The only other poet of importance who shewed an interest in cosmology during this period was John Dryden (1631–1700). He was, for a short period (1662–67), a Fellow of the Royal Society though he never engaged in scientific research. There are occasional references to scientific themes in his writings but the few mentions of the Copernican system give no indication whether he accepted it or not.

On the whole, therefore, English literature in the 17th century does not manifest any great interest in cosmological questions. The poets and essayists were aware of their existence but did not feel particularly concerned with them. It has often been asserted in the past that the Copernican theory was deeply disturbing to many people because it displaced man from his central position in the universe and thus made him appear less important in the scheme of things. A. O. Lovejoy pointed out, many years ago, that this was not, in general, the case[135]. For the popular preachers and moralists the centre of the universe was not the most important but the least and lowest place of all – the place where the 'dregs of the universe' (*faeces mundi*) were to be found. Terrestrial matter was the most imperfect; it must therefore be at the greatest possible distance from the empyrean heaven. The lesson they drew from this was that man in himself, and except in so far as he is ennobled by the love of God, is a vile and worthless creature. Gilbert, Carpenter and Wilkins[136] all found it necessary to refute the traditional argument that the earth must be in the centre because it is composed of the most imperfect material. They met it by maintaining that the earth is not unimportant and that its matter does not differ essentially from that of the other planets.

The Copernican system undoubtedly did disturb some people. This, however, was not because it reduced man's importance but because it undermined his confidence in the power of reason. The essential features of the traditional Aristotelian-Ptolemaic system had seemed for many centuries to be established with complete certainty. But then, suddenly, the ordinary man found himself confronted with three rival systems, none of which was capable of proof. Instead of demonstrative certainty he had to be content with doubt and hypothesis. It was this sense of uncertainly which, both for Burton and Donne, was conducive to melancholy and which led Milton to tell Adam that he should not seek to unravel the secrets of the heavens. Eventually the work of Descartes and Newton would help to restore men's faith in the power of reason but, until then, the loss of intellectual security was a cause of discomfort to many.

II. THE COPERNICAN SYSTEM IN SCOTLAND

Scotland until the end of the 17th century was politically, economically and culturally independent of England. The first step towards union was taken in 1603 when James VI of Scotland succeeded to the throne of England as James I. Thereaf-

---

[135] A. O. Lovejoy, *The Great Chain of Being* Cambridge, Mass., Harvard University Press, 1936.

[136] W. Gilbert, *De Mundo* (1651), pp. 116–7; N. Carpenter, *Geography Delineated* (1625), p. 100; J. Wilkins, *That the Earth may*), Prop. VI.

ter the two nations had the same king but in other respects continued to be politically independent. It was not until 1707 that a full political union was effected.

The intellectual climate in Scotland also differed widely from that of England. The country was economically poor and, especially in the 16th century, politically unsettled. As a result, many of the more enterprising young men went abroad for their studies. They were to be found as students, and later as teachers, in many of the European universities. A few of them are known to have played an active part in promoting or opposing the Copernican system.

Perhaps the most interesting of these expatriate scholars was Duncan Liddel (1561–1613). He received his early education in Aberdeen but left his native country at the age of 18 to study at Frankfort a. O. and at Breslau where he learnt of the Copernican system from Paul Wittich. He then lectured on mathematics at the universities of Rostock, Frankfort and Helmstedt, finally returning to Scotland in 1607[137]. In 1613 he endowed a professorship of mathematics at Marischal College, Aberdeen, which was not, however, inaugurated until 1626 owing to financial difficulties. He also bequeathed a fine collection of mathematical, scientific and medical books to the same college, which included copies of both the first and second editions of the *De Revolutionibus*. In the second edition he transcribed a complete text of the *Commentaries* of Copernicus[138].

His friend John Caselius tells us that Liddel, in his mathematical lectures at Rostock (1588), expounded the Ptolemaic, Copernican and Tychonic systems side by side (without, it would seem, making any definite choice between them) and he added that, so far as he knew, Liddel was the first in Germany to treat of the Copernican system in this way. Tycho Brahe had explained his (not yet published) system to Liddel when the latter had visited him at Hven in 1587. Unfortunately the friendly relations between the two were ruptured a few years later when Tycho began to suspect, without any good evidence, that Liddel was claiming to have invented the Tychonic system himself[139]. There is, in fact, no reason to think that he ever made such a claim.

Liddel, while a student at Frankfort, had been taught by a fellow Scot, John Craig, who also spent much of his life in German universities. Craig, however, was a traditional Aristotelian with little sympathy for contemporary developments

---

[137] Liddel's academic career in Germany and hist attitude to the Copernican system were described by his friend John Caselius in a letter to John Craig reproduced in his preface to Liddel's *Ars Medica* (Hamburg, 1608). Further information is to be found in a letter of Daniel Cramer to Holiger Rosencrantz, 31st March, 1598, printed in T. Brahe, *Opera Omnia* (Ed. J. L. E. Dreyer), VIII (1925), 37–43.

[138] See W. P. D. Wightman, *Science and the Renaissance. An annotated Bibliography of the Sixteenth-Century Books relating to the Sciences in the Library of the University of Aberdeen*, Edinburgh and London, Oliver and Boyd, Vol. II. (1962).

[139] See letters by and to Brahe, published in his *Opera Omnia*, VIII, 37–43. 56–59, 205–206.

in astronomy. He maintained the immutability of the heavenly bodies and entered into a controversy, very ineffectively, with Tycho Brahe on this subject[140].

Two Scotsmen are known to have been actively propagating the new astronomy in Europe during the early 17th century. These were Thomas Seggeth (c. 1569–1627) and John Wedderburn, or Wodderburn (1583–1651). Seggeth went to Padua in 1597 where he became friendly with Galileo. In 1611 he was helping Kepler with his observations on the satellites of Jupiter in Prague. When Kepler published his *Narratio de observatis quatuor Jovis satellitibus*, Seggeth added a set of epigrams, one of which included the famous saying *Vicisti Galilaee!*, in praise of Galileo[141].

Wedderburn was a pupil of Galileo's at Padua and wrote, in 1610, a treatise defending the latter's *Sidereus Nuncius*[142] against an attack by Martin Horky. He continued to live on the Continent but made occasional visits to Scotland. It is not known whether he attempted to interest his fellow countrymen in the new astronomy.

A few other names may be briefly mentioned. James Cheyne, of Arnage[143], was born about 1545, graduated at Kings College, Aberdeen, in 1566, and was later ordained priest in the Catholic Church. He lectured in astronomy and other subjects at the university of Douai from 1572 and later at Paris. In 1575 he published a work on spherical astronomy[144] which was substantially based on the De Sphaera of Joannes Sacrobosco. It made no mention of Copernicus.

John Barclay (1582–1621), of Scottish parentage, was born and educated in France and came to London after the accession of James I. In 1614 he published *Icon Animarum*[145] in which the Copernican system was briefly referred to and decisively rejected.

The most competent work on astronomy published by a Scotsman before 1660 was probably Hugh Semple's *De Mathematicis Disciplinis* (1635)[146]. Semple was born at Craigevar, entered the Society of Jesus at Toledo in 1615 and became rector of the Scots College at Madrid where he died in 1654. As was to be expected under the circumstances, he rejected the Copernican theory. The main reason he gave for doing so was the absence of any observable stellar parallax but he added

---

[140] Brahe, *Opera Omnia* IV (1922), 415–476; 477–488. See also the correspondence between Brahe and Craig in Vol. VII, pp. 175–182 and subsequently.

[141] J. Kepler, *Narratio de observatis quatuor Jovis satellitibus*, Frankfort, Zacharias Palthenius, 1611.

[142] *Quatuor problematum quae Martinus Horky contra Nuntium sidereum de quatuor planetis novis disputanda proposuit. Confutatio per Ioannem Wodderbornium scotobritannum*, Padua 1610.

[143] See W. P. D. Wightman; *James Cheyne of Arnage*. In: *Mélanges Alexandre Koyré*, Hermann, Paris, Vol. 2 (1964), pp. 587–601.

[144] *De Priore Astronomiae Parte seu de Sphaera libri duo*, Douai, 1575.

[145] *Ioannis Barclaii Icon Animorum*, London, J. Bill, 1614. An English translation was published ander the title: *The Mirrour of Mindes*, London, J. Norton for T. Walkley, 1631; 2nd edition 1633.

[146] *Hugonis Sempilii Craigbaitaei Scoti e Societate Jesu de Mathematicis Disciplinis libri duodecim..*, Antwerp 1635.

that there were also philosophical and scriptural objections. He accepted Tycho's world system with slight modifications and shewed himself to be acquainted with the work of Kepler, Galileo and other modern astronomers. He discussed opposing theories with a moderation and respect not always to be found among the traditionalists of his time. Semple wrote also a more detailed work — *Dictionarium Mathematicum* — which was never published.

If we turn now to writers who studied or worked in Scotland, we find very few who wrote on astronomy during the 16th and 17th centuries. Setting aside, for the moment, the university thesis lists which were in a special position and were read only within a very limited circle, the available material is scanty indeed.

George Buchanan (1506—1582), a famous scholar and Latin poet in his own day, was born at Killearn, Stirlingshire, and educated partly at St Andrews, partly in Paris. He wrote a long astronomical poem, *De Sphaera*, published posthumously at Geneva in 1584[147] in which he discussed and rejected the possibility of a moving earth.

One of the most vigorous but least competent opponents of the new learning in the early 17th century was Alexander Ross (1591—1654). He graduated at King's College, Aberden, and later migrated to England where he was appointed master of the free school at Southampton. In 1634 he published a treatise on the immobility of the earth[148] in which he attacked the Copernican and semi-Copernican works of Philip Lansberg and Nathanael Carpenter[149] respectively. This was answered by John Wilkins in his *Discourse concerning a new planet* (1640)[150] to which Ross replied in another work: *The New Planet no Planet* (1646)[151]. Ross's books were violent in tone but very superficial in content. It is unlikely that they would have had much influence with educated readers.

A more moderate and intelligent defence of the Aristotelian cosmology was to be found in David Person's *Varieties* (1635)[152]. He held that celestial bodies are incorruptible and did not believe that new stars were natural phenomena. Instead, they were "extraordinary workes of the great maker, threatening mortalls by their frownings" (p. 7). Comets in general were sublunary but he recognised that some of them had been shewn by recent astronomers to be above the moon. He rejected Copernicus on the ground that the universe, as it revolves, must have an immovable centre which is the earth.

---

[147] It was included in a collection of his poems: *Franciscanus et Fratres*, Geneva 1584, and was published separately in 1585. It went through many editions.

[148] *Commentum de terrae motu circulare: duobus libris refutatum. Quorum prior Lansbergii, posterior Carpentarii argumenta ... refellit*, London, T. Harper, 1634.

[149] See p. 210.

[150] See p. 222.

[151] London, L. Young, 1646.

[152] *Varieties: Or, a Surveigh of rare and excellent matters, necessary and delectable for all sorts of persons...*, By David Person, of Loghlands in Scotland, London, Richard Badger for Thomas Alchorn, 1635.

I have been unable to find any other books in which the question of a moving earth was discussed before 1663, the year when James Gregory published his *Optica Promota*[153]. This marked the beginning of a new epoch in Scottish science. It was primarily a treatise on optics in which Gregory described the reflecting telescope which he had invented. At the end he added an appendix in which he discussed a number of astronomical problems. He fully accepted the heliocentric theory together with Kepler's first law of planetary motion. Instead of the 2nd law, however, he used Seth Ward's approximation in which the empty focus of the ellipse was treated as an equant point.

James was the first of a distinguished family of Gregorys who for many years played a leading part in establishing the vigorous scientific tradition which has characterised the Scottish universities ever since. He took his degree of M. A. at Marischal College, Aberdeen, in 1657. In 1668 he became the first professor of mathematics at St Andrews University whence he moved in 1674 to Edinburgh to become the first occupant of the corresponding chair in that university. He was succeeded at St Andrews by his nephew, James Gregory the younger. The Edinburgh chair remained vacant after the elder James's death in 1675 until 1683 when it was filled by another nephew, David Gregory (1661–1708) who in 1691 became Savilian Professor of Astronomy at Oxford. David published, in 1702, *Astronomiae Physicae et Geometricae Elementa*[154]. This was based on Newton's *Principia*, which he fully accepted.

Only one more work remains to be noticed: *Physiologia Nova Experimentalis*[155] by James Dalrymple, first Viscount Stair (1619–1695). Dalrymple had taken his degree at Glasgow University in 1637 and taught there from 1641 to 1647. Afterwards he had a distinguished career as a lawyer and a politician but found time to write the *Physiologia* during a period of exile in Holland. A large section of it was devoted to astronomy. In it he discussed the systems of Copernicus, Tycho, Descartes, Duhamel and others. His own theory was a somewhat implausible modification of Descartes' vortex theory in which the earth was at the centre of the universe and was surrounded by a vortex which gave it a diurnal rotation. This terrestrial vortex was embedded, without mutual interference, in a larger vortex which carried the sun and the superior planets around the earth. Circulating around the sun was a secondary vortex in which were embedded the inner planets, Mercury and Venus.

It will be clear from what has been said so far that the published literature of the period gives us a very incomplete picture of the history of the Copernican sys-

---

[153] *Optica Promota, seu abdita radiorum reflexorum & refractorum mysteria, geometrice enucleata: cui subnectitur Appendix subtillissimorum astronomiae problematon resolutionem exhibens*, London, J. Hayes for S. Thomson, 1663.

[154] Oxford 1702.

[155] Leyden 1686. The term *Physiologia* was here used, as in some other 17th century works, to cover all branches of natural science.

tem in Scotland. We can, however, obtain much more detailed information from the graduation theses of the Scottish universities during the 17th century. I will conclude, therefore, with a brief summary of these documents.

Scotland differed from England in having five universities — two in Aberdeen: King's College and Marischal College (which were combined into a single University of Aberdeen in 1859), and one each in Edinburgh, Glasgow and St Andrews. Each of these was, of course, much smaller than Oxford or Cambridge but taken together they must have provided a good higher education for a much larger proportion of the population than was available in England.

The teaching system differed from the English in several respects. (1) The Arts course lasted for 4 years only and led directly to the degree of Master of Arts. (2) In general, the Scottish universities throughout the 17th century adopted a system of circulating regents, i. e. there were no specialist teachers but each lecturer (or regent, as they were called) took a particular class for the whole 4 year course, lecturing in all subjects to the same set of students: Greek in the 1st year, logic in the 2nd, ethics in the 3rd, natural philosophy with some mathematics and astronomy in the 4th. At the end of the four years he would start again with a new class in its 1st year, and so on. The Arts Faculty consisted therefore of precisely four regents, working in rotation.

There were some exceptions to this system. Marischal College until about 1640 had specialist regents, each of whom lectured to all students in one subject only. King's College had a similar system between 1616 and 1638. Thereafter both reverted to the normal system of circulating regents. In addition Marischal College had a professor of mathematics from 1626 who lectured to all the students in that subject. St Andrews founded a professorship of mathematics in 1668 and Edinburgh in 1674. Mathematical astronomy was included in the mathematics course but cosmological questions concerning the structure of the universe, the nature of the heavenly bodies etc. were generally treated in natural philosophy. Towards the end of the century, however, this distinction tended to disappear.

The 4 year course culminated in a graduation ceremony which included a public disputation. The regent responsible for the graduating class would publish beforehand a long list of "theses", or topics for debate, covering substantially the whole of the course. The candidates were then called upon, in turn, to defend one or other of the theses against objections propounded by regents, students or visitors. The theses expressed the opinions of the regent himself and the students were bound to defend his view whether they agreed with it or not. Towards the end of the century, however, this requirement was progressively relaxed. Many of these published theses are now no longer extant but where they have survived it is possible to reconstruct the contents of the lecture courses in considerable detail. The most complete series is available for Edinburgh University. Of the 113 years from 1596 (the earliest known list) until 1708 when this form of disputation was abandoned, the lists for at least 68 years are extant. The other universities are less well represented but, except for Glasgow, enough lists have survived to give a general indication of the teaching[156].

The value of these lists for our present purpose is that they enable us to determine the views of the different regents on the Copernican system and to discover, at least approximately, when it was first taught to the students. Not all the lists contain explicit reference to the system but nearly all contain some cosmological theses from which the attitude of the regent to it can be inferred. I shall concentrate mainly on the teaching in Edinburgh since the information for this university is so much more complete than for the others.

From 1596 till 1616 the cosmological teaching in Edinburgh was predominantly Aristotelian, following closely the theories propounded by Aristotle in his *Physica* and *De Coelo*. The only regent during this period who did not conform was James Knox who, in 1601, asserted, against Aristotle, that the heavenly bodies are changeable. The year 1616 marked the beginning of a 10 year period during which the teaching was progressively modernised. William King, in this year, referred to the comet of 1577 as evidence that celestial bodies are mutable. Eight years later, in 1624, he also mentioned the new stars of 1600 and 1604 in support of this conclusion. However, he was not altogether happy about his rejection of tradition. He concluded his discussion with a warning that we cannot hope to reach certainty on this question; it is safer simply to admit that we are ingorant.

Another regent at this time, Andrew Young, was apparently making astronomical observations with his students between 1617 and 1620. In his theses for 1617 and 1621 he published data for the elevation of the sun at Edinburgh at the vernal equinox of 1617 and the summer solstice of 1620 respectively. He also reported, in 1621, observations on the comet of 1618 which proved that its diurnal parallax was undetectable; hence it must have been situated far beyond the sphere of the moon. He therefore rejected the theory of solid celestial spheres. Young died in 1623 and there is no evidence that the observations were continued by his successors.

A third modernising regent during the same period was James Reid. In 1618 he defended the Ptolemaic system but four years later, in 1622, he asserted that many different hypotheses, including the Copernican, can explain the planetary paths equally well. He concluded from this that it is impossible to reach any certain conclusions in cosmology. Finally, in 1626, he suggested, very cautiously, that we may hold at least as a supposition, that the earth rotates on its axis while remaining at the centre of the universe. He defended himself against any accusation of religious unorthodoxy by pointing out (a) that he was speaking only hypothetically – *ex suppositione* – and (b) that Scripture does sometimes describe phenomena as they appear to the senses rather than as they are in themselves.

---

[156] For a list of the extant theses see H. G. Aldis: *A List of Books printed in Scotland before 1700*. Reprinted with additions..., Edinburgh, National Library of Scotland, 1970.

Unfortunately these theses of 1626 raised a theological storm[157]. The ministers of the Scottish Church immediately insisted that Reid should be dismissed from his teaching post. The Edinburgh Town Council, which at that time controlled appointments to the university, resisted their demands for several months but eventually, under mounting pressure, gave way and asked Reid to resign, which he did. The ostensible reason for this attack was very trivial: in one of his theses Reid had mildly criticised a minister who, a few months previously, had delivered a violent attack upon philosophy in one of his sermons. This was seized upon as a *casus belli* by the ministers but it was clearly only a pretext; the real target was evidently the University philosophical teaching as a whole.

Whatever may have been the precise motives of the theologians, their victory did in fact crush the progressive movement for many years. The immediate consequence, clearly shown in the theses from 1627 to 1632, was a reversion to strict Aristotelian orthodoxy, with its solid planetary spheres and immutable celestial bodies.

No theses have survived from the years from 1633 to 1640 inclusive but by 1641, when the series resumes, the solid spheres had apparently been tacitly abandoned since there was henceforward no further mention of them. Between 1641 and 1650 some at least of the regents still regarded celestial bodies as unchangeable and none explicitly rejected this principle. During this period the Copernican system was never mentioned but the generally traditional tone of the theses makes it unlikely that any of the regents could have taught it.

From 1651 to 1658 there is another gap in the series. By 1659 the immutability of the heavenly bodies was no longer being taught. The Copernican system was now beginning to receive more attention: between 1659 and 1669 it was discussed but rejected in four out of the seven extant lists; the others did not refer to it. At least one regent — John Wishart in 1661 — was teaching the Tychonic system and it is quite possible that the others were doing so too, though they did not refer to it explicitly.

In 1670, John Wood, allowed his students for the first time to accept the Copernican system if they wished, but he did not himself express any opinion on its truth. The same option was given in 1674 and 1675 but the system was still being rejected by more conservative regents in the two following years. The first to accept a moving earth unreservedly and definitively was Gilbert McMurdo in 1682. This year marked a turning point in the university teaching. A completely new team of regents took over, all of them firmly committed to the philosophy of Descartes including, with some reservations, his philosophy of nature. All accepted a Cartesian heliocentric system.

When Newton's *Principia* was published in 1687 it was taken up with remarkable alacrity at Edinburgh. As early as 1688 there was a favourable mention of

---

[157] The incident is described by Thomas Craufurd, *History of the University of Edinburgh from 1580 to 1646*, Edinburgh, 1808, p. 106–7.

it in the theses of Alexander Cockburn. Herbert Kennedy in 1690 and again in 1694 expounded and accepted its basic principles. In the surviving theses of the early 18th century, the cosmological teaching was firmly based upon it.

It is not possible to follow the development of cosmological theory in the same detail for the other Scottish universities since not nearly so many of their thesis lists have survived and they are particularly scarce for the important years between 1650 and 1680. The evidence suggests, however, that the Copernican system may have been taught at the Aberdeen universities and St Andrews at least ten years earlier than at Edinburgh. Of the lists which are extant special mention may be made of (1) Andrew Cant's theses for Marischal College (1654), the first (apart from the ill-fated Edinburgh theses of 1626) which admitted the possibility of a moving earth, though Cant himself preferred the Ptolemaic theory; (2) William Sanders, at St Andrews, 1674, fully accepted the Copernican system and quoted Kepler's first two laws of planetary motion in its support; (3) James Gregory the younger, St Andrews 1690, gave a detailed exposition of Newton's physics and cosmology which he unhesitatingly accepted.

Finally, something must be said about the Professors of Mathematics, whose views were not always represented in the thesis lists. The first Chair, as we have said, was inaugurated at Marischal College in 1626. The founder, Duncan Liddel, had laid it down that the professor should be "well versed in Euclid, Ptolemy, Copernicus, Archimedes and other mathematicians", and that he should teach mathematics and astronomy to the students in their 3rd and 4th years. The first professor was William Johnston who held the post from 1626 until his death in 1640. There is, in the Aberdeen University Library, a MS text of his dictated lectures for 1633–34. recorded by James Dounie, one of his students[158]. It includes a treatise on the theory of the planets, in which the systems of Ptolemy, Copernicus and Tycho Brahe were discussed, together with the semi-Copernican theory that the earth is at the centre of the universe but rotates on its own axis. Each theory was treated objectively and sympathetically; each was apparently regarded as tenable. Johnston himself seemed to favour the semi-Copernican theory but his final conclusion was that since it is impossible to attain to any certainty on the question it is perhaps better to rely on the evidence of our senses and to follow Ptolemy. This may have been said simply in deference to his philosophical colleagues who, at this time, would certainly have regarded a moving earth as metaphysically unacceptable. On one point, however, Johnston was quite definite: we must abandon the theory of solid celestial spheres. He made effective use of Tycho's arguments to establish this conclusion.

During Johnston's tenure of office, astronomical theses were not included in the annual disputations at Marischal College so we do not know whether his views changed as time went on. It seems clear, however, that during the 1630's the students of this university received more competent and up to date instruction in astronomy

[158] MS. Aberdeen University Library, M. 181.

than at any other British university with the possible exception of Oxford. The professorships in mathematics at St Andrews and Edinburgh haxe already been mentioned. The Gregorys who were the first occupants of these Chairs — James the elder, James the younger and David — played a leading part in the rapid and complete modernisation of the science teaching which began about 1670.

Conclusion. There is no evidence, either from books or university records, that there was much interest in the Copernican system in Scotland before the 1660's. The first to accept it publicly was James Gregory in 1663. In Edinburgh University it does not seem to have become academically respectable until about 1670. It was given sympathetic treatment at Marischal College, Aberdeen, in the 1630's by William Johnston and again in 1654 by Andrew Cant but the latter, at least, did not accept it and the former, in his lecture course, avoided any firm conclusion. It is impossible to fix any precise date when it was first accepted in any of the universities apart from Edinburgh since the extant thesis lists are too few. It was presumably taught by James Gregory at At Andrews from 1668 and at Edinburgh in 1674–75. It may have been cautiously admitted in some of the missing Aberdeen theses of the 1660's but probably not earlier. The earliest known lists in which a moving earth was accepted are A. Alexander's at Marischal College, 1669 (with the proviso, however, that the earth, though a planet, is stationary with respect to its vortex); William Sanders's, 1674, at St Andrews, George Middleton's, 1675, at King's College, Aberdeen, and Gilbert McMurdo's, 1682, at Edinburgh.

The Scottish universities in the first half of the 17th century were predominantly Aristotelian and, in general, hostile or indifferent to the new astronomy. But from 1670 onwards they modernised their teaching with exceptional speed and throughness. By 1680, and for many years afterwards, they compared favourably with any other university in Europe for their interest in modern science and for their readiness to incorporate its more significant findings into their teaching syllabuses.

BIBLIOGRAPHY

I. The Copernican System in England. General Studies

Johnson, Francis R., *Astronomical Thought in Renaissance England*, Baltimore, Johns Hopkins Press, 1937 (reprinted: New York, Octagon Books, 1968). This is much the most complete study of the subject, up to the year 1645, that has ever been made. It includes detailed discussions of all the important sources with extensive quotations. It is essential reading for all students of the period, and, as a source book, is thoroughly reliable. Johnson's enthusiasm for the achievements of English scientists does, however, lead him to exaggerate the extent to which the Copernican system was actually accepted, both by some of the individuals whom he discusses and in the country as a whole. But he presents his evidence in such detail that the careful reader can generally make the necessary corrections for himself.

Applebaum, Wilbur, *Kepler in England: The reception of Keplerian astronomy in England, 1599–1687*. (Thesis). Diss. Abs. Int. *30*, 1969, 2440A. I have not had an opportunity of seeing this thesis.

Jones, R. F., *Ancients and Moderns. A study of the rise of the scientific movement in seventeenth-century England*, 2nd ed. St Louis, Washington University Press, 1961 (reprinted: California University Press, 1965). The cosmological debates of the 17th century are examined in the light of the more general conflict between new science and old Aristotelianism at the time.

Kocher, Paul H., *Science and Religion in Elizabethan England*, San Marino, California, The Huntington Library, 1953.

Nicholson, Marjorie, *English Almanacs and the 'New Astronomy'*, "Annals of Science", 4, 1939, 1–33. A comprehensive survey of the English almanac-makers of the 17th century and their attitude to such questions as the Copernican and Tychonic systems, comets, novae, etc.

Universities and Colleges. Cambridge

Costello, W. T., *The Scholastic Curriculum in Early Seventeenth Century Cambridge*. Cambridge, Mass., Harvard University Press, 1958. A good account of the formal scholastic training given at the university, but has little to say on the state of scientific and mathematical instruction.

Hoskin, Michael A., *'Mining All Within': Clarke's Notes to Rohault's "Traité de Physique"*, "The Thomist", 24, 1962, 353–363. Describes the introduction of Newtonian science into the Cambridge teaching syllabus, under the guise of annotations to Rohault's work.

Nicolson, Marjorie, *The Early Stage of Cartesianism in England*, "Studies in Philology", 26, 1929, 356–374. Chiefly concerned with Henry More: his attitude to Descartes and his influence on the growth of Cartesianism at Cambridge.

Webster, C., *Henry More and Descartes: Some New Sources*, "Brit. J. Hist. Sci.", 4, 1969, 359 377. Supplements and to some extent supersedes Marjorie Nicolson's study.

Gresham College

Johnson, Francis R., *Gresham College: Precursor of the Royal Society*, "J. Hist. Ideas", 1, 1940, 413–438.

Hill, Christopher, *The Intellectual Origins of the English Revolution*, Oxford, Clarendon Press, 1965. Chapter 1 includes a detailed discussion of the influence of Gresham College and of the London scientists generally. Good background material but is inclined to exaggerate the influence of Puritanism on English science.

Ward, John, *The Lives of the Professors of Gresham College*, London, 1740. Reprinted by The Johnson Reprint Corporation, New York, 1967.

The Copernican System and English Literature

Coffin, C. M., *John Donne and the New Philosophy*, New York, Columbia University Press, 1937. Shews that Donne had a good general knowledge of the work of Copernicus, Tycho, Gilbert. Kepler and Galileo. He was interested in the new astronomy but he himself remained essentially Ptolemaic in his outlook.

Meadows, A. J., *The High Firmament. A survey of astronomy in English Literature*, Leicester University Press, 1969.

Nicolson, Marjorie, *Science and Imagination*, New York, Cornell University Press, 1956. A collection of articles published from 1935 onwards. The best general treatment of the influence of the new science on English literature during the 17th century.

Svendsen, Kester, *Milton and Science*, Cambridge, Mass., Harvard University Press, 1956. A comprehensive survey of the popular encyclopedias and similar sources available to Milton, and the influence which they exerted on his writings.

Studies on Individual Scientists

Gunther, R. T., *The Uranical Astrolabe and other Inventions of John Blagrave of Reading*, "Archaeologia", 79, 1929, 55–72.

Johnson, Francis R., *The Influence of Thomas Digges on the Progress of Modern Astronomy in Sixteenth-Century England*, "Osiris", 1, 1936, 390–410.

Johnson, Francis R., *Thomas Digges, the Copernican System, and the Idea of the Infinity of the Universe in 1576*, „The Huntington Library Bulletin", No. 5, 1934, 69–117.

Lawton, H. W., *Bishop Godwin's 'Man in the Moone'*, "Rev. of English Studies", 7, 1931, 23–55.

McColley, Grant (Ed.), *Francis Godwin's 'The Man in the Moone'*, Smith College Studies in Modern Languages, 19, No. 1, 1937. A reprint, with introduction and notes, of the first edition of Godwin's book, London, 1638.

Kelly, Sister Suzanne, *The 'De Mundo' of William Gilbert*, Amsterdam, Menno Hertzberger, 1965.

Roller, Duane H. D., *The 'De Magnete' of William Gilbert*, Amsterdam, Menno Hertzberger, 1959.

Stevens, Henry, *Thomas Harriot: The Mathematician, the Philosopher, the Scholar*, London, 1900.

Tanner, Rosalind C. H., *The Study of Thomas Harriot's Manuscripts*, I. *Harriot's Will*, "History of Science", 6, 1967, 1–16.

Pepper, Jon V., *The Study of Thomas Harriot's Manuscripts*, II. *Harriot's Unpublished Papers*, ibid. 17–40. Very little of Harriot's astronomical work has yet been published or investigated adequately. These two articles provide a preliminary survey of the unpublished material.

Quinn, David B. & Shirley, John W., *A Contemporary list of Hariot References*, Renaissance Quarterly, 22, 1969, 9–26.

McColley, Grant, *Nicholas Hill and the "Philosophia Epicurea"*, "Annals of Science", 4, 1939, 390–405. An analysis of this work, with special reference to the scientific ideas propounded in it.

Barocas, V., *Jeremiah Horrocks (1619–1641)*, "J. Brit. Astron. Assoc". 79, 1968, 223–226. A brief general survey of his life and work.

Gaythorpe, S. B., *Jeremiah Horrocks and his 'New Theory of the Moon'*, "J. Brit. Astron. Assoc." 67, 1957, 134–144. A technical exposition of Horrocks's theory of the moon's orbit.

Plummer, H. C., *Jeremiah Horrox and his "Opera Posthuma"*, "Notes and Records of the Royal Society", 3, 1940, 39–52. The "Opera Posthuma" contains Horrocks's defence of Kepler and the Copernican system.

Whatton, A. B., *A Memoir of the Life and Labours of the Rev. Jeremiah Horrox... To which is appended a translation of his celebrated discourse upon the Transit of Venus across the Sun*, London, 1859.

Johnson, Francis R. & Larkey, Sanford V., *Robert Recorde's Mathematical Teaching and the Anti-Aristotelian Movement*, "The Huntington Library Bulletin", No. 7, 1935, 59–87.

Patterson, L. D., *Recorde's Cosmography, 1556*, "Isis", 42, 1951, 208–218. Examines Recorde's attitude to the Copernican system. A useful corrective to Johnson's rather exaggerated estimate of the extent to which Recorde was a Copernican.

Bowen, E. J. & Hartley, Sir Harold, *The Right Reverend John Wilkins, F. R. S.*, "Notes and Records Roy. Soc.", 15, 1960, 47–56.

McColley, Grant, *The Debt of John Wilkins to the 'Apologia pro Galileo' of Tomasso Campanella*, "Annals of Science", 4, 1939, 150–168.

McColley, Grant, *The Ross-Wilkins Controversy*, "Annals of Science", 3, 1938, 153–189. A discussion of the controversy between Alexander Ross and John Wilkins referred to on p. 230 above. Gives extensive references to other 17th century works which illustrate the relations of science and religion at this time.

Shapiro, Barbara, *John Wilkins (1614–1672). An Intellectual Biography*, Berkeley, University of California Press, 1969.

Parsons, E. J. S. & Morris, W. F., *Edward Wright and his Work*, "Imago Mundi", 3, 1939, 61–71.

II. The Copernican System in Scotland

Anderson, P. J., *Notes on Academic Theses, with a Bibliography of Duncan Liddel*, "Aberdeen University Studies", No. 58, 1912.

Favaro, Antonio, *Galileo Galilei a Padova*, Padova, Editrice Antenore, 1968. There is a short biography of John Wedderburn on pp. 279—285 and a note on Thomas Seggett on pp. 215—216.

Johnstone, James F. Kellas, *The Lost Aberdeen Theses*, Aberdeen University Press, 1916. Reprinted, with additions, from "Aberdeen University Library Bulletin", Vol. 2, No. 12. It includes a general description of the Scottish graduation theses and their place in the university teaching system.

Naiden, James R., *The Sphera of George Buchanan*, Philadelphia, W. H. Allen, 1952. A translation of Buchanan's astronomical poem with notes, bibliographical data etc.

Rosen, E., *Thomas Seget of Seton (1569 or 1570 to 1627)*, "Scottish Hist. Rev." 28, 1949, 91—95.

Turnbull, H. W. (Ed.), *James Gregory Tercentenary Memorial Volume*, London, Bell, 1939. Includes many unpublished letters and papers of Gregory.

Turnbull, H. W., *Early Scottish Relations with the Royal Society*, "Notes and Records Roy. Soc.". 3, 1940, 22—38. Chiefly concerned with James Gregory.

HENRIK SANDBLAD
Göteborg University

THE RECEPTIONS OF THE COPERNICAN SYSTEM IN SWEDEN

1.

At the time of the publication of Copernicus' *De revolutionibus orbium coelestium*, the learned and scientific culture of Sweden was in a profound state of decay. The Lutheran Reformation had resulted in confiscation by the State of the ecclesiastical property upon which all learned education rested. Medieval monastery culture had disappeared and all its libraries were scattered, the educational system was totally disorganized, and the University of Upsala, founded in 1477, ceased to exist for more than half a century. That university, or *studium generale*, was the only one within the State of Sweden, which at that time (and till 1809) included Finland as well. The few Swedes, who acquired any higher education and had any close contact with international science during this period, had to go abroad, above all to universities in Protestant Germany.

In the 1570's,a modest attempt at reviving the University of Upsala was made, and the astronomy was taught as well. One of the teachers, Olaus Jonae Luth, left an astronomical text-book from 1579 in a manuscript which is probably based on lectures given at the university[1]. Characteristically enough, the picture of the world represented there is entirely the Ptolemaic one, in the form it was given by Sacrobosco and his commentators, including Hartmann Beyer, the disciple of Melanchton, and it lacks any trace of knowledge of Copernicus, in spite of the fact that Luth studied at three German universities and obtained his master-of-arts degree.

In fact, it is as late as 1588 that the name of Copernicus is mentioned in Swedish literature for the first time, in a large Calendarium duplex, published by the otherwise unknown Georgius Olai (d. 1592)[2], who had spent some time in Germany where

[1] Olof Luth, *Nogre stycker aff thenn frije Konst Astronomia* (Cod. Holm. D 77). With an introduction ed. by H. Sandblad, *Skrifter utg. av K. Hum. Vetenskaps-Samfundet i Uppsala*, 29:4. Upsala & Leipzig, 1935.

[2] Georgius Olai Upsaliensis, *Calendarium duplex, Christianorum et Iudaeorum, Cum prognostico astrologico*..., Stockholm, 1588.

he probably acquired his vast knowledge of astronomy. There he probably learned also about the current astronomical discussion concerning the Copernican system, although there are no obvious traces of that in his work. In the introduction to his calendarium however, he points out that there are two different calculations of the orbits of the planets: one by Johannes Stadius who follows the method of Nicolaus Copernicus and Erasmus Reinhold's *Tabulae Prutenicae*, based on that method, the other by Cyprianus Leovitius who follows the old *Tabulae Alphonsinae* and the Tables of Johannes Schonerus. Since the difference between these two is often great, Georgius wants to quote them both, to enable examination "thereby in the changes of weather" as to which is the most correct. Furthermore, in the calendarium itself he often gives information "secundum Copernicos". He does not go further than that in using Copernicus' calculations.

During the last decades of the century, more and more Swedish students went to universities of the continent, and, of course, in doing so many of them learnt about the results of the new natural research. And one also finds that in their dissertations at foreign universities, many of them had reasons to deal with Copernicus. Thus a certain Johannes Laurentii Gevaliensis defended a thesis *De elementis*[3] in Greifswald in 1597, briefly citing the Copernican theory of the three motions of the earth: the diurnal one around the earth's axis, the annual around the sun, and the annual motion of declination which would explain precession. However, basing his argument upon the motion concept of the prevailing Aristotelian physics he denies the possibility of terrestrial motion. For the future, that concept was to be one of the main obstacles for the Copernican system. Defending a theses *De stellarum natura et proprietatibus*[4] in Rostock in the same year, Jacobus Simonis Stockholmensis uses the same argument to prove the immobility of the earth in the centre of the universe, polemising with dissentient opinions but without mentioning the name of Copernicus. Johannes Schroderus, later famous as a statesman and a leading personality in Swedish cultural politics under the name of Johan Skytte, defended a dissertation on various philosophical and scientific problems in Marburg in 1598[5], mentioning Copernicus with admiration and recognition along with Ptolemy, though adopting without hesitation the system of the latter. In surprising contrast with that, two years later a collection of theses was discussed at the same university by Jonas Petrejus Upsaliensis[6]. This dissertation is a strikingly sharp defence of the Copernican doctrine of the motion of the earth, and the author does not even hesitate to maintain that it is supported by Holy Scripture, introducing some rather daring interpretations of the *Bible*. That dissertation should be regarded as a very

[3] The praeses and perhaps author was Prof. Jacobus Seidelius.

[4] Praes. Johannes Sturmius.

[5] *Problemata ex artium liberalium suavissimis et amoenissimis fontibus*, praes. Rodolphus Goclenius.

[6] *Centuria thesium physicarum de coelo et terra*, praes. Nicolaus Chesnecopherus, a Swede who was professor of mathematics in Marburg for a short time.

unusual phenomenon, however, and it seems that to some extent the sharp phrasing of the theses may have been due to the dialectic exercise at the public discussions. In the same year, the same scholar defined another dissertation in which some theses about the immobility of the earth in the middle of the world are presented for discussion[7]. It is this dialectic purpose of the academic literature of that time which often makes it difficult to estimate the meaning of the opinions stated, especially as regards views as delicate and controversial as those of cosmology.

At that time, however, scientific activities within the boundaries of Sweden had entered a new phase with their richer possibilities, since the University of Upsala had been re-established in 1595. The first Professor of Astronomy of the renewed educational centre was the young Laurentius Paulinus Gothus (1565–1646), later a bishop and finally archbishop, philosophically and pedagogically an ardent follower of Petrus Ramus. Even during his student years in Rostock, where his most prominent teacher of astronomy was Heinrich Brucaeus, friend of Tycho Brahe, he published an almanac and an astrological practice for the year of 1592[8]. There he deals with a forthcoming eclipse, mentioning its time according to the Alfonsine calculation. But at the same time he gives his own, dissentient estimation, adding that it corresponds to "Calculo Copernicano", which, in fact, means only that he used Copernicus' Tables. However, in Rostock no doubt he was given the opportunity to study the Copernican system thoroughly. It was presented in lectures there, although the teacher, Brucaeus, adhered principally to the Ptolemaic cosmology[9].

About his work as professor of Astronomy in Upsala (till 1600) Laurentius Paulinus himself several years later states that he was not only teaching the celestial motions according to the Tables, but also lectured on the three different planetary hypotheses: the Ptolemaic, the Copernican and the Tychonic. This information is fully confirmed by a large number of detailed notes, made by Paulinus himself in a copy of Georg Peurbach's famous work *Theoricae novae planetarum* from the middle of the 15th century, still used in academic education all over Europe many years later[10]. The notes are made in 1599, obviously in connection with lectures at the university[11]. They contain detailed accounts of Copernicus' planetary system, illustrated with a large number of figures, with mathematical comparisons to the Ptolemaic system and also that of Tycho Brahe. It is obvious that Paulinus had

---

[7] *Εγκυκλοπαιδεια sincerioris & Socraticae philosophiae*, praes. Rodolphus Goclenius.

[8] *Almanach Och Practica wppå thett M.D.XCII. Åhret...*, Greifswald, 1591.

[9] So he does in his principal astronomical work, *De motu primo*, Rostock, 1573, 5th ed. 1604, which was used in Swedish academic education for a long time.

[10] Laur. Paulinus has used the edition Basileae, 1573. His copy now belongs to the library of the Upsala Observatory.

[11] The Latin original is published with a Swedish translation by N. V. E. Nordenmark, *Laurentius Paulinus Gothus föreläsningar vid Uppsala universitet 1599 över Copernicus hypotes*, [in] "Arkiv för astronomi", vol. 1, No. 24, Stockholm, 1951.

a great interest in the Copernican theory, but he does not express his own standpoint.

Thus it is evident that the students in Upsala at this time obtained detailed knowledge of Copernican astronomy. Copernicus' doctrine was, as yet, however, hardly considered so dangerous and repudiable as would be the case during the following period, when Aristotelian scholastic philosophy with ecclesiastical orthodoxy took command of university learning, in Sweden as well as in many other European circles.

Those Swedes showing a more definite interest in the Copernican system in the period immediately following are found outside university life.

One of them is Sigfrid Aronus Forsius (d. 1624), a clergyman of Finnish extraction, natural philosopher and an astrologist of strong apocalyptic inclination. During the early 17th century, he published almanacs with prognostics, and in one of them, that of 1610[12], he talks about the reasons for the uncertainty of astrological calculations, one of which is that the foremost astronomers have offered such dif ferent hypotheses about the planetary motions. In doing so, he briefly describes the three great world systems, those of Ptolemy, Copernicus and Tycho. This account is particularly interesting in that it is the first actual description of the Copernican system printed in Swedish literature. Forsius does not express his own opinion; he only emphasizes that another long period of observations is needed in order to ascertain the truth. He returns to the question in his large, principal work of natural philosophy, *Physica*, completed in 1611, the printing of which was prevented by ecclesiastical orthodoxy because of certain heretical, Paracelsic parts[13]. In its chapter on cosmology, Copernicus' heliocentric view is mentioned very briefly, with the addition that it is universally disapproved of, because it is incompatible with *Holy Scripture*. For the future, Forsius explicitly adopts the geocentric principle, and it seems as if he inclines to accept Tycho Brahe's system. He does not give any scientific arguments against Copernicus, and it may be regarded as characteristic that he, consequently, disregards the strong objections raised by Aristotelian physics, from which Forsius diverges in several passages of his work[14].

At this time, Sweden had its first — also for long the only — Copernican, Johannes Thomae Bureus (1568—1652). Bureus was a polyhistor with a very strange personality, ana antiquarian, a linguist and a religious mystic on a neo-Platonic basis with strong cabbalistic and Paracelsic elements. His work and thinking are entirely separated from the academic circles and their official doctrines, apart from the fact the he, in the 1590's, was matriculated at the University of Upsala, where he may

---

[12] *Prognosticon astrologicum... Till thet Åår... MDCX*, Stockholm, 1609.

[13] *Physica Eller Naturlighe tings Qualiteters och Egendomars Beskrijfuelse*, first published by J. Nordström, *Sigfridus Aronus Forsius, Physica* (Cod. Holm. D 76), Upsala, 1952.

[14] Forsius's natural philosophy has been studied especially by S. Lindroth, *Paracelsismen i Sverige till 1600-talets mitt*, Upsala, 1943 (Lychnos-Bibliotek, 7), pp. 391 et seq.

have heard the teaching of Laurentius Paulinus. Anyhow, his peculiar views are the products of his own studies and his own speculation[15].

Like so much connected with this man, the documents illustrating his astronomical views are not always unambiguous. But it seems that he entertained Copernican notions while still quite young. In his diary there are astronomical notes from 1601, where Copernicus' theory of the motion of the earth and the immobility of the sun is mentioned in a way indicating that it was not unfamiliar to Bureus himself. His printed publications are scarce, and his written heritage largely consists of miscellaneous notes, often somewhat chaotic, which makes it impossible to get a clear picture of his views on astronomical subjects[16]. It can hardly be regarded as remarkable that he used Copernicus' planetary Tables, because so did many astronomers who otherwise adhered to the old system. It is far more interesting that he, in his notes from the beginning of the 17th century, drew a model of the Copernican planetary system. Furthermore, one finds that at the same time he was occupied with extensive cosmological speculations, which were very close to a Copernican view but above all influenced by Nicolaus Cusanus, whose doctrine of a cosmos that is infinite and accordingly has no fixed centre — notions which were of the greatest importance for Giordano Bruno's cosmology — is obviously behind Bureus's statements that the centre of the earth is everywhere, i.e. in all its bodies, and its circumference nowhere; thus no fixed centre can be determined. As a logical consequence, Bureus maintains that all celestial bodies are mobile. Probably also influenced by Cusanus, Bureus arrived at a notion no less daring and heretical — it was also cultivated by Forsius — that all celestial bodies are inhabited by living creatures.

Thus it seems that, as early as this time, Bureus had not only arrived at a Copernican standpoint, but that he was also familiar with thoughts reaching much further. During his later years, when on the whole he had reached a more circumspect and consequent way of thinking, he expressed himself more clearly in these matters, above all in a printed natural-philosophy publication in 1641[17]. In that work, he very schematically presents his peculiar, biblically inspired conception of the origin and structure of the cosmos. The sun is at rest in the centre of the universe, and outside the earth revolves among the other planets, while the innumerable fixed stars are situated farthest out under the firmament. Thus, in this work Bureus stands out as essentially a Copernican.

Some words should also be said about another prominent polyhistor, Georg Stiernhielm (1598—1672), best-known as the greatest Swedish poet of the century but also a linguist, a natural scientist and a philosopher of very radical, anti-Aristo-

[15] Lindroth, *op. cit.*, pp. 82—252, contains a thorough monograph on Bureus and his opinions.

[16] These astronomical notes in his diary and his various manuscripts are dealt with by H. Sandblad, *Det copernikanska världssystemet i Sverige, I. Aristotelismens tidevarv*, [in] "Lychnos" 1943, pp. 156 et seq.

[17] *Hebraeorum philosophia antiquissima*, Upsaliae, 1641. Cf. Lindroth, *op. cit.*, pp. 218 et seq.

telian views on the neo-Platonic tradition. In many ways, he was a Burean scholar, but whether that applies to cosmology as well may be regarded as uncertain, although the speculations of the two in that domain have much in common. In his cosmological conception, Stiernhielm was particularly influenced by Giordano Bruno. Together with him, he embraces the idea of an infinite world-space, with the consequence that it is futile ot talk of a fixed centre of the universe; it is everywhere and nowhere. The earth is a planet, which on the other hand the sun is not, and all stars and planets are suns and earths in an infinite number of worlds. Thus, the firmament is not the heaven of our earth but belongs, says Stiernhielm, to every star or material body which is part of it, like another earth or another world; accordingly, this thesis means that from any other star the cosmos looks the same as from our earth. In fact, these notions include Copernicus' system but they reach much farther. To the predominant way of viewing nature at that time they had indeed a revolutionary significance. Stiernhielm never published them, however, and in fact his Swedish contemporaries would have regarded them as absolutely intolerable.

In their cosmological speculations, Bureus and Stiernhielm are isolated phenomena, very interesting on their own merits but insignificant for the future progress of the Copernican system in Sweden. One obvious reason for that is the fact that they were printed or otherwise propagated to such a small extent. The road along which the new picture of the world had to struggle, came up against conservative university science.

2.

During the first part of the 17th century, the State of Sweden carried through that political and military expansion, which for almost one hundred years made it one of the Great Powers of Europe. Simultaneous with this expansion, a magnificent cultural policy with a large-scale expansion of all higher education was started under King Gustav Adolphus. The University of Upsala was now given the organization and the material resources which enabled activities tolerably comparable to those of the larger continental centres of learning. By and by, new universities arose in different parts of the Baltic. Upsala remained the leading university, and the mathematical sciences received there three professorial chairs, the holder of that of astronomy having the significative title of "professor Ptolemaicus". In the general educational regulations, issued in 1626, Copernicus is mentioned among a number of other, mostly older, "authores probatissimi". It would be rash, however, to assume from that the Copernican system had met with approval in Swedish university learning, although, now as before, the students were introduced to it. The special instructions for the professorship of astronomy mention only three text-books, all based on the classical picture of the world: Sacrobosco's *Sphaera*, Peurbach's *Theoricae* and Brucaeus's *De motu primo*.

The military enterprises were also used for direct support of national cultural life. An extensive number of books taken as spoils of war on the continent were brought home, above all to Upsala. Thus it occurred in 1626 that the Swedish forces in East Prussia took those books and manuscripts by Copernicus that had been bequeathed to the Cathedral of Frombork. They are still kept in Swedish libraries and archives, mainly the Upsala University Library[18].

The astronomic standpoint dominating the learned life of Sweden during this era, is found at its best in the academic dissertation literature. The discussion about Copernicus and his system carried on there, will now be summarized. It can be said at once that in this literature one hardly finds any contribution of international significance. Not only does it lack true originality — which indeed it did not claim — it is also conspicuous that only occasionally does it contain any work showing a deeper and farther-reaching study of the cosmological questions of principle so intensely and vitally debated during this period. It can also be said that in the main the development in Sweden follows the general European and continental development, that only here and there does the Swedish material show any direct or close contact with the current exchange of opinions in the culturally progressive countries[19].

The first author we meet here in Upsalian literature is Martinus Erici Gestrinius (1594—1648). After studying at various German universities he became Professor of Mathematics in Upsala in 1621. He seems to have devoted quite as much time to astronomy, however, and his activities as a teacher for 27 years were of great importance for the introduction of the new astronomy into Swedish science. As one might expect, he adheres to the geocentric principle, but in many details he accepts the results of up-to-date studies, which is shown as early as 1622 in a thesis *De stellis*. A dissertation *De sole*, ventilated in 1632, is more interesting than any other under his patronage. The respondent, and perhaps the author, was Simon Kexlerus, later to become the first Professor of Mathematics at the University of Turku, Finland, founded in 1640. The author sets out to deal with the annual motion referring to the sun, of which many hypotheses — three in particular – had been proposed. Of those he wants to confine himself to "observationes eruditissimi viri & excellentissimi Astronomi Nicolai Copernici". He at once dismisses Copernicus' theories of the earth's multitudinous motion and the immobility of the sun in the centre of the universe; without further argumentation he declares that they are too remote from the physical reality and completely absurd. However, he says, they have led to important observations worth of closer examination and explanation, especially concerning certain irregularities of the motion of the sun. The author wants to explain them in a way that makes Copernicus' theories of the motions of the earth superfluous. He does it by means of complicated epicycle and deferent theory, in

[18] The Upsala University Library now owns three copies of the original edition of *De revolutionibus*. One of them, having Georg Joachim Rheticus's personal autograph, was captured at the same time together with the other books of the Jesuit College in Braniewo.

[19] A detailed investigation fully documenting the discussion about Copernicus in the Swedish academic dissertations of this period is made by Sandblad, *op. cit.*, pp. 160—188.

the main coinciding with the Ptolemaic explanation of the irregularities of the planetary motions. To this treatment of certain special problems in the Copernican system some corollaries have been added, in which the author states his general attitude to Copernicus' fundamental ideas. The questions as to whether the sun is at rest in the centre or is moving as the central one of the planets in the ethereal region, both receive the answer that both alternatives are astronomically possible, but only the latter physically. The question as to whether the earth, situated in or outside the centre, is mobile or immobile, is answered thus: if it is situated outside the centre of the universe it is mobile, but if it is situated in the centre, both alternatives are possible.

Kexlerus's dissertation, undoubtedly bearing the impress of the opinion of his teacher, Martinus Gestrinius, is the first Swedish work in which Copernican lines of thought are subjected to any real discussion. The work evinces a thorough study of *De revolutionibus* and a marked admiration of Copernicus' adhievements. The author's attitude to the Copernican system seems on the whole benevolent; in rejecting it, he gives physical reasons only, whose content is, however, not even suggested.

In a hand-written astronomical text-book by Gestrinius[20] of the same year, 1632, the question of the place of the earth in the universe is treated somewhat more exhaustively. For three different reasons, all of them purely astronomical, the author assumes that the earth is immobile in the centre of the universe: (1) of the signs of the zodiac six are always above and six below the horizon; (2) the size, grouping and relative distance of the fixed stars appear identical everywhere on earth; (3) everywhere, the time from sunrise to noon is the same as from noon to sunset. With these arguments, some of the most common ones in the more popular criticisms of Copernicus, the author rests content. One special chapter shortly relates the most importan- hypotheses that have been given through the ages, of the structure of the universe and the planetary motions. Of course, the Copernican system is also described; it is observed that its fundamental idea had been presented already by Aristarchus, long before the birth of Christ. Gestrinius does not take a stand for or against any hypotheses; he only points out that hitherto the Ptolemaic one has been followed "in the schools" – i.e., university education because, knowing that, anybody may easily initiate himself into the others on his own. This statement is probably meant to justify the fact that he made the Ptolemaic hypotheses the basis of his own work, which was presumably meant for junior university studies. The whole of this exposition of the different systems later re-appears in Gestrinius' printed text-book *Urania*[21].

Judging by his extant works, this is what was written about the Copernican system by Martinus Gastrinius, the foremost representative of the mathematical sciences in Upsala and Sweden during the time up to the middle of the century. It is true

---

[20] MS No. A 502: 10, Upsala University Library (untitled).
[21] *Uraniae libri IV*, Upsaliae, 1647.

that he rejects the heliocentric principle, but he has not made any detailed refutation, and on the whole his position seems to be cautious and rather divergent. As a teacher of astronomy he undoubtedly influenced the following generation considerably.

Several contemporaries of Martinus Gestrinius in Upsala briefly touched on different occasions upon the Copernican system. One of them is Johannes Franck (1590–1661), Professor of Medicine and in certain respects influenced by Paracelsus, though as a natural-philosopher he was foundamentally an Aristotelian. He devoted several dissertations to astronomical and physical subjects, and one of them, *De calore solis* in 1625, was earlier regarded as indicating strong Copernican sympathies. That, however, was due to a misunderstanding of a polemic passage, where Franck talks about the famous astronomer Nicolaus Copernicus, who thinks that the earth is moving, while the sun is immobile, a view nowadays zealously "a quam plurimis" defended as correct. That statement did not imply any direct support of Copernicus, although Franck greatly admired him and was obviously interested in his system, whose success abroad he used as an argument in that part of the dissertation where the question of the heat of the sun was dialectically treated. In later dissertations he also stands out as a spokesman of a conservative cosmological conception. In one of them, *De orbium coelestium realitate*, 1627, the question is discussed whether the different planetary spheres really exist, as classical physics taught, or whether these spheres, in accordance with the view common among more modern astronomers, are purely fictions, heaven being one great etheral space, where the planets of their own force run in their set orbits, "like fish in the water and birds in the air". Strictly formally, Franck defends the former standopint, but it is obvious that this is only a dialectical gambit, and that he is actually espousing the opposite opinion. The most important thing, however, is that Franck's discussion of the question is wholly and entirely based upon the geocentric standpoint. That is also clearly shown in a corollary concerning the question as to whether the earth is moving in a circle, as Copernicus and his supporters allege, and heaven is immobile instead. This question is answered in the negative. It seems clear that Franck's principal conception is geocentric – perhaps rather concurring with the of Tycho Brahe, whom he calls "Astronomorum Aquila & Atlas".

Franck's statement that at this time the Copernican system was widely spread abroad, should be interpreted with a certain caution. But no doubt it had established its ground at several German universities, even if it was not always taught publicly; and there it could influence the numerous Swedish students still going to the continent. One example of that is a philosophical dissertation[22] presented in 1626 in Greifswald by the Swede Olaus Moraeus. In one of the corollaries the respondent agrees to defend the opinion maintained by Copernicus, Kepler, Origanus and others, that heaven is at rest while the earth is moving with the sun as a centre.

---

[22] *De ente eiusdem conceptu*, praes. Johannes Segerus.

Such bold theses, however, were not vindicated by those students staying in Sweden. Nor are there, in the Upsalian literature of the next decades, any great number of passages mentioning Copernicus' name or even discussing the cosmological matters of principle. In several passages, Copernicus' system is mentioned very briefly and dismissed equally briefly with theses out of Aristotelian physics, sometimes with an acknowledgment of those modifications caused by Galileo's sensational telescopic discoveries. One of the most interesting and able contributions to this literature is a dissertation *De sole*, ventilated in 1649 for the Professor of Optics and Mechanics, Martinus Olai Nycopensis (1596–1657), who probably was the principal author himself. Having given an account of the Ptolemaic system, the author points out that it was entirely changed by Copernicus and the Copernicans, who put the sun, immobile, in the place of the earth, while the earth takes that of the sun and is subject to a threefold motion: the diurnal, the annual and that of libration. This view, it is said, is old enough, because even Philolaus and the Pythagoreans espoused it. Thus, here is a reference to the Pythagorean doctrine of the central fire of the universe, around which the earth and the other celestial bodies move. In later times, particularly during the discussion of the Copernican system, this central fire was incorrectly interpreted as identical with the sun – i. e., a misconception re-appearing in Martinus Nycopensis' dissertation. Finally, the author sums up the system formed by Tycho Brahe, with which, he says, most modern astronomers agree. To him that hypothesis seems the most correct. As expressed by Longomontanus, the reason is that it is some kind of a middle course between the old Ptolemaic and the admirable Copernican hypothesis: on the one hand it excludes the complicated epicycle theory, used by Ptolemy in an attempt to explain the irregularities of the annual planetary orbits, and on the other it excludes Copernicus' absurd theory of the annual motion of the earth. Consequently, Tycho's hypothesis is the best and most expedient way of explaining the symmetry of the cosmos.

In this quotation from Longomontanus – i. e. Christen Sørensen of Lomborg, the great Danish mathematician and astronomer, a disciple and assistent of Tycho – the criticism of Copernicus concerns only the annual motion. In fact, Longomontanus deviated from Tycho's system in so far that he accepted the Copernican theory of the diurnal motion of the earth around its axis. It is not clear whether the Upsalian author agrees with this so-called semi-Tychonic standpoint, given final shape by Longomontanus in his principal work *Astronomia Danica* (1622), from which the quotation is taken[23]. Anyhow, here as in other dissertations by Martinus Nycopensis, one finds that he principally accepts Tycho's system. On the whole, he is a typical representative of numerous scientists all over Europe at this time; they all found the Ptolemaic system untenable, but at the same time they were incapable of accepting a heliocentric conception and therefore they espoused the Tychonic or at most the semi-Tychonic.

---

[23] *Astronomia Danica*, Amsterdami, 1622, *Pars posterior*, lib. I, cap. I, p. 21.

Clearly, those supporters of the geocentric principle so far mentioned have not to any major extent given the motives for their stand, and the astronomical arguments against Copernicus are very meagre. The question presenting itself is: what was the fundamental cause of the opposition so commonly offered to Copernicus by the representatives of academic learning?

The answer seems evident. The conclusive arguments came from Aristotelian natural philosophy which almost totally leavened university culture during this period, in Sweden as in most places in Europe. As long as that dominating position remained unshaken, the obstacles to Copernican system were insurmountable. There is no doubt that these traditional scholastic physics lie behind the criticisms of the new doctrines even in cases where no real attempts at refutation are made.

In the year 1625, a teacher of philosophy in Upsala, Aeschillus Petraeus (1593–1657), later Professor of Theology and bishop of Turku, presented a Collegium physicum in a series of dissertations. In one of them[24] he enters upon the question of the motion of the earth, the possibility of which he denies. His polemics are mainly directed against the Pythagoreans but also against Copernicus and William Gilbert, who deduces evidence of the rotation of the earth out of his pioneering research concerning terrestrial magnetism[25]. Petraeus refers to biblical passages: in general, that occurred very rarely during this early phase of the Swedish discussion of Copernicus. He also cites one of the astronomical arguments formerly encountered in Martinus Gestrinius. These are only passing objections, however, and the author adds that strictly astronomically the motion of the earth is possible as a hypothesis to explain the phenomena, but from a physical point of view it cannot be treated as real. We have met this opinion already, though without any further indication of the physical arguments. They are given here, however, and to Petraeus himself the physical arguments are vital in repudiating the Copernican system. He thoroughly relates the Aristotelian doctrine of elements and mechanics, according to which the motion natural to the elements is rectilinear; of necessity, that makes the earth immobile in the centre of the universe. Along with this natural motion a motus violentus can occur, too, but since such a motion lately decreases, that could not be the one in question either.

These Aristotelian mechanics are the basis of the most thorough settlements of account with the Copernican system occurring in Swedish scientific literature during this period. Not unexpectedly, these contributions come from the representatives of philosophy in Upsala. Thus, a dissertation *De terra* was ventilated in 1632 under the presidency of the professor of logic Laurentius Stigzelius (1598–1676), one of the most skilful and devoted representatives of Aristotelianism in Sweden. There, the question of the position of the earth and the questions of its alleged motions are

---

[24] *De aqua et terra.*

[25] *De magnete, magneticisque corporibus, et de magno magnete tellure, Physiologia noua*, Londini, 1600, lib. VI.

exhaustively treated, with reference to various Aristotelian authorities, mainly the famous Aristoteles commentary complied by the Jesuit College of Coimbra[26]. The author's opinion that the earth owing to its nature is immobile in the centre of the universe, can be confirmed, he says, by several biblical passages but above all physically proved. Making the concepts and definitions of Aristotelian mechanics his starting-point, he maintains that if the earth is not immobile its motion must be either natural or violent or possibly praeternaturalis. If it is natural, it has to be either rectilinear or circular. In the same way as Petraeus, but more exhaustively, he shows, strictly logically, that a terrestrial motion can not be natural in either way. Nor can it be violent, because such a motion can not be permanent. Finally, one can not think it praeternaturalis either, because then the earth would depend on another body, which can not be proved. Thus, from these and other "firmissimis argumentis" it can be concluded that the earth is immobile in the centre of the universe. Therefore, no other cause is recognized here than that referring to the nature of the earth, inspired by God at the beginning of the world. Besides, it is said, much that is obviously false is used in astronomical hypotheses to explain phenomena more conveniently; but it simply can not be accepted as physically real, which even the famous astronomer Nicolaus Copernicus seems to want to allege as an excuse in his preface. Thus, here as so often before, Osiander's anonymous preface of *De revolutionibus* is used as an argument against Copernicus.

None of this detailed confutation of the fundamental thoughts of the Copernican system is in the least original. On the contrary, it is highly typical of the criticism emanating from traditional Aristotelian philosophy. It is precisely because of that, that its arguments out of traditional physics are of such great interest: they are identical with the arguments that were repeated again and again all over Europe at this time, and for a long time yet preserved the dominance of the geocentric picture of the universe.

Stigzelius soon became a Professor of Theology, and finally archbishop of Upsala. His successor in the professorship of logic, Olaus Unonius (1602–66), was not as prominent; however, he was also a firm Aristotelian of a battling spirit. Among his unusually numerous dissertations from the 1640's, several are of interest in this context. The reason for that is the vehement polemics they direct, mainly with the same physical arguments, against the doctrine that the earth is moving, a doctrine espoused by ancient philosophers and roused out of its ashes by Nicolaus Copernicus of Toruń, giving him many echoers "conjurati in verba Magistri". In a dissertation from 1663[27], passages of the *Holy Scripture* are quoted along with the old, perpetually repeated physical arguments and to some extent the astronomical ones. With greater reliability than any mathematical evidence, says the author, these passages maintain that the sun and the other stars are moving, while the earth is immobile. The author

---

[26] *Commentarii Collegii Conimbricensis Societatis Jesu in IV. libros de coelo Aristotelis Stagi ritae*, Lugduni, 1594, especially lib. II, cap. XIV (pp. 363 et seq.)

[27] *Novem themata philosophica*, esp. them. VIII and IX.

also strongly emphasizes that in these matters as well as others, the *Bible* must be interpreted literally.

This ardent defence of the authority of the *Bible* in cosmological matters merits the greatest interest: for from the whole nature and wording of the argumentation it seems obvious that it is directed against the Cartesian philosophy, which at this time had started to gain ground in different quarters of Europe and through which the Copernican notions rapidly established themselves. Since the foundation of the scholastic conception of nature has started tottering before the Cartesian attack, biblical arguments, as we shall soon see, were more than ever used against the heliocentric conception; they remained as the last resource of the traditional picture of the universe. Thus, at the same time, and mainly, the battle of Copernicus' system became a battle of the jurisdiction of the letter of the *Bible* in natural science.

We have seen what stand the repesentatives of philosophy in Upsala adopted towards the Copernican system. Nor is any positive attitude to be seen in those of the mathematical sciences even in the 1660's. As late as 1669 a dissertation on Theoria Maris was ventilated before the Professor of Astronomy, Jonas Fornelius; there, an account of the three great world systems is given; in doing so the author does not explicitly take a stand for either of them, but the whole exposition starts from the geocentric principle. It can also be noted that Kepler is mentioned among the followers of Copernicus, but this dissertation in no way touches upon has investigation of the motion of Mars which formed the basis of his first two laws and which was indeed published a full 60 years earlier.

Consequently, it can be established that not one true follower of Copernicus had appeared in academic Upsala, the scientific centre of Sweden, as late as the end of the 1660's; on the contrary, from certain quarters Copernicus' doctrines were now as vehemently attacked as ever. But during the following decade new, refreshing winds finally began to force their way into the lecture-rooms of scholasticism.

Before we take a look at that process, the conditions at the other Swedish seats of learning will be somewhat touched upon. After Upsala, the University of Finnish Turku was the most important; it was highly active soon after its foundation in 1640. In the disserations from its first decades one finds several contributions to cosmological matters of principle. Constantly a geocentric conception is maintained, though supported mainly by the traditional arguments from Aristotelian physics and the biblical passages that were generally used against Copernicus. An astronomical text-book is of greater interest. It was written by Andreas Thuronius (1632–65), Professor of Philosophy and later of Physics as well, during the early 1660's[28]. Thuronius cannot accept Copernicus' doctrine of the annual motion, because it is incompatible with the *Bible*, but he does accept the diurnal motion and unequivocally adopts the semi-Tychonic system. In doing so, Thuronius took a very advanced position in contemporary Swedish learning.

---

[28] MS No. A 301, Upsala University Library (in Swedish, untitled).

Under the presidency of the first Professor of Mathematical sciences in Turku, the already mentioned Simon Kexlerus, a dissertation was ventilated in 1661; this is the most important Swedish contribution to the criticism of Copernicanism so far. The author is Petrus Laurbecchius (1628–1705), a teacher of philosophy and later a professor and bishop. It is entitled *De circuli quadratura et vero mundi systemate, adversus Copernicum Redivivum* and is especially directed against the famous learned German Daniel Lipstorpius, who had appeared as a devoted follower of Descartes' natural philosophy; in his work *Copernicus Redivivus, seu de vero mundo systemate*, 1653, he zealously advocated the Copernican system in its Cartesian form. Laurbecchius's dissertation is the first sign that the Swedish university world observed the threat of Cartesian philosophy, under the cloak of which Copernicanism sought to intrude. Laurbecchius himself espouses Tycho Brahe's system, asserting its superiority to the Copernican one with both astronomical and physical arguments in a very penetrating criticism of Lipstorpius, evincing both erudition and dialectical keenness. The details must be passed over here; it can hardly be said, however, that the Aristotelianly trained Laurbecchius does justice to the argumentation of his adversary in his polemics. It is not possible for him to understand the Cartesian physics on which Lipstorpius bases his lines of thought; these two men speak totally different languages.

How strong was the position of the geocentric system in Turku even in the 1670's mad be seen from the astronomical dissertations they presented: they are all entirely traditional as regards cosmological matters of principle. Accordingly, as late as that decade the Copernican system had secured no follower in Turku, anyhow none appearing publicly.

The University of Lund came into being in 1668, to meet the neads of the Southern provinces of Sweden, captured from Denmark ten years earlier. In its first, rather miscellaneous collection of teachers the holder of the Professorship of Astronomy, Anders Spole (1630–99), was one of the best, and probably the most skilful astronomer Sweden had so far. He took his chair immediately after a long stay abroad, during which he studied in Holland, England and Italy, but above all for many years in Paris. There he could not avoid getting close contact with the current vivid discussion of Cartesian natural philosophy, which did not, however, leave any deeper impress on his output. Instead, his great astronomical authority was the Jesuit anti-Copernican Giovanni Battista Riccoli whom he met in Bologna; in his extensive work *Almagestum novum*[29] Riccoli presented his own system which was a simplification of Tycho Brahe's. This dependence is clearly displayed in the only dissertation by Spole in Lund worth mentioning in this connection – i. e., *De terra* in 1674, which as a matter of fact shows an insight into recent astronomical literature that is unusual for Swedish circumstances.

---

[29] *Almagestum novum astronomiam veterem novamque complectens*, Bononiae, 1651.

The activities at the University of Lund were soon interrupted by a new war and a Danish invasion, and shortly afterwards Spole received a call to the astronomical chair in Upsala. And immediately after his arrival there, he was drawn into a dramatic process which marks the beginning of a new era in Swedish natural science research.

3.

As is well known, Descartes himself visited Sweden, having received a summons to the court of Queen Christina in Stockholm where he died in 1650. That visit did not, however, leave any clear traces of his doctrines. The earliest definite influence of Cartesian natural philosophy appears in the beginning of the 1660's when it turns out thait it has ganed a firm footing in the faculty of medicine in Upsala. At this time, it consistedi of two brilliant and young men, Petrus Hoffvenius (1630—82) and Olaus Rudbeck (1630—1702); the latter is the most famous, although in this case he kept more in the background[30]. They were both eminent anatomists and had been studying in Holland, the second home country of Descartes and his doctrines, and there they were strongly impressed by the new way of thinking. Hoffvenius's public spread of Cartesian opinions soon caused violent conflicts with orthodox theology, which regards this new-fashionedness as a threat to the true faith[31]. At the end of the decade, the contentions temporarily ceased, but the Cartesian notions were still spread, mainly by Hoffvenius. In the 1760's, he taught Cartesian physics for a succession of years to students of all faculties, and during that period he wrote a number of small dissertations meant for exercise concerning various natural-philosophical problems, which he collected and published in 1678 under the title of *Synopsis physica.* It is a mystery that this work managed to escape the theological censorship, because it is entirely a text-book of Cartesian physics, though discreetly enough the name of Descartes is not mentioned.

This work, essentially based on works by the eminent German Cartesian Johann Clauberg, became fundamental to education in the natural sciences in Upsala for a long time to come. There, Hoffvenius briefly and very instructively presents the new celestial mechanics, radically different from the Aristotelian one and to which Descartes gave final shape in Principia philosophiae[32], with its fundamental principles

---

[30] That nevertheless Rudbeck was a convinced Copernican seems obvious from a remarkable treatise on comets which he wrote in 1665 as a letter to Nicolaus Heinsius, the Dutch humanist. It was sent by Heinsius to Stanislaus Lubienietzki, the learned Polish exile, and published in his *Theatrum cometicum*, Tom. I, Amsterodami, 1668; see esp. pp. 355—359.

[31] These and later conflicts about Cartesianism are related in detail by R. Lindborg, *Descartes i Uppsala*, Upsala, 1965 (Lychnos-Bibliotek, 22); their special astronomic aspects by H. Sandblad, *Det copernikanska världssystemet i Sverige*, II. *Cartesianismen och genombrottet*, [in] "Lychnos" 1944/1945, pp. 79 et seq.

[32] *Principia philosophiae*, Amsterodami, 1644, pars III.

of matter and motion and its strange vortex theory. The planetary doctrine based on this theory of the motions of universal matter, Hoffvenius presents in a special dissertation (No. X) among those included in the *Synopsis*, entitled *De Planetis eorumque Phenomenon causis*. The planets all come into existence in the same way and they constitute dark bodies which obtain their light from the sun, in the vortex of which they are carried around. Among them only two kinds can be distinguished, however: simplices seu primarii and compositii seu secundarii. Except for the motion around their own axis, the former kind have a single motion together with the vortex into which they are drawn; they are Mercury, Venus, Tellus—i. e., our earth—Mars, Jupiter and Saturn. The latter kind perform several whirling motions, on one hand their own and on the other those related to the bigger vortex absorbing them; this is true of the moon which was first drawn into the vortex of the earth and then along with the earth into the big vortex of the sun. For that reason, it follows the earth in the diurnal motion of its axis, and at the same time it is brought along around the sun in the annual motion. The same is true of the four Jupiter satellites: first they were drawn into the vortex of Jupiter and then with Jupiter into that of the sun. The single planetary motions in the vortex of the sun are faster or slower the nearer or the farther they are from the sun. The distance of a planet is not only due to its size but also to its solidity; the bigger and more solid the planet is, the greater force it must possess to remove itself from the centre, according to the law of nature. Accordingly, that is what decides the fixed relative order, which of old it has been possible to observe within our heavens[33].

Thus, in the greater vortex including our planetary system, the earth is carried around the sun; and as it is always situated between the same parts of the celestial matter it can in fact be considered immobile, "as a ship, which is not driven by wind or oars and is not stopped by anchors, is at rest at sea, even if the body of water carries it along in a current"[34]. The latter argumentation is based on the Cartesian definition of motion: motion is the moving of a body from the vicinity of the bodies being in its immediate proximity to the vicinity of other bodies[35]. In this way Descartes, of course, was able formally to adhere to the biblical doctrine of the immobility of the earth, at the same time as he, in the main, accepts the principal idea of Copernican cosmology, even though from natural-philosophical starting points quite different from those of earlier Copernicanism. And in this form, through Hoffvenius, Copernicus is taught for the first time in the academic education of Upsala, forcing his way into Swedish scientific literature.

One of the students taking part in the exercise, in which was discussed the above-mentioned dissertation by Hoffvenius on the planetary system, was Nils Celsius (1658—1724), not yet 20 years old and the son of the Professor of Mathematics,

---

[33] Cf. ibidem, p. 147 et seq.

[34] Ibidem, p. 26.

[35] *Principia philosophiae*, pars II, p. 25.

Magnus Celsius[36]. That young man would soon make an important and sensational contribution to the history of Swedish Copernicanism.

Having moved to the astronomical chair in Upsala in 1679, Anders Spole was shortly afterwards approached by young Celsius who wanted to defend a dissertation *De principiis astronomis propriis* under his presidency. With his customary cautiousness, Spole consented, after making Celsius modify some excessively bold expressions. But his cautiousness did not suffice. As soon as the dissertation was published, it turned out that it contained several statements that were intolerable from an orthodox point of view. The faculty of theology immediately sounded the alarm, and thus, in the cathedral itself, was started one of the fortunately few heresy trials in the Swedish history of learning. Celsius escaped with serious admonitions from his theological guardians, but the public discussion had to be cancelled and further distribution of the dissertation was forbidden[37].

No doubt, the Copernican sympathies clearly shown by Celsius in his dissertation are mainly a result of his participation in Hoffvenius's education in natural-philosophy. But another influence also, is traceable — i. e., that of his father, Magnus Celsius, who died earlier the same year and who obviously approved of the dissertation, intending to preside over it himself. Magnus appears to have been strongly attracted by the Copernican system, although he did not find it possible to declare his opinion publicly.

The anger of the theologians at Nils Celsius's dissertation was not caused principally by the special question of the motion of the earth, but by the fact that it vindicated certain leading natural science questions of the greatest consequence. Celsius starts by pointing out that it is the experience obtained through observations that must be the foundation of a science devoted to exploring celestial phenomena. Without observations, nothing of value can be performed in astronomy. In its efforts to acquire true knowledge of nature, astronomy is hampered by various kinds of prejudices: on the one hand the preconceived opinions of the masses, often based on habitual ways of thinking without regard to the experience to be obtained from the evidence of the senses — and even that evidence can be deceptive, unless verified by careful experiments. Another kind of prejudice to be fought by the natural scientist are those wrongly understood and cunningly adjusted passages of the *Holy Scripture*, which "spread a cloak of alleged piety over the truth".

The latter question, the validity of the word of Scripture in natural science, Celsius discusses thoroughly, obviously in order to forestall objections anticipated as regards his cosmological conception. In doing so he uses statements by great teachers of the Church, mainly Augustine, who sharply denounces those who allow matters

[36] The Celsius family is one of the most prominent dynasties in the Swedish history of science. Nils Celsius became the father of Anders Celsius (1701—44), the constructor of the centograde thermometer, Professor of Astronomy in Upsala.

[37] This incident is related in detail by Sandblad, *op. cit.*, pp. 84—91, with complementary additions by Lindborg, *op. cit.*, pp. 183—187.

of nature and things celestial to be decided by faith, contrary to reason and experience. When deciding matters of natural science, Celsius continues, one should use the biblical texts with discretion. Because it is obvious, as can be pointed out by ecclesiastical authorities, Jerome in particular, that in matters not concerning salvation itself, the sacred writers use expressions adapted to the view generally held by the Jewish people at the time when the biblical scriptures came into being; they were not meant to express the exact natural reality. Therefore, correctly understood and interpreted, the *Bible* by no means necessarily comes into collision with natural science.

These Celsian arguments against the biblical letter-worship were no new in themselves: earlier, often with a reference to the same ecclesiastical authorities, they had been used by Galileo and many subsequent supporters of the new, empirical science. They had been used especially by the Cartesians, when they wanted to prove to their adversaries the compatibility of their way of thinking with the *Bible*. Thus for instance, Daniel Lipstorpius in his *Copernicus Redivivus* referred to above. But in Sweden these views are openly presented here for the first time in the dissertation of the young Celsius.

A Cartesian strain is also discernible in Celsius' statement that we must not quite literally believe the evidence of our senses; only a well-conceived and well-performed experiment can give the corret answer to the questions we ask nature. Furthermore, he points out the importance of technical means, simpler or more complicated, and the laborious work required in their handling if the results of experiments, are to be reliable. By way of examples of this he mentions the rich instrumental, experiences obtained recently by practical astronomy through Johannes Hevelius the selenographer, G. D. Cassini, Robert Hooke and others.

In his very skilfully and logically composed exposition, Celsius from this point enters on those hypotheses of the celestial processes that can be proposed on the basis of phenomena observed by the means of instruments. On these hypotheses are based those general theories of the structure of the universe, of which mainly three have been modelled in the course of the ages. As the first of these Celsius mentions the "Philolaic", and discreetly he altogether avoids mentioning the name of Copernicus. Thus he links himself with the above-mentioned, old conception established and mainly spread by the French Copernican Ismaël Boulliau's large work *Astronomia Philolaica*[38], that Philolaus – and the Pythagoreans on the whole – espoused a heliocentric system. Thus it could be maintained that this system – i. e. in reality the Copernican – had in fact the oldest tradition, the authority of antiquity, on its side. – The three systems have, says Celsius, many common features and each of them has a great deal in its favour. But they cannot all be valid; one of them must be nearer the truth, although indeed several can be usable and possess

[38] Ismaël Bullialdus, *Astronomia Philolaica. Opus novum, in quo motus Planetarum per nouam ac veram Hypothesim demonstrantur*, Parisiis, 1645. Earlier though anonymously, Boulliau had presented his opinion in *Philolai sive Dissertationum de vero systemate mundi*, Amstelodami, 1639.

a truth which may by hypothetical though not absolute. The actual astronomical content of the different systems and the differences between them are not touched upon by Celsius; consequently, the motion of the earth is not discussed or even explicitly mentioned.

In the final point of the dissertation, Celsius finally declares his position in the most cautious words imaginable. It should be allowable, he says, to use any system that can present a calculation corresponding in any way to observations and phenomena. Astronomers believe that system should be accepted as the best, however, which proves best adapted to the carrying out of the calculations — i. e., that which by the brevity of its wording is more advantageous than the others — which has a disposition that makes it simpler and easier to learn and use, which more accurately presents the phenomena and indicates their causes and, finally, that which is based on more reliable evidence. In these respects, the Philolaic system seems to take priority over other systems hitherto put forward. Consequently, all the most renowned and prominent contemporary astronomers espouse that system; they may not consider it an absolutely true system but they do consider it the most advantageous, the most convenient and that which in practice is the most applicable, while being at least hypothetically true.

Such modest statements, then, were not allowed to be delivered and discussed in Upsala 136 years after *De revolutionibus orbium coelestium*. So far, theologians could restrain the new opinions through their external means of enforcing their will, by suppressing the printing of the dissertation. What they wanted to defend was above all, as the hearing in the cathedral shows, not only the jurisdiction of the *Bible* in natural science, but also their own exclusive right to interpret the *Bible*. Naturally enough, the special question of the validity of the heliocentric system was overshadowed by this principal controversy, which had to be solved before a new conception of nature could come through in astronomy as well as in other fields. It should be sufficiently clear from the foregoing that Nils Celsius' attack is linked with the emergent Cartesianism and this whole episode shows how greatly the fate of Copernicanism is associated with the general development of the history of ideas. The triumph of the Copernican system could not become possible without a defeat for the allied forces constituted by scholastic physics and orthodox *Bible* faith.

It would soon be evident that the opinions represented by Celsius, and Cartesian philosophy as a whole, could not, once they gained a footing at the university and in the intellectual life of Sweden, be prevented from further circulation by any censorship. From the material extant it cannot be established in detail how the new thoughts were spread, but it is certain that it happened very rapidly. Because when the struggle of Cartesianism blazed up afresh in the middle of the 1680's, it turned out that, as the faculty of medicine had done earlier, almost the whole faculty of arts and science in Upsala took sides with the new way of thinking. By now Aristotelian philosophy outlived itself, and on the whole the theologians had to restrict themselves to trying to vindicate the traditional supremacy of the biblical word in science. But even that was more and more difficult for them, and in its Cartesian form the Copernican world system now rapidly entered the different fields of learning.

Leading Cartesians during this period were Johan Bilberg (1646–1717), a Professor of Mathematics and also a learned philosopher and theologian, and Andreas Drossander (1648–96), successor of Hoffvenius in a medical chair and a pioneer in experimental physics in Upsala; both were very prominent teachers and skilful in polemics. Both also had been studying for many years abroad. Bilberg is the most renowned of them; he became the foremost apostle of Cartesianism in Sweden, and through a fighting spirit and a talented presence he incurred the genius hatred of his adversaries.

The work that can be considered the principal document of Swedish Copernicanism comes from the circle of Bilberg and Drossander. It is written in Swedish in a rather popular, unacademic style, entitled *En diskurs mellan de opponerande Ptolemaei och de responderande Cartesii adhaerenter angående Systema Mundi.* It was printed for the first time as a kind of serial story in Nils Celsius's almanacs 1722–24, but it was obviously written in the 1680's and preserved in a great number of handwritten copies showing that at an early date it was widely spread, without being hampered by the censure which hung over printed publications. No doubt, it is in this very form, in which the heliocentric system is presented here, and with the arguments used here, that Copernicanism – if that term may be used – definitely broke through in Upsala towards the end of the 17th century and after that elsewhere in Sweden as well. The question of the authorship of the Discourse is somewhat uncertain. At any rate it is obvious that it came into being among leading Cartesians, and it was probably prepared in some kind of co-operation, with Bilberg as the real author[39].

The work has the form of a kind of dialogue, where the supporters of the geocentric world picture direct their objections against the heliocentric system and are answered point by point, mainly with Cartesian arguments. All this exchange of opinions is very clearly and instructively presented, giving an excellent insight into the cosmological discussion at that time, demonstrating the arguments used and the grounds on which it was held.

At first, Bilberg – if we may regard him as the author – touches upon the different obstacles, apart the poor reasoning power of man, raised against man's correct investigation of nature: on the one hand the prejudice ingrained since childhood; on the other, a proneness to remember false opinions early acquired better than any later knowledge; furthermore, the false names and terms that are used but which do not agree with the matter itself (this apparently refers to scholastic science); and finally that some people do not take sufficient trouble to examine nature but rest content with what the external sences indicate to begin with. Thus, many dis-

[39] Extant manuscript copies in various libraries are recorded by Sandblad, *op. cit.*, p. 93, where the problem of the authorship is also examined.

putes have arisen among scientists, the dispute concerning the motion or orbit of the sun not being the smallest. In fact, Nils Celsius had introduced his thesis in 1679 with similar views, and they were not uncommon in those representing a modern natural science. And the model is the introduction of *Principia philosophiae*, where Descartes gives an account of the epistemological foundations of his way of thinking, a presentation re-appearing in Hoffvenius's *Synopsis*[40].

In the Discourse itself, Bilberg first of all brings up the biblically based criticism of the heliocentric system, and he does that from the principal standpoint already declared by Celsius. Starting from a number of special examples, he shows the untenability of the practice of following the *Bible* word for word in matters not concerning Christian faith itself, and especially in matters of natural science. In doing so, he points out that in certain cases the Bible demonstrably cannot mean anything but approximate information, and furthermore it is obvious that in the *Holy Scripture* God uses expressions adapted to the weak and insufficient intelligence of man, which demands a certain concretion; that does not mean that they correspond to the actual state of things. Next, Bilberg discusses some of the biblical passages, preferably those used against Copernicus by supporters of the old picture of the world, showing the obvious absurdities resulting from a word for word interpretation. For instance the story in the *Book of Joshua* 10: 12—13 that is perpetually used in anti-Copernican polemics; furthermore, the word "absurdity" applies to passages likewise frequently used where it is said that the Lord fixed the earth so that she neither totters, nor moves; here Bilberg, along with the Cartesian motion concept, can maintain that the earth is not moving when it is carried in its vortex around the sun, since it is never moved from those bodies in the ether that are most near to it.

Having thus met the biblically based criticism, Bilberg brings up some objections of an astronomical character made by the Ptolemaics, refuting them by means of Cartesian mechanics. Then he examines Tycho Brahe's system, finding it even more untenable than the Ptolemaic from the standpoint of vortex mechanics. He goes on to give reasons which, in his opinion, are those especially favourable to the third, heliocentric system — which he calls the Cartesian — emphasizing that it is an improvement on Copernicus' system. Having stressed that in worldly matters one should not assume anything contrary to divinely inspired reason, he points out that even with superior instruments it is impossible to estimate the distance of the fixed stars; from the parallax of Saturn, which is small or non-existent, one can conclude, however, that this distance is extremely great. For that reason, the universe has a circumference so immensely great that the fixed star heavens, performing the kind of circular motion supposed in the geocentric system, would move with a velocity so enormous that it is incompatible with our reason, which demands a plausible relation between the time and the distance covered[41]. Furthermore one can in-

[40] *Principia philosophiae*, I: 1 ff.; *Synopsis physica*, I.

[41] This argument was not unusual in Copernican literature: see e. g. Lipstorpius, *Copernicus Redivivus*, Lugduni Batavorum 1653, p. 85.

deed, despite the fact that the fixed stars are so enormously distant, observe that they have a gleaming light which is more vivid than that of the less distant planets and never darkened. From that Bilberg maintains, it can be assumed, still in accordance with Cartesian physics, that they have their own light and like the sun consist of a "fire-flowing" matter, which could not possibly retain its round shape if it moved with the fantastic velocity that the old system would imply.

Bilberg also points out that the Cartesian system gives a simple explanation of the irregularities of the velocity of the planets, which forced Ptolemy to take refuge in his complicated epicycle and deferent theory. He maintains that those irregularities are only apparent, illustrating his very popular argumentation with a typically Cartesian image. When you travel by boat, seeing another boat right in front of you, it appears not to move; if your own boat is going twice or three times as fast, so that you not only overtake it but even pass it, it appears to be going backwards, because you are looking more at the boat that is not able to keep the pace than at your own boat; but if you meet another boat, it appears to go fast even if it is going slower than your own, because you do not so much consider what your own is doing as the fact that the other one is getting out of sight. In a similar way do the motions of the other planets appear from the earth. "This is so clear that nobody following this reason can have doubts about it".

The last part of the Discourse deals with certain physical questions. The usual Ptolemaic objection is quoted: if the earth is revolving, man will come under her, and large houses and towers must fall and be destroyed. Bilberg replies that it is known that the earth is fixed in its atmosphere, which holds everything together with its *pressio*, so that if this pressure did not exist, the animals could not live either; as a proof of that he refers to air-pump experiments with animals. Finally, in Bilberg's account, the traditional argument re-appears that if the earth is moving, a ball thrown up vertically would not fall down at the same spot, which in fact it does, but some distance from the thrower. Bilberg refutes this by emphasizing once again that according to Descartes the earth does not move but is at rest in its heaven, as when you shake an apple and do not hear the pip rattle, you say that the pip is immobile although the apple itself is shaken. During the motion of the heavens with the earth, its atmosphere by means of its pressure holds everything together towards the centre, so that the earth and the atmosphere should be regarded as one simple body, and accordingly the ball must fall back to the point from which it was thrown. To be sure, this explanation closely agrees with that given to the same phenomenon by Copernicus, from other physical starting-points: the inner part of the air element is so mixed with water and earth that it has the same nature as the earth, and accordingly follows the earth in its orbit without resistance. When the Discourse was written, it was still unknown in Upsala, as in most quarters, that a few years earlier Newton had proposed the hypothesis that a body thrown up vertically in fact does not return to the same point, but must deviate somewhat in the motional direction of the earth, the centrifugal force making it move faster than the earth; this, however, is a theory that was not experimentally proved until much later (J. F. Benzenberg). Thus, by contrast with previous argumentators, the old Aristotelian experiment provided direct proof of the rotation of the earth.

In his final words, Bilberg brings up an objection that he had probably heard only too often: you should not give offence — but the heliocentric doctrine does give offence to many. Using their reason to some extent, many of those opposing this opinion, he says, clearly confess that the opinion is founded on good reasons. Yet they do not want to follow it so that they may not offend any simple man. But is it more offensive to give offence with false doctrines than to be offended by the truth? God has given us reason; if we do not use it properly, it is a sin, and we are not worthy of the gift of God. "Thus without any doubt the greatest offence is to know the truth and still not confess it".

4.

Thus, in this form and with arguments of the kind which is found in the Discourse, the heliocentric system now rapidly met with support in Upsala, and soon also in those wider circles to which the relatively popular character of the argumentation was adapted. Copernicanism, in its Cartesian shape, now took over outside the theological quarter, one point of support of the classical world picture after an other.

The rapidity of its success in scientific life can be read in the dissertation literature. That it is shown in the astronomical theses ventilated for Bilberg is less remarkable. But in the other dissertations also, from the 1690's and the first years of the 18th century, the new situation is clearly seen: the validity of the heliocentric system often enough being regarded as rather obvious, any real refutation of the old cosmology is hardly considered necessary[42].

But, naturally, the change could not come everywhere at once. That is particularly true about the aging Spole who held the astronomical chair in Upsala till 1699, still adhering to a cautiously conservative view, at least publicly. Many of the numerous dissertations ventilated under his presidency bear witness to that. In some of them it is said apropos of the different planetary systems that philosophical reasons and a certain experience speak for the Copernican system, but the Ptolemaic is better in keeping with the *Holy Scripture*, and therefore it is safer to adhere to it[43]. It is interesting to see that the Copernican system is assigned superiority from a philosophical point of view: consequently, the author does not accept the arguments of traditional Aristotelian physics. Surely this is due to the influence of Car-

---

[42] A detailed investigation of the Swedish dissertations in question during the following pe riod is made by Sandblad, *op. cit.*, pp. 101—125.

[43] *De sole*, 1685, and *Contemplatio planetarum*, 1691. The question is also discussed in *Solis contemplatio*, 1692, where the Tychonic system receives a certain preference.

tesianism. Spole's attitude to that is rather vague in his printed works, and it seems that he tried to avoid expressing a definite opinion, but he certainly was not uninfluenced.

The most important dissertation by Spole, however, are those two which constitute the first two parts of the astronomical text-book which he never completed[44]. Here, time after time, the religious argument re-appears as decisive concerning the world system, and writing about the planets Spole in the main follows traditional astronomy – remember that the work in question is a text-book, the character of which is usually rather conservative – although on many individual points Spole inserts the latest scientific results. Thus, on one point he refers to Newton; that is one of the very first times Newton is even quoted in Sweden[45]. Against the heliocentric system and Copernicus he directs a number of objections. Those he was able to find in ample quantity in his great master Riccoli, who in *Almagestum novum* gives what is probably the most extensive criticism of Copernicanism ever to see the light[46]. Spole quotes the concluding judgement in which Riccoli summarizes his settlement of accounts with Copernicus: the view of the earth and the sun which is opposed to the Ptolemaic conception is also opposed to physics and to the letter of the *Bible* and is thereby condemned heretical by the holy cardinal congregation, chosen by Pope Paul V and Pope Urban VIII (i. e. those popes under whom *De revolutionibus* was put on the Index and Galileo's trial was conducted). Undoubtedly, it is strange to see this truly papistical argument, altogether based on the authority of the Roman church power, cited in an academic thesis in the Sweden of Lutheran orthodoxy. Nor does Spole find that, against the biblical evidence of the immobility of the earth, he is able to accept "effugium Renati des Cartes" that the earth is at rest in its atmosphere, which in its turn is moving and in doing so carries the earth with it. That does not prevent him from adducing Cartesian authorities in other contexts, as in the final chapter on the theory of the sun and its motions. Choosing between the different systems, however, he ends by recommending the Ptolemaic one, because one should rather pay reverence to the *Holy Scripture* than vindicate another opinion, so that one does not give offence to others, and the common man in particular, since the astronomers have not yet presented arguments supported by reliable observations. Here one may recall those final words of Bilberg's Discourse, quoted above, as to what should be regarded as offence.

So much for Spole's text-book, and so much for Spole in his printed publications. But among his unprinted and extant ones, which were lectures, there are passages somewhat changing the picture of his opinions, especially a rather extensive manuscript entitled *Prooemium Cosmographiae*[47]. There, Spole gives space

[44] Published in 1694 and 1695 with different, very circumstantial titles.

[45] It concerns the irregularities of lunar motions, *Principia mathematica*, lib. III, prop. XXII et seq.

[46] Lib. IX, sect. IV.

[47] Included in MS. No. A 504, Upsala University Library; undated.

to the different opinions of the Planetary system without himself adopting the biblical standpoint. He has also inserted a detailed account of Descartes' cosmology according to *Principia philosophiae*, and apparently he values it highly. Furthermore: his presentation of general physics is essentially based on Descartes, and he presents Cartesian celestial mechanics as the almost self-evidently valid: thus, the earth is only one planet among the others in the vortex of the sun. Of course the difference between Spole's printed publications and this manuscript may to some extent be due to the fact that he changed his opinion. But the essential explanation must be that out of caution, in order not to give "offence" he avoided making his real standpoint publicly clear. By his restraint he thus assisted in maintaining the traditional picture of the world, but in his teaching, which did not get much publicity, he simultaneously contributed to furthercoming among his students heliocentric cosmology in its Cartesian form. And indeed, such a double game is far from usual in the history of Copernicanism.

Spole's successor, Petrus Elvius Sr. (1660–1718), was not an important man and did not quite have the qualifications to utilize the extremely rapid development of astronomical science towards the end of the 17th century and during the beginning of the new century, above all characterized by the name of Newton. Elvius, fundamentally a Cartesian, in many respects represented a decidedly more modern outlook than Spole, however. In the astronomical dissertations during his time as professor (till 1718), the tone is quite different from what it was previously; when the question of the different world systems is mentioned, Copernicus and his successors, starting with Kepler, are often richly used and mentioned with unmistakable respect, even when the Ptolemaic view is simultaneously referred to. A *Historia astronomiae ellipticae* in 1703 is typical of the new situation; entirely basing itself on Copernican astronomy, it deals with Kepler's laws and their development through newer experiences in celestial mechanis. In that context, Newton's *Principia* is also mentioned and the theory of gravitation is briefly referred to, but for his part the author prefers the explanation of planetary motions given by the Cartesian vortex theory. Accordingly, in this thesis we are already far from the classical picture of the world, and the geocentric principle is here totally dismissed. The same typical features are found in, for instance, the theses *De causis motuum coelestium* in 1716 and *De planeta Venere* in 1717. Ideed, in the latter an account of the Tychonic system is initially given along with the Copernican, but the presentation itself starts from Copernican astronomy, mainly Kepler and Boulliau, which appears to be self-evident to the author.

Thus it is clear that it was during Elvius's time that the geocentric picture of the world was finally abandoned in Upsalian astronomy, even though it was less due to the scientific ability of the official representative at the university than the power of the general Cartesian tide of the time. Another man supporting it, more influential than Elvius as a scientist, was Bilberg's successor in the mathematical chair, Harald Vallerius (1676–1716). He was a marked Cartesian, a disciple of Bilberg and Drossander, and by largely devoting himself to astronomy as well, he also considerably contributed to the strengthening of heliocentric cosmology in its Cartesian form, as is shown by his dissertations around the turn of the century[48]. As a matter of fact, so far as is known it is in one of those that Newton and his *Principia* are mentioned for the first time in the scientific literature of Sweden[49]. And not until this time – in conjunction with Newton – are the results of Kepler's research really used in Swedish science.

Even if Elvius and his contemporaries and disciples lacked the qualifications to understand Newton, dismissing his theory of gravitation in a causal way – which indeed they were not alone in doing in the learned world of Europe at that time – it still means a step forward that Newton and his successors received attention and were brought into the discussion at all.

As regards the question of the different world systems, it is conspicuous how it slips out of the field of interest more and more at the beginning of the 18th century. When Cartesian philosophy has conquered the area of natural sciences, and the biblical and theological authority has been broken there, new problems of lesser principal consequence come to the fore. The conflict between a geocentric and heliocentric system is no longer able to attract the main interest; they can even be said to be as they were in a dissertation in 1710, about "dominium exigui puncti"[50]. At this time the dominion of Cartesianism is so strongly fortified that for a time, another couple of decades to come, it raises insurmountable obstacles to the penetration of the new Newtonian celestial mechanics. In that respect, the conditions in Upsala in Sweden do not differ from those in many quarters of continental Europe.

Upsala was still the intellectual and scientific centre of the realm of Sweden, and the development there tends to be well reflected at the other seats of learning. But the special conditions of the latter deserve to be mentioned briefly for this period also. It should, then, be pointed out that during the period dealt with here, the activities and output of any note in Swedish natural science were almost entirely connected with the universities; at any rate that is true of the disciplines discussed here.

The university of Turku, also visited by many students from the Swedish mother country, was still very lively. It had the closest contact with Upsala and was the university that was most sensitive to the singals from there; it also obtained many of its teachers from there. In Turku Cartesianism started to assert itself more in the scientific discussions from the middle of the 1680's especially in a number of dissertations under the presidency of the Professor of Physics, Petrus Hahn (1650–1718), in which opinions both pro and contra were presented. Though giving his disciples great freedom in their dissertations, Hahn was not himself a convinced

[48] Most emphatically in *Parallelismus microcosmi et macrocosmi*, 1711.
[49] *De centro terrae*, 1693.
[50] *De praecessione aequinoctiorum Copernicana* (praes. Elvius).

Cartesian; that was clear from his standpoint in 1697 when the new ideas finally caused an open conflict in Turku. The vice Chancellor of the university, Bishop Johan Gezelius Sr., then appeared as the main adversary of Cartesianism, supported by the faculty of theology, but it was defended by almost the whole faculty of arts and science. The conflict was soon settled, and in the end the theologians here were not more successful than in Upsala in their endeavour to suppress the academic freedom of teaching.

It seems that the cosmological matters of principle were not directly brought up during the conflict. But it is significant that just at the time of the Cartesian breakthrough in Turku, they start to teach the Copernican system openly there. That is done by Magnus Steen (d. 1697), Professor of the Mathematical Sciences. In a thesis *De placitis astronomorum praecipuis* in 1694, ventilated under his presidency, the Ptolemaic, Copernican, Tychonic and semi-Tychonic systems are discussed, and in particular the Copernican and Tychonic ones are weighed against each other. The traditional objections from the followers of the geocentric principle are cited and met with the usual Copernican counter-arguments; the biblically based objections to Copernicus are also refuted in terms in which one seems to discern an echo of Bilberg's Discourse. Finally, it is conciliatorily pointed out that everyone may consider the earth immobile or mobile without risking his salvation, and that from and astronomical point of view it is, insignificant whether a hypothesis is physically true or false, so long as it is astronomically true — i. e., suitable for determining phenomena. The question is brought up again in a thesis *De hypothesibus astronomicis Copernici & Ptolemaei* in the spring of 1697, when the Cartesianism conflict was in full swing. The author here shows that he is very familiar with Cartesian celestial mechanics, and evidently, although not saying so, under that influence he makes his choice between the two systems mentioned in the title. The author concludes by cautiously saying that although he does not want to insist that Copernicus' hypothesis is physically true, it is no doubt preferable because of its infallible truth in astronomical argumentation and its remarkable consistency with "nature and phenomena". The biblical counter-arguments are altogether left aside here; the author cleverly limits himself to treating the problem from an astronomical point of view; nevertheless, he leaves no room for doubt about the strenght of his Copernican conviction.

Thus, at this time, the heliocentric system has apparently gained a firm footing in Turku and should be rather generally spread outside the circle of theologians. A strange relapse into old ways of thinking manifests itself, however, under Steen's successor Lars Tammelin (1669—1733). Immediately before the turn of the century he went abroad for purposes of study, during which, in Leiden, he came in contact with the extremely lively Cartesianly inspired natural research, obviously without imbibing any of its spirit, however. He had a strong inclination for theology, later turned a clergyman and ended as bishop in Turku. In several cases, the astronomical theses over which he presided are on a surprisingly low level. In part, they might as well have been written a century earlier[51]. The idea that the earth moves around the sun is mainly contested with biblical arguments, and moreover those arguments out of Aristotelian mechanics are cited that were repeatedly used against Copernicus during the preceding century; even in a dissertation in 1712 the Copernican system is repudiated as still being insufficiently proved, and Riccoli is cited. None of these dissertations is further removed from the classical picture of the world than Tycho Brahe's system. Yet they can hardly be considered representative of the Turku faculty at this time.

When the last-mentioned thesis was published, the Great Scandinavian War had long been raging near the gates of the university, hampering education. Soon the activity had to be cancelled altogether. The reorganization at the beginning of the 1720's was made in the field of natural sciences under the sign of a new era. These ways of thinking, the languishing of which can be seen in Tammelin's dissertations, were forever gone. For decades to come, the mathematical chair was held by Nils Hasselbom (1690—1764), under whom the new scientific research — in which the most prominent names are Huygens and Newton — had its real introduction in Turku.

A university even worse struck by the war was that of Dorpat (now Tartu) in the Baltic province of Livonia (today Esthonia). It was founded as early as 1632, but the activity during the following century was frequently interrupted for longer periods by the wars and the Russian invasions, and hence the scientific standard was rather low at times or at least uneven. Little is known of the penetration of the new natural research and Cartesianism in Dorpat; the material extant seems to give no closer information. As in Turku, the general development there mainly followed that of Upsala, but in certain respects direct contact with Germany also played its part. Only one concrete circumstance, of the greatest interest in this connection, will be touched upon here. It concerns Sven Dimberg (d. 1731), a disciple of Bilberg and Drossander, who held the professorship of mathematics in Dorpat 1690—98, no doubt one of the most prominent teachers during the whole Swedish period of this university, active in the modern spirit in mathematics, astronomy and physics. His position in relation to his Swedish contemporaries is apparent from the fact that in 1695—96 he lectured on mathematics according to Newton, and for his last year in Dorpat he announces "an even deeper penetration into the analysis of Newton's *Principia* and into his higher mathematics"[52]. This is undoubtedly a fascinating inside picture: in that poor little village on Livonian soil, while there is a Russian threat of war, Dimberg is in the chair, explaining to a few students, who no doubt had inferior previous knowledge, scientific lines of thought by far not reached at the much better equipped seats of learning in the Swedish mother country.

---

[51] For a closer account, see Sandblad, *op. cit.*, p. 120 et seq.

[52] G. von Rauch, *Die Universität Dorpat und das Eindringen der frühen Aufklärung in Livland 1690—1710*, Essen, 1943, pp. 384 et seq.; cf. Sandblad, *op. cit.*, p. 122 et seq.

In Lund, the dependence on Upsala was generally less noticeable, partly because of the close relations with Germany, and from the very start many of the professorships of the university were filled with Germans. One of them became the most skilful representative of Cartesianism in Lund, when it first reached there at the beginning of the 1670's – namely, Christian Papke, (1634–94), who at first taught physics and philosophy, later theology, and finally became a bishop in Lund. With such a man among the theologians, the resistance of conservative theology was naturally considerably weakened. On the whole, the development in Lund in the field of natural sciences seems to have come about rather quietly but also very slowly, the personal forces being rather insignificant since Spole had left for Upsala. No obvious trace of Copernican ways of thinking are to be found in the meagre Lundian literature of astronomical nature until some time after the beginning of the 18th century. Then there was a vital change, when the chair of astronomy was taken over by Conrad Quensel (1676–1732), an exceptionally eminent teacher. He represented the Copernican system in its Cartesian form, as he shows in a polemic in 1720, where he answers an anonymous anti-Copernican pamphlet[53]. His argumentation is of the same nature as that of Bilberg's *Discourse*, and several of his examples appear to be taken from there[54]. On the whole, it appears as if Quensel's youthful impressions from Upsala of the 1680's, where he became a student very early, formed the basis of his way of viewing science as revealed by him in this polemical pamphlet and, to be sure, passed on by him to his numerous disciples in Lund.

Thus, in the end one finds that at long last, around the year of 1720, the heliocentric idea of the universe has gained a firm footing at all the Swedish seats of learning, although it would be some considerable time before it was generally accepted even in the learned world outside the rank and file of the scientists themselves. And, as already mentioned, at the same time Nils Celsius, once the youthfully bold pioneer of Copernicanism in Upsala, starts printing in his almanacs Bilberg's Discourse on the world systems. Copernicanism beeing deeply rooted in the scientific world, its dissemination in wider circles follows accordingly, and then one of the most widely spread publications is skilfully used: the almanac. The subsequent gradual penetration into the broader strata of society of Copernicus's doctrine of the relation between the sun and the earth is a complex process, difficult for historical research to grasp; and moreover, even today that process is hardly completed.

BIBLIOGRAPHY

A detailed examination of the penetration of the Copernican system into Sweden, mainly based on the academical dissertation literature and thoroughly documented, is made by H. Sandblad, *Det copernikanska världssystemet i Sverige*, I. *Aristotelismens tidevarv*; II. *Cartesianismen och genombrottet*, [in] "Lychnos", Annual of the Swedish History of Science Society, resp. 1943 and

[53] About these polemics, see Sandblad, *op. cit.*, p. 125.

[54] Above, p. 260.

1944—1945 (English summary: *The Copernican System of the Universe in Sweden*, I. *The Aristotelian Era*; II. *Cartesianism and the Triumph of Copernicanism*). Important completions have been made on one particular point by N. V. E. Nordenmark, *Laurentius Paulinus Gothus föreläsningar vid Uppsala universitet 1599 över Copernicus hypotes*, [in] "Arkiv för astronomi", ed. by the Royal Swedish Academy of Sciences, vol. 1, No. 24, Stockholm, 1951, and concerning the Cartesian period by R. Lindborg, *Descartes i Uppsala, Striderna om 'nya filosofien' 1663—1689*, Upsala, 1965 (Lychnos-Bibliotek, 22; English summary: *The Contentions about Cartesianism in Uppsala 1663—1689*). One special aspect of the development is treated by H. Sandblad, *Galilei i Sveriges lärda litteratur till Magnus Celsius*, [in] "Lychnos" 1942 (French summary: *Galilée dans la littérature scientifique Suédoise jusque vers 1680*), completed by O. Walde, *Nicolaus Granius, Galilei och Kepler*, in the same volume.

Several of the persons mentioned in this essay, and their cosmological views, have been treated in monographs: Johannes Bureus and S. A. Forsius by S. Lindroth, *Paracelsismen i Sverige till 1600-talets mitt*, Upsala, 1943 (Lychnos-Bibliotek, 7), chaps. II & IV; Georg Stiernhielm by J. Nordström in his introduction to Stiernhielm, *Filosofiska fragment*, I, Stockholm, 1924, and also by Lindroth, *op. cit.*, chap. VII; further N. V. E. Nordenmark, *Anders Spole*, in the "Annual of the Royal Swedish Academy of Sciences", Stockholm, 1931, and the same author, *Svensk astronomi och svenska astronomer 1700—1730*, [in] "Arkiv för matematik, astronomi och fysik", ed. by the same Academy, vol. 24, A, No. 2, Stockholm, 1933. The two latter works are, however, not altogether reliable in details.

The general history of astronomy in Sweden (and Finland) during the period in question is treated by N. V. E. Nordenmark, *Astronomiens historia i Sverige intill år 1800*, Upsala, 1959, with two bibliographical supplements by J. Nordström, Upsala, 1960 & 1965 (Lychnos-Bibliotek, 17: 2), and in a more concentrated form in English by P. Collinder, *Swedish Astronomers 1477—1900*, *Skrifter rör. Uppsala universitet*. Ser, C, No. 19, Upsala & Stockholm, 1970. The earlier Swedish astronomy, before Copernicanism, is also treated by H. Sandblad in his introduction to Olof Luth's astronomical text-book (see above note 1). Various conditions, treated in the present essay, concerning the position of astronomy at the universities, are treated in the histories of C. Annerstedt, *Upsala universitets historia*, I—II: 2, Upsala, 1877—1909; K. F. Slotte, *Matematikens och fysikens studium vid Åbo universitet*, Helsinki, 1898 (*Åbo universitets lärdomhistoria*, 7), and M. Weibull & E. Tegnér, *Lunds universitets historia*, II, Lund, 1868; these older works are all minutely documented.

About Copernicus himself there is only one monograph in Swedish: H. Sandblad, *Nicolaus Copernicus*, Stockholm, 1962, a book that is popular and that also gives a suvey of the gradual acceptance in Europe of the Copernican system. Copernicus's library and manuscripts, brought to Sweden in 1626 and kept in different libraries there, mainly in Upsala, have been treated by scholars from various countries in different connections. Among accounts made by Swedish scholars P. Högberg, *Copernicus-minnen i Uppsala*, in the journal "Populär astronomisk tidskrift", 1943, may be mentioned.

JUAN VERNET
University of Barcelona

COPERNICUS IN SPAIN

The introduction of Copernican ideas into Spain, has, until now, been the subject of a brief article by Father Antonio Romañá Pujo S. J.[1] (noted in the *Bibliografia Kopernikowska 1509—1955*[2]) in which the author follows Ernst Zinner[3] in his basic ideas. Previously, José Gavira[4] and A. Frederico Gredilla in their *Biografiá de José Celestino Mutis con relación de su viaje y estudios practicados en el Nuevo Reino de Granada*[5] had dealt with this topic fairly extensively — not to mention the continuous allusions, rather more rhetorical than scientific, to be found throughout all polemics on the subject of Spanish science[6]. Subsequently, after the date of the completion of the *Bibliografia Kopernikowska*, Vicente Peset Llorca[7] has published a first--class document which basically contains the study in Latin carried out by Gregorio de Mayans y Siscar (1699—1781)[8] in 1773 at the request of the Polish Samuel Luther Geret, of Toruń. This monograph, now two hundred years old, is based on the remains of the expurgatory Indices of the Inquisition. It will be duly appreciated in the lines to follow. Complementary materials can be found in the different works of J. M. López Piñero[9], which will be referred to at the appropriate place.

---

[1] *La difusión del sistema de Copérnico*, "Euclides" 4, 35—6 (Madrid 1944), 23 pages.

[2] H. Baranowski, *Bibliografia Kopernikowska*, Warszawa, 1958, p. 237, number 1775.

[3] *Entstehung und Ausbreitung der Coppernicanischen Lehre*, Erlangen, 1943.

[4] *Aportaciones para la geografia española del siglo XVIII*, Madrid, 1932.

[5] Madrid, 1911.

[6] One may consult the excellent anthology of texts on this matter which has just been published by Ernesto and Enrique García Camarero in "El Libro de Bolsillo", number 260 (Madrid, 1970) of the Alianza Editorial.

[7] *Acerca de la difusion del sistema copernicano en España*, "Actas del II Congress de Historia de la Medicina Española" 1 (Salamanca, 1965), 309—324.

[8] A great Spanish scholar who became librarian at the Royal Palace and maintained correspondence with Voltaire, Muratori and other learned men of the age.

[9] In passing, we point out his little book *La introducción de la ciencia moderna en España* (Barcelona, 1969) which contains a good description of the cultural atmosphere of late seventeenth century Spain.

Obviously, in Copernicus' work, one must distinguish between two aspects which were the subject of very different treatment by erudite Spaniards. First, the parts susceptible to mathematical application for the calculation of ephemerides were appreciated from an early date through the Prutenic tables, by the astronomers and astrologers of the peninsula, since the acceptance of their numerical values did not imply the admittance of their heliocentric concept of the universe. This heliocentric concept of the universe, which is the basic subject for discussion here, is, even nowadays attacked by some eccentric authors. We believe that in the introduction of Copernican theory into Spain, two periods can be sharply distinguished: the first (1543–1633) lasts until the definitive condemnation of Galileo by the Roman Inquisition, the second is from this date (1633) until the middle of the eighteenth century; as the inquisition lost its power and its penalties became inconveniences rather from mortal, enlightened members of society gradually admitted the new concept of the universe. It is logical, as we shall see, that both periods should lack monolithic consistency, and that in both, strong characters can be found who "believe" in Copernicus, but who, in the second period and for many decades, spin elaborate hypocritical tales (the *taqiyya* of the Moslems, or the *mental restriction* of the Christians) so as not to betray themselves in front of others but still remain loyal to the dictates of conscience. Many of the fundamental texts of this polemic can only be appreciated in the original tongue in which they were written, and they lose force in translation. From the nineteenth century, which, for obvious reasons, we will dispense with here — the acceptance of the heliocentric system is absolute and can be openly admitted without reproach.

For us to be able to understand the attitude of the Spanish authorities in the two periods alluded to, we must bear in mind the dominating position achieved by the Iberian peninsula in the sixteenth century, the uninterrupted decadence throughout the seventeenth century, and the abandoning of all dreams of power in world politics of the eighteenth century. Although on the appearance of printing (1480) the Catholic Monarchs did, to the best of their ability, motivate and facilitate freedom of expression, they soon changed their tune as can be seen from the texts of the royal edict promulgated on July 8th, 1502. They had realised the full importance for propaganda of the new invention and they decided to prevent any such activity by potential rebels (The Moors, converted Arabs etc.). Among the pages of this decree (which does not seem to have been rigourously applied) the following paragraphs stand out[10]:

> Don Fernando and doña Isabel... because we have been informed that you, the said publishers and printers at the said presses, and merchants and their agents have been, and are accustomed to printing and bringing for sale to these our kingdoms many printed books on all subjects both in Latin and in Romance, and that many of them contain errors in the content therein, and others are immoral and others apocryphal and fictitious, and others recently made up of vain superstitious

---

[10] See Antonio Sierra Corella, *La censura en España. Indices y catálogos de libros prohibidos* (Madrid, 1947), pp. 79–84.

things which have given rise to some harm and certain disturbances in our kingdoms [...]. We command and prohibit – that henceforward you will not dare, directly or indirectly to make nor print from type, any book of any faculty, or reading, or work, be it short or long, in Latin or in Romance without first seeking our permission and special license to do so. Be not so daring as to sell in our kingdoms any books you might bring from abroad – whatever their quality or subject – without first submitting them for examination by those said persons[11], or whosoever should be appointed to see and examine them by special license [...] under penalty of losing all such books and works which should be burnt publicly in the square of that city or town or place where you had them made or printed or where you sold them or had them sold. You will also forfeit all the income that you would have received or gained. And in further penalty you will also pay a fine equal to the value of the books burned on your account. And furthermore, for the same offence you will not be allowed to continue your office. And we decree that henceforth no one shall dare to sell any book nor any reading matter – lengthy or brief – whether it be from within our kingdoms or without, before first submitting it for examinations and approval.

It is such circumstances that Sebastian Kurz[12] wrote (21 March 1543) to Charles V of Germany (Charles I of Spain):

Nicolas Copernicus, the Mathematician, has written six books on *Revolutionibus orbium caelestium* which have been printed these past few days. And as the subject matter is no less marvellous than it is novel; and as it has never been seen, heard nor thought that (the) sun is the centre of everything and does not have the motion of its own, contrarily to what has been thought until now by all writers; and that the world has its motion through the Zodiac in the same manner as we believed the sun to have its motion. I have been so bald as to send this to your Majesty, as I know that your Majesty is a keen mathematician, and that you will be pleased to see and hear the opinion and fantasy of this author, so praised and approved by many mathematicians, for according to his theory, all the motions of the heavens can be discovered much more easily than was previously done by the belief in the motions of the Sun. I humbly beseech Your Majesty to receive this as a small service.

The eventual fortune of this copy of the *De revolutionibus* is unknown to us. Philip II in his decree of September 7th 1558 strengthened and reinforced the previous censorship implemented by the Catholic Monarchs.

And whosoever should print or submit for printing a work or book, in an unauthorised manner, without first having submitted the text to undergo the said examination for approval and license in the said form, is liable to suffer the death penalty; and to lose all his possessions. And such books and works should be publicly burned[13].

This decree was followed by that of 1559:

We decree the henceforward: none of our subjects and inhabitants, of whatsoever civil status, condition or order, ecclesiastic or secular, friars, clerics or any other should leave these Kingdoms to go and study or teach or learn, or go and reside in universities, centres of learning or colleges outside these Kingdoms; and that those who up to the present time should reside or find themselves in universities, centres of learning or colleges should leave such places, and no longer be found

[11] The decree goes on to enumerate the royal commissioners for the task of censorship; they are all Bishops of Castille and Andalusia. In A. Corella, *op. cit.*, pp. 274–275, one may see an example of the means adopted to prevent the smuggling in of books.

[12] Cf. M. Bataillon, *Charles Quint et Copernic* B. H. 22 (1923), pp. 256–258.

[13] A. Sierra Corella, *La censura...*, p. 97.

there within four months of the publication of this decree; and that the persons who against the content of our decree travel and depart to study and learn, read, reside or remain in those universities, centres of learning or colleges, outside these kingdoms; or those that already being in them and not leave, travel and depart within the said time, without returning or revisiting them, be they ecclesiastics, friars or clerics, of whatever estate, dignity and condition, let them be held for foreign ers and aliens of these Kingdoms, and let them lose and be deprived of all their titles that they had in these lands, and let these laws relate and apply to them, on penalty of loss of all their possessions, and perpetual exile from these kingdoms[14].

We do not know[15] of any case where the death penalty was in fact carried out; possibly, as in modern times, self-discipline was sufficient to avoid greater offences· However, we are aware that the decree of 1559 did produce ominous results in the development of Spanish science since Philip II forbade even the Jesuits to leave Spain to study abroad; also that the Professors of mathematics and astronomy at the Imperial College (founded in 1603, transformed into a Royal centre of studies in 1625[16], and as such, in competition with the universities) were nearly always foreigners, or sometimes of foreign descent. Foreigners were the only people with access — not to a liberal education (for nowhere in the Europe of this period did academic freedom exist as it does today) but to the teachings of different schools of thought.

Neither was the activity of the individuals, mostly Spaniards, grouped round the Academy of Mathematics[17] (1582—1634) founded by Philip II, particularly brilliant.

Despite these coercive circumstances, one must recognise that the Copernican doctrines were not essentially affected. It could be that Philip II did not see any danger in them or that he remembered his father's inclinations and hobbies, or that he wished to retain his Royal perquisites in the face of the clerical powers[18]. As a result, Copernicus' doctrine did not meet with resistence[19], and very soon his name appeared in books with Spanish authors: Pedro Núnez Salaciense and Pablo de Alea mention him; the tabular part of his work is made use of in so early a year as 1582 by Vasco de Pina[20]; Juan de Herrera, director of the above-mentioned Academy of Mathematics, wrote in 1584 to Cristóbal de Salazar, the provisional ambassador in Venice, asking for the following books, which give a good idea of what was considered as suitable reading in the Academy:

---

[14] Text from C. Sanchez-Albornoz, *España, un enigma histórico* (Buenos Aires, 1962) II, 553—4.

[15] A. Sierra Corella, *La censura...*, p. 99.

[16] Cf. Jose Simon Diaz, *Historia del colegio Imperial en Madrid.* 1 (Madrid, 1952), p. 85 and 121.

[17] Cf. J. Simon Diaz, *Historia...*, p. 47—52.

[18] Cf. A. Sierra Corella, *La censura...*, p. 113, 153 and 212 in which one may see how the King could authorise the circulation of books banned by the Church.

[19] Cf. A. Romañá, *La difusión...*, p. 6—8.

[20] Cf. Acisclo Fernández Vallín, *Cultura científica de España en el siglo XVI* (Inaugural speech for his admission to the Royal Academy of Science, Madrid 1893), p. 63 and 64.

Proclus, on the first books of Euclid, in Latin; Pedro Montanno (*sic*) on the tenth book of Euclid; the *Spheres* by Theodosius, in Juan Penna's translation; the *Sphere* of Father Clavius, recently compiled; the two books of Heron on *Spiritualibus*, rendered by Comandino; all the books which were printed by Guido Baldo Marchinoni, Marquis del Monte, on mathematics; and among them one of the *Mechanics* in Italian; the *Mechanics* of Aristotle (sic) in Italian by Picolomini; all the works that exist in Italian by Mercurius Trismegistus; if there are any translations of theoretical works on the planets — besides the one by Picolomini, and the introduction to them — for they can be found here; *if Copernicus has been translated into vulgar Italian send me a copy*; if the *Machines* are in translation, they can be sent[21].

From the account we think one may infer that:

1. Herrera, a good Lullist, and as such, acquainted with Latin, was trying to get hold of two kinds of books: a) for his pupils, in the vernacular; b) for himself or other professors, in Latin.

2. Herrera held in his possession a *De revolutionibus* and that what he wanted was to make the book accessible to the students.

That this could well have been so is proved by the text of the influential Constitutions of Salamanca University[22]:

"Title XVIII. Concerning the Chair of Astrology ... in the second year, six books on Euclides and Arithmetic up to square and cubic roots; and the *Almagest* by Ptolemy, or his *Epitome* by Regiomontanus, or Geber or Copernicus, by the vote of the auditors on the substitution of the *Sphere*"[23].

In view of these precedents, there is no cause for amazement at the fact that Andrés Garcia de Céspedes (d. 1611) wrote a commentary on the *Theorica* of planets by Peurbach which "consists of three parts ... in the first, the theories, according to the doctrine of Copernicus; in the second — according to our observations, the causes for the inequalities in the movements of the sun and the moon both in Copernicus and in King Alfonso are revealed; and in the third, the stations of the planets are dealt with in a treaty on parallaxes[24].

But the most interesting case of all is that of the Augustinian Monk, Diego de Zúñiga (1536—1597), who in his early years, used his father's name, Arias[25].

In 1573, he held the Chair at Osuna University and in 1584 he published his *Commentary on Job*[26] in Toledo, the preparation for which went back at least to

[21] Cf. Felipe Picatoste y Rodriguez, *Apuntes para una biblioteca científica española del siglo XVI* (Madrid, 1891), p. 148.

[22] *Constituciones* (Salamanca, 1584). The copy we have used also contains, with its own little page, *Estatutos hechos por la muy insigne Universidad de Salamanca* (Diego Cusio, 1595) and the text we transcribe is placed under the rubric of the year 1561.

[23] That is to say that Salamanca evaded the traditional text of the epoch — the one by John Holywood or Sacrobosco, in the most democratic way possible — through the vote of the pupils.

[24] J. A. Sanchez Perez, *La matemática* (in "Estudios sobre la ciencia española del siglo XVII", Madrid, 1953), p. 607. According to this author, it is kept in manuscript 26, lst on the right No. 20 in the Real Academia de la Historia.

[25] Biographies in Picatoste, *Apuntes*, p. 339—344; Marcial Solana, *Historia de la Filosofía española*, vol. III (Madrid, 1941), p. 221—266 and bibliography quoted in this work.

[26] *Didaci a Stunica... in Job Commentaria.*

1579, as this is the date given for the royal license to print. On page 205 (article 9, 5) – later expurgated by the Inquisition – and which was largely crossed out by the pious readers who possessed editions produced before the prohibition, we have the following passage, preserved for by Mayans[27]:

You wish to know, magnificent, illustrious and most learned Sir, what has been the progress of the Copernican system in Spain. So that I may most briefly and clearly indicate everything pertaining to this matter, to omit mention of the ancient philosophers to whom it was agreeable for the sun to be the centre of the World, and the earth moving round it; there was also your most celebrated colleague Nicolaus Copernicus, the outstanding astronomer and author of six books on the revolutions of the heavenly orbs, who in a short time gained distinguished adherents to his new-and-old doctrines. Among these latter, to answer your inquiry, I should mention the most learned Spanish theologican Diego de Stuñica, and Augustinian Eremite; he followed the Copernican system in his most erudite commentary on Job, Chapter 9, Verse 5; page 205 of the edition published at Toledo in 1584 with four plates, by the press of Joannis Rodericus. The testimony of Stúñica is worth discussing, it runs thus:

«He lays down an additional power belonging to God in order to show his supreme power oined to his infinite wisdom. [For those to whom this particular passage seems difficult, it might be illustrated by the Pythagorean doctrine imagining the earth to move by its own nature; by no other means can we explain the motion of the planets so greatly differing in speed and slowness. This opinion was held by Philolaos and Heraclitus Ponticus, as relates Plutarch in his book *de Placitis Philosophorum*. Also in agreement was Numas Pompilius, and more important, the Divine Plato in his later years. For he said that any other opinion was most absurd, as Plutarch in his *Numa* narrates; and Hippocrates in his book on currents of air τῆς γῆς ὀχημα (sic), he said the earth to be a vehicle.

In our time, Copernicus announced a motion of the planets, in accordance with this ancient opinion. Nor is there any doubt, that from this doctrine are derived far better places of the planets than from the *Almagest* and other writings. For it is certain that Ptolemy could not explain the motion of the equinoctial points, nor to establish a definite, stable initial point for the year. He himself said so in *The Almagest*, Book 3, Chapter 2 and left the discovery of these things to those later astrologers (sic) who could compare observations over an interval greater than that available to him. And although both the Alphonsines and Thabit Ben Core tried to explain these, still they achieved little. For as Ricius proves, the positions according to the Alfonsines are mutually inconsistent. The theory of Thabit is, I grant, better and from it can be established a stable initial point for the year, as Ptolemy desired. However, the equinoctial points appear to have progressed further than his theory could allow for. And finally the sun is now recognised as closer to us by at least forty thousand stades, than it was previously, and neither Philolaos nor the other astrologers knew of this motion.

In summary, the reasons of all these things are by Copernicus most expertly described and demonstrated in terms of the motions of the earth; while all the others fit better.

This doctrine of his is not at all contradicted by the saying of Solomon in Ecclesiastes, "But the earth is fixed for eternity" (*Terra autem in aeternum stat*) [in the margin: *Ecclesiast*]. For this means only that however various may be the succession of epochs and the works of man on the earth, still that earth is one and the same and maintains itself unchanged. For the passage itself shows this; in full it reads:

"Generation passes-away and generation comes-to-be, but the earth is fixed in eternity" (*Generatio praeterit, et generatio advenit: terra autem in aeternum stat*). [In the margin: "The motion of earth is not against *Scripture*".] Therefore it does not fit with the context of the passage, if it is explained (as the philosophers generally do) in terms of the immobility of the earth.

---

[27] Peset's edition: *Acerca de la difusión...*, p. 316–317.

As for the argument that this chapter of Ecclesiastes, and many others in sacred scripture mention the motion of the sun, which Copernicus wished to fix in the centre of the universe; this is not at all contrary to his doctrine ast he motion of the earth is commonly attributed ot the sun in ordinary speech, even by Copernicus himself and those who follow him; they will frequently refer to the earth's course as the sun's course. Indeed, there is no passage in the sacrosanct writings which speaks so clearly of the immobility of the earth, as he proves it moves. Along these lines that the passage we are discussing is easily explained to show the marvellous power and wisdom of God, who can put and maintain in motion the earth, since that earth is so heavy by nature.

It says "and its pillars tremble" so as to show from a positive assertion the earth moved from its foundations. Those who do not regard this opinion of ancient and modern philosophers as established — although not altogether convincingly — may be won over by the fact that it tallies with the earthquakes by which the earth is sometimes shaken».

Thus far was the writing of Diego do Stúñiga, who put in his margin "The motion of the earth is not contrary to *Scripture*", and since he was an outstanding theologian, he aroused no disturbances, and a deep silence ensued on this matter.

This passage must be placed in relation to what Zúñiga himself says in his *Physics*, Book four, Chapter five: "De totius mundi constitutione" which appears in his work *Philosophiae prima pars*[28]. It is important to see that Zúñiga shows that there is no contradiction between Copernican doctrine and the *Holy Scriptures*, and as López Piñero[29] notes, its great merit lies in knowing how to interpret those passages of the *Bible* which are not theological, in the light of scientific knowledge of the time.

But the application of the telescope to astronomy, realised by Galileo, and, more than anything, his joyous manifestations on behalf of the irrefutable proof which his observations brought to the heliocentric system (which were soon suppressed by the Roman Inquisition), produced a much more drastic attitude in the governors of Spanish scientific policy. They did not stop publications of the ephemerides, despite the fact that Copernicus, is mentioned in these Suárez Argüello, 1608; Freyre de Sylva, 1638; Lázaro Flores 1663; Father Zaragoza, or the Argentinian Buenaventura Súarez (1706—1739) in his *Lunario*[30], etc.). However, they did apply the prohibition made by the Sacred Congregation of the Index on 5th March 1616, which Mayans[31] transcribes as follows:

Ex eo tempore conciliari coeperunt maximi tumultus: nam, ut videri potest in Riccioli *Almagesto*, libro 9, sect. 4, cap. 40, num. 2, pag. 496, ibi legitur Extractus Decreti Congregationis Eminentissimorum S. R. E. Cardinalium sub Paulo V. editi V. Martii 1616. qui sic se habet.

Et quia etiam ad notitiam praefatae Congregationis prevenit, falsam illam doctrinam Pythagoricorum, *Divinae Scriptura* omnino adversantem de mobilitate Terrae, et immobilitate Solis, quam Nicolaus Copernicus *De Revolutionibus Orbium coelestium*, et Didacus Astunica in Job etiam docent; jam divulgari, et a multis recipi, sicut videre est ex Epistola quadam impressa cujusdam

---

[28] Ed. Toledo, 1597.

[29] *Galileo en la España del siglo XVII* — "Revista de Occidente" 40 (July 1966), 99—108.

[30] Cf. Guillermo Furlong, *Matemáticos argentinos durante la dominación hispánica* (Buenos Aires, 1945), p. 58.

[31] Text *apud* Peset: *Acerca de la difusión...*, p. 318. Summary by Pierre Aubanez, *Le génie sous la tiare. Urbain VIII et Galilée* (Paris, 1929), p. 38.

Patris Carmelitae: Lettera del R. P. Maestro Paolo Antonio Foscarini, Carmelitano, sopra l'opinione de i Pittagorici e dell Copernico della mobilita della Terra e stabilita dell Sole, e il nuovo Pittagorico Sistema del Mondo: in Neapoli per Lazzaro Scorrigio, 1615; in qua dictus Pater osten, der conatur, praefatam doctrinam de immobilitate Solis in centro Mundi, et mobilitate Terrae, conconam esse veritati, et non adversari *Sacrae Scripturae*. Ideo ne ulterius hujusmodi opinio in perniciem Catholicae veritatis serpat; censuit dictum Nicolaum Copernicum *De Revolutionibus Orbium* et Didacum Astunica in Job suspendendos esse, donec corrigantur, librum vero P. Pauli Antonii Foscarini Carmelitae omnino prohibendum, prout praesenti Decreto omnes respective prohibet, damnat, atque suspendit. In quorum fidem praesens Decretum manu et sigillo illustrissimi, et Reverendissimi D. Cardinalis S. Caeciliae Episcopi Albanensis signatum et munitum fuit die 5. Martii 1616. Romae ex Typographia Cam. Agost. anno 1616.

This resistance to accept the Copernican system is already sketched out by the cosmographer Rodrigo Zamorano (1542–1620), who in the prologue of his *Arte de navegar* shows his acquaintance with Copernicus, whom he must have followed in the tabular part but not in the theoretical part[32]. The condemnation of 1616, with the prohibition of these passages by Zúñiga relative to the heliocentric system[33], appears in the *Index* published in Seville in 1632[34] and in those of 1640, 1707 and 1747. The final stroke to these incidents was made with the condemnation of the whole system by Urban VIII (22nd June, 1633).

From this date, the heliocentric system would not be publicly defended again until well into the eighteenth century. And this panorama remained unchanged, in spite of the fact that from the reign of Charles II onwards, there appeared such erudite men as Crisóstomo Martinez (1687) who went abroad, subsidised by the Sovereign himself (a tacit derogation of the 1559 decree[35]) to study; and in spite of the fact that in Zaragoza and Madrid, round Juan José of Austria, the bastard brother of the King and universal minister, a group was formed of scientific innovators who read all that was published in the rest of Europe (a tacit derogation of the 1558 decree), and despite the establishing in Barcelona of the embryo of one of the first academies of science in the country, the Academy of the "Desconfiados" (i. e. Distrusting) (1700), which later (1754) was to be transformed into the Royal Academy of "Buenas Letras". But these innovations, as we have said, must have caused sufficiently great upheavals to their instigators. We know that the chief sailing master of the Casa de Contratación in Seville, Juan Cruzado (d. 1692) travelled through Holland, Germany and England in the company of the Cardinal Infante don Fernando, between 1633 and 1644. There is no doubt that he knew of the scientific progress of the countries he visited, since in 1679 he relates in a piece of information how he had spent a lot of money in "having mathematical books and instruments which have been produced in Europe, brought to Spain to be shown to and

[32] J. Pulido, *El piloto mayor. Pilotos mayores, catedráticos de cosmografia y cosmógrafos de la Casa de Contratacion de Sevilla* (Sewilla, 1950).

[33] The part of the above passage which appears in brackets was supressed.

[34] Cf. Peset, *Acerca de la diffusión...*, p. 319; A. Sierra Corella, *La censura...*, p. 267 onwards.

[35] Cf. Lopez Piñero, *La introducción de la ciencia moderna en España*, "Revista de Occidente" 35 (February, 1966), 133–156.

discussed with the sailing masters who come to be examined; and not failing in my watchfulness to notice anything which may be of use, I have gone so far as to exchange letters ... with the Academy in London on matters concerning navigation, and other subjects which I profess. It seems that they approved, for they replied to me, and printed my letters in books which come out every year[36] which have come into my possession; and continuously, I am studying books produced in the six languages that I understand and speak". This individual (who, in some of the reports, is alleged to be of French nationality) was seized by the Inquisition in 1691 and we do not know for sure on what grounds, although one suspects that his relation with foreign countries would count against him[37].

In this respect, Admiral Antonio de Gaztañeta y de Iturribálzaga (1656–1728) was more fortunate. He was able to introduce the latest discoveries made in nautical matters in France and England into his works.

The Copernican system, therefore, remained, throughout this period, subject to the opinion expressed by José Vicente del Olmo[38], secretary to the Holy Office, who, in his *Nueva descripción del orbe de la tierra* (Valencia, 1681) refers to Manilius and Ovid to demonstrate the central position of the Earth; makes comments on the Psalm quoted by Zúñiga (without quoting Zuñiga) and assures us that such a movement refers to earthquakes. But in Chapter V we read:

This opinion, besides the fact that it was not necessary to introduce it to save the appearances of the haevenly bodies, is condemned by the congregation of the most Eminent Cardinals formed by his Holiness Paul V for the disposition of the Index of forbidden books, on March 5th 1616; and afterwards, on June 22nd 1633, by the decree of his Holiness Urban VIII of the same Congregation, it was declared to be erroneous in belief, absurd and false in Philosophy.

The condemnation of 1633 gave rise to three different attitudes among Spanish thinkers:

1. Absolute subjection, with or without refutation in principle of the heliocentric system;
2. Acceptance of the systems of Tycho or Descartes;
3. Presentation of the Copernican system under a hypothetical form. This last current, as we shall see was to gather moment from the early eighteenth century, to triumph in the end as the word *hypothesis* became *theory*.

1. Subjection to the condemnation

José Gavira[39] has quite clearly established the ideological line of this school of thought: Ferrer, who makes an apology of the ptolemaic system, maintains that the crystalline ring that separates the *primum mobile* of the sky from the stars, is formed by a mass of the purest waters – "so that the blessed can enjoy

---

[36] I think he is referring to the *Philosophical Transactions.*

[37] Cf. J. Pulido, *El piloto...*, p. 869–905.

[38] Cf. J. Gavira, *Aportaciones...*, p. 45.

[39] *Aportaciones...*, passim.

themselves in these seas of crystal, so as not to miss a mode of diversion". Hurtado de Mendoza, who reproaches Copernicus for "resurrecting the dead and forgotten opinion of Philolaos", and although realising that the heliocentric doctrine fits in well with his own observations, adds: "it is absurd, devoid of philosophy, and erroneous in faith [...] notwithstanding the efforts made by Galileo, Bullialdus, Hevelius and other Copernicans to try and prove the contrary". González de Urueña and Cansino defend similar positions; Fr. Murillo admits that as there is no rational proof to establish which of the two systems — Ptolemaic or Copernican — is the true one — one must submit oneself to the indications of biblical texts: the immobility of the Earth; Hualde tries to prove it with the following examples: (1) a stone thrown from the top of a tower always comes to rest at the foot of the tower; (2) when two cannons are fired in opposite directions, the cannon ball fired to the East does not travel further than the other one. The Cordovan astrologer Serrano says in the verse prologue of his *Tablas Philippicas*:

> Then was born that illustrious man Copernicus Torinense (*sic*), the promoter of that elegant system dreamt up in ancient times by Aristarchus of Samos and Philolaus; the one to which the name of hypothesis is allocated – an imaginary supposition – for it says the immobile sun is the centre and that the Earth is mobile in circulation.

This same Serrano elucidates his opinion in his *Astronomía Universal*:

> This system ingeniously explains all that we can see in the sky and observe in the stars, but since this idea of the movement of the earth and immobility of the sun has been condemned by the Congregation of Cardinals it is expressly permitted merely as a hypothesis or supposition; that is, it must be held as false that the Earth really does have the disposition of the Copernican system; but if it did have it, one would easily be able to explain the celestial movements and appearances of the heavenly bodies; and so for this purpose we consider it and deal with it only as a hypothesis.

And Father Rodriguez is of the same opinion in his *El Philoteo en conversaciones el tiempo* (p. 316):

> "The calculations with which one attempts to adjust its movements and anomalies (i. e. of the Newtonian system) by a magnetic action or central-attraction are morally impossible, (although one may suppose they are sound mathematically) even though they may occur in this fashion physically".

And of Newton he holds the opinion that:

> "Even if he did discourse on many physical theorems and discover many things in light and in colours, despite all this, as to the causes and order of movement by his beloved magnetism, no one remaining unprejudiced and disinterested will swear by his words".

And Fr. José Cassani (1673–1750), the author of an excellent *Tratado de los Cometas*[40] finds himself in a similar position.

It is curious to note that within this retrograde line, one finds the two most

[40] Cf. Armando Cotarelo Valledor, *El "Tratado de los cometas" del P. Cassani (1703)*, in: "Anales para el Progreso de la Ciencias" 1 (Madrid, 1934), 485–520; and Constantino Eguía, *El P. José, Cassani, cofundador de la Academia Española*, in: "Boletín de la Academia Española" 22 (1935), 7–30.

traditional universities: (I) Salamanca, which in 1770 refused to allow the study of Newton, Gassendi and Descartes, because their principles "do not accord so well with the revealed truths as do those of Aristotle"[41], forgetting that Salamanca was the first university to inaugurate the study of Copernican theory and which had among its staff in the sixteenth century Jerónimo Muñoz, one of the greatest antiaristotelians of the century as seen from his study of comets of 1572[42]. And (II) Alcalá hardly went much further: in 1772 the *Sphere* of Sacrobosco; the *Use of the Astrolabe* by Gemma Frisius, the *Theorica* of Peurbach and *The Alfonsine Tables*, were all taught, and one reads that the 'theories' of Peurbach will each be interpreted 'by Ptolemy'.

2. Acceptance or invention of other systems

Among the most peculiar authors to be cited are the Franciscan Francisco Angeleres (c. 1680) who made a picturesque synthesis of Copernican doctrines with astrology and physiognomy[43], and José Santiago de Casas, who published his hypothesis in Madrid (1758) with the title *Relox universal de péndola y en él nueva idea del sistema del universo*. The work is retrograde for the date when it appeared, and on account of the puerility of the author — a puerility shared by the long list of censors and references connected with the work — which tries to explain the celestial appearances whilst keeping the Earth in the centre of the universe, and giving it a simple oscillatory movement along the meridians to explain the variation of the solar declination throughout the year. He denies the movement of terrestrial rotation because "such a violent revolution of such an extremely heavy body as is the earthly globe on its imagined axis, whilst maintaining its centre immobile, just cannot be conceived; it would be against nature which is characterised by tranquility in the works of the Author of all things [...] Let us see what would happen if the earth rotated as the Copernicans maintain. Are the wild or hurricane winds the inner part of the atmosphere or outside the atmosphere? If I am to be told that the atmosphere runs together with the globe, it is essential to believe then that they originate within; and if the atmosphere is unchanging, from whence does the calm come when hurricanes cease?"

Worthy of more serious consideration are those who, abandoning fantasies, tried to reconcile faith and science in the adoption of some acceptable system. In this vein, Carlos de Sigüenza y Góngora (1645—1700) adopted the Cartesian system of the vortices or "tourbillons"; Juan Bautista Corachán (1715) in his *Tractatu de Cosmographia* follows Tycho's system modified by Riccioli "Whom I find

[41] Cf. J. Sarrailh, *La España ilustrada de la segunda mitad del siglo XVIII* (Madrid, 1957), p. 484 and 101.

[42] *Libro del nuevo cometa y lugar donde se hacen y como se verá por la paralaje cuán lejos están de la tierra* (Valencia, 1572).

[43] Cf. J. Mª. Lopez Piñero, *La introducciòn...*, p. 142.

more up to date and in line with experience"[44], as do the majority of Spanish astronomers of his time. The doctor and philosopher, Andrés Piquer[45] tells us in his *Lógica* (p. 166):

> Copernicus says that each day the earth gives a complete turn on its axis, and that in one year, it turns around the sun, which we assume to be in the centre of the universe. And when human understanding tries to grasp that this idea of Copernicus does not conform itself to the truths of Sacred Scripture nor to the truths we acquire with experience; when, on the contrary, Tycho Brahe says that the Earth is in the center of the world, that the Sun and all the planets revolve every day around the Earth, and that Mars, Jupiter, Saturn, Mercury and Venus in their yearly movement circle around the Sun; realizing that all this concurs with experience and the reason, it adopts the system of Tycho Brahe as the likely one.

González Cañaveras, in his treatise, describes four astronomical systems: Ptolemy, Copernicus, Tycho Brahe and Martianus Capella, and he adds one more of his own invention, calling it "Physical-Astronomical-Solar" which is nothing else, in fact, than Copernicus — although adapted (so he says) only to facilitate calculations[46].

These discussions on the systems of the world were already published by the mid eighteenth century and they appear in the programme of normal academic activities with or writhout an indication as to which system should be defended. For example, in *Conclusiones Matemáticas, Seminario Real de Nobles* (Madrid, 1748) we read:

> They will explain that the system of Philolaus, Pythagoras and Aristarchus of Samos, renewed by Nicolas Copernicus, as well as the censorship which it earned from the Congregation of Cardinals, the Inquisitors of Rome against the person of Galileo de Galileo and in which way it can be used[47].

In the contest which the Imperial Royal Seminary of Nobles of Barcelona celebrated around 1760, we read exclusively specified: "Description of the systems of Claudius Ptolemy, Tycho Brahe, Nicolas Copernicus and René Descartes; critical opinion and judgement of any of the said systems"[48], and in the *Programa de ejercicios de examen para los caballeros seminaristas del Real Seminario de Nobles de Madrid* (1775)[49] we read: "About the system of Copernicus (which we admit as a hypothesis and not as a thesis)".

These dissertations occur as a result of the scientific and critical spirit which came to birth with the rise to the throne of Ferdinand VI. Likewise, Francisco Gius-

---

[44] Peset, *Acerca*, p. 313 and 322. One may see the opposite case expressed in Gavira, *Aportaciones*, p. 45, and J. Sarrailh, *La España ilustrada*, p. 493.

[45] Cf. Alejandro Sanvisens Marfull, *Un médico-filòsofo español del siglo XVIII: el doctor Andrés Piquer* (Barcelona, 1953).

[46] Cf. Gavira, *Aportaciones...*, p. 48.

[47] Cf. Sarrailh, *La España ilustrada*, p. 497.

[48] Cf. the text in the Central Library of Catalonia, Folletos Bonsoms, Cat. 4, No. 4743.

[49] Cf. Gavira, *Aportaciones...*, p. 50; Sarrailh, *La España...*, p. 497.

triniani published an *Atlas Universal* in Lyons (1755) in which he says with reference to Copernicus[50].

Unfortunately his book came out at a time when all new opinions were regarded with suspicion. Everywhere people spoke only of Reformation: under the pretext of reforming customs of the clergy there was a swarm of people who attacked the faith of the Catholic Church; and those who still defended it were frightened by that host of particular opinions which were being advanced in the place of the Church's Dogmas, themselves threatened with suppression...

And much more symptomatic, on account of its apparent lack of coherence— — *et pourcause* — are two of the *Cartas Eruditas* which Father Benito Feijóo dedicated to the theme under the title *Sobre el sistema copernicano* (Vol III letter 20) and *Progresos del sistema filosófico de Newton en que es incluido el astronómico de Copérnico* (Vol IV, letter 21). As a priest, he found himself obliged to defend—or at least not to challenge directly — the decrees of 1633, and his reaction, as Gavira summarises it (p. 52), consists in establishing a heretical syllogism: "the only hindrance to the acceptance of the admirable Copernican system is its opposition to the *Sacred Scriptures*; the persistence of the ptolemaic system among our astronomers can only be explained by our scientific backwardness, *ergo* our excessive respect for the sacred scriptures was the cause of scientific backwardness". And to lessen the impious consequence, Feijóo adopts to Tycho's system.

3. The hypothesis-theory of the heliocentric system

One may deduce, from all that has been explained so far that the adopting of the Copernican system in Spain was accomplished by the way of hypothesis and with much precaution. It seems that one of its first partisans was father José Zaragoza S. J. in his *Esfera en común celeste y terráquea* (Madrid, 1674), who affirms as Mayans indicates[51], in his book 2, 1, 8, that "this belief, although ingenious, is condemned by the Cardinals, the Inquisitors, as contrary to the divine writings, although in the guise of a hypothesis or a supposition, everyone may make use of it in calculations of the planets: thus only the actual reality of this composition is condemned, but not the possibility of it being so [...] The belief of Copernicus has two parts: firstly that the sun does not move; secondly that the earth has an actual and daily movement and is not a centre. The first was censored on the grounds of *formaliter haeretica*, as being expressly in opposition to the *Divine Writings*: the second on the grounds of theological teaching, for it is at least erroneous *in fide*. Consult Ricciolo. Vol 2. p. 496".

Let us continue: one cannot deduce decisive consequences from this affirmation with regard to his real opinion, for Zaragoza is also the author of an *Astronomia nova methodo iuxta Lansbergii hypothesim ad meridianum Matritensem accomo-*

---

[50] Cf. Gavira, *Aportaciones...*, p. 49.
[51] Cf. Peset, *Acerca...*, p. 321–322.

data (1670)[52] which permits one to suppose that he tacitly admitted the heliocentric system from the moment he followed Lansberg, a disciple of the new system.

Armando Cotarelo y Valledor, the author of an excellent monograph on Zaragoza[53], has gleaned the ideas which have come forth — sometimes involuntarily — on the theme from the writings of our author; the fact that "Venus and Mercury can be mutually eclipsed, because either can be inferior to the other" (*Esfera*, p. 161) indicates that he was inclined towards the theory of Tycho modified by Riccioli[54]. When he analyzes the immobility of the Earth (*Esfera*, p. 196) he says that despite the fact of it being evident to the senses, "one may reply that a sense is deceived because the immensity of the heavens is such that the distance of the Earth and Centre is like a point in comparison with the immensity of the heavens, and therefore sight cannot distinguish the differences of stars, days, etc." He is also acquainted with the ellipticity of the planetary orbits (p. 78) which gives a good result in the calculation of the ephemerides, but he prefers to admit a spiral movement. This polemic on the shape of the orbits of the celestial bodies was then in vogue and father Juan de Ulloa S. J. (1722) admitted that the planetary orbits were oval shaped[55] as Azarquiel and Alfonso the Wise had thought of Mercury as well as Kepler, initially, in his *Astronomia Nova* on Mars; and Vicente Mut (d. 1687) in his observation on the course of the comets, of 1664, had asserted that the former followed a curvilinear orbit[56], similar to a parabola.

The Cistercian Juan Caramuel y Lobkowitz (1606—1682)[57] also concerned himself with these questions in his *Mathesis Biceps. Vetus et Nova* (Campania, 1670). He shows himself to be a partisan of Tycho, but he adds "that God omnipotent can bring about, this nigh, that the Moon, or Mercury, or Venus, or the Sun, or Mars, or Jupiter, or Saturn, or any of the immobile stars should rest in the centre of the world without an astronomer being able to notice, tomorrow, that anything has been moved by God"[58].

One step further on stands Father Tomás Vicente Tosca (1651—1723)[59]. He alludes to our problem in different places of his works. In the *Compendio Mathematics* he closely follows the work of Father Milliet Deschales. He devotes volume VII

---

[52] Cf. A. Cotarelo, *El P. José de Zaragoza y la Astronomía de su tiempo*, in: "Estudios sobre la ciencia española del siglo XVIII" (Madrid, 1953), p. 218.

[53] Cf. A. Cotarelo, *El. P. José de Zaragoza...*, p. 65—223.

[54] The modification consists in supposing that Mercury and Venus rotate round the sun — according to the Egyptians, and also that Mars does likewise — according to Tycho; but leaving the Moon, Sun, Jupiter and Saturn concentric to the Earth.

[55] A. Romañá, *La difusión...*, p. 15.

[56] A. Cotarelo, *El. P. José de Zaragoza...*, p. 103.

[57] Cf. Ramón Ceñal S. J.; *Juan Caramuel. Su epistolario con Atanasio Kircher S. J.* "Revista de Filosofía" 12, 44 (1953), 101—147.

[58] Cf. A. Romañá, *La difusión...*, p. 19.

[59] Upon which you should consult Roberto Marco Cuellar, *El "Compendio Mathematico" del padre Tosca y la introduccion de la Ciencia moderna en España.* "Actas del 2° Congreso de Historia de la Medicina Española 1 (Salamanca, 1965), 325—57; Peset, *Acerca...*, p. 322—24.

and part of volume VIII to astronomy. Tosca makes Mercury and Venus and Mars revolve around the Sun, and speaking of the movement of the Sun he adds:

> All that we have said about the solar ellipse, according to the common hypothesis that gives the Earth stability and the Sun movement, is verified also in Copernicus' hypothesis which attributes stability to the sun and movement to the Earth; for the only difference between these two theories is that in the latter — according to the Copernicans the Earth moves around the periphery of the ellipse, and the sun is in that focus or centre of the real movements, where common hypothesis would place the Earth.

When speaking of the superior planets—Mars, Jupiter and Saturn—he claims that the *hypothesis* of Copernicus explains with such simplicity its second inequality that "it seems impossible to improve on it". When in the twenty fourth thesis of *De la Geografia* (book 1, chapter 2) he faces up to the problem we are interested in he notes that "Copernicus and his followers attributed two movements to the Earth; the first that the Earth, as though it were one of the seven planets, travels through the annual orb or ecliptic;... the other movement is daily and it moves around its centre... making a complete circle in twenty four hours. It is not necessary for us to pause here on this subject, since, in different texts on Astronomy, these have been explained extensively, where the facility with which the celestial phenomena or appearances are explained with these two movements is evident. This system may be considered in two ways: the first is as a hypothesis, and the second as a real fact: as a hypothesis, there is no doubt that it is one of the best that have been debated..."[60].

Evidently he realises that if he does not recognise the rotation movement, the heavenly bodies—given the distance at which they lie from our globe—would have to travel at a fantastic velocity to complete their circuit in twenty four hours, but "there is no doubt that the creator could, for his high purposes, put this movement in the stars and not in the Earth"; although he also confesses that "the philosophical arguments with which the movement of the Earth is defended are not conclusive either, arguments which consist in the experiments of throwing a ball in an easterly direction, and another to the west (the distances covered should be different), and in dropping a stone from the top of a tower it should deviate from the vertical position of its fall. All this to conclude: "since there is no evident reason, nor experiment that concludes such a movement, it is necessary to say that the Earth is immobile and movement is in the Sun, according to several texts of *Holy Scripture*".

Tosca expresses similar ideas in his *Compendio Philosophico* (Valencia, 1721) which were copied by Mayans and transmitted to Geret[61].

The Triumph of Copernicus

The heliocentric system can be considered as definitively accepted by the Spanish astronomers from the moment the Inquisitor Francisco Pérez de Prado gave the

---

[60] Text in Peset, *Acerca...*, p. 323—24.
[61] Text in Peset, *Acerca...*, p. 314—16.

imprimatur without reticence to the works which exposed Copernicanism by way of hypothesis, as shown in the letter which Mayans[62] wrote to him on April 29th, 1747.

Sir: although I now have great peace of mind, and I live with good faith, I am retaining the prohibited books which I showed your Excellency in virtue of the license which has been verbally conceded to me; however, desiring to secure this favour in a written form (if you have no objection to this) I beg your Excellency to concede this.

At the same time, I am explaining to your Excellency the enjoyment I have at knowing that we have a General Inquisitor who examines the evidence personally, and who carries out Justice, leaving the reputation of the writers quite unharmed as far as possible. I say this on account of the controversy of previous times, concerning the manner in which don Jorge Juan should explain himself and his ideas on the system of the world which, after many ancient philosophers, Copernicus followed and strengthened.

He goes on with the history of the problem and adds:

If we enter into a consideration of the causes of one or another condemnation, we will find that they were not founded on absolutely sound philosophical reasoning, as Fr. Tosca has proved very well [...] I am of the opinion that Copernicus' system is reproved by *Sacred Scripture* (leaving aside other texts), especially in v. 5 Ps. 103 which says "Qui fundasti terram super stabilitatem suam: non inclinabitur in seculum seculi". But this and other authorities can have an explanation which is not necessarily absurd and which can fit in well with the opinion of Copernicus; and great theologians of Christianity attribute this explanation to it, fitting in with the Hebrew text and the version by St Jerome; from this arises the argument that it is absolutely unnecessary to consider the statement of Copernicus as heretical, as many writers have affirmed. And that it should not be held as heretical, besides what is argued in an erudite way, by my friend Luis Antonio Muratori, in his most learned work *De ingeniorum moderatione*, it can be said that in the present case this condemnation is not a theological conclusion, stricly speaking, because it is not derived from theological premises; and we have even seen, following Fr. Tosca, that these are not even sound premises.

Following on from what has been said, we arrive at the conclusion that if what I say has any probability, what Fr. Zaragoza says is sound at least, (bk 2 of The *Sphere*, prop. I n. 8) and according to Fr. Tosca (bk 2 ch. 20) there is no danger in following this system hypothetically.

And even though don Jorge Juan does not approve it, he may refer to it and use it as a hypothesis. And I rejoice very much that Your Excellency should have resolved the matter in this manner, as I have been told — so that everyone may know that Your Excellency does not remove all freedom that may be useful to Science, from those inspired with ideas, in your capacity as promoter of Science.

This state of opinion permitted Jorge Juan y Santacilia (1713—1773) to publish his *Observaciones astronómicas y físicas* (Madrid, 1748) although he had to admit in the prologue that the Copernican system was a hypothesis. This accounts for the fact that Vaquette d'Hermilly translated Juan into French saying "The author of this work does not speak as a mathematician when supposing the opinion of those who affirm the earth revolves round itself to be false, but as a man who writes in Spain, that is to say, where the Inquisition exists"[63].

[62] Text in Peset, *Acerca...*, p. 314—16.

[63] Cf. Sarrailh, *La España...*, p. 497.

The victory — and I stress that I mean the victory among scientists — of the Copernican system, was achieved some time between the episode we have just related (1747) and the publication of the last work by Jorge Juan: *Estado de la astronomia en Europa* (1774). In a passage quoted by Juan Sempere Guarinos[64] he says:

> To wish to establish that the earth is immobile is the same as wishing to overthrow all the principles of mechanics, physics and even astronomy, without leaving help nor strength in the human sphere to be able to satisfy us.
>
> These reflections have already been made throughout almost the whole of Europe. There is no kingdom that is not "Newtonian" and as a result, Copernican; but, not even in imagination there is any attempt of offend the *Holy Scriptures* which we should venerate so much. The spirit in which the *Scriptures* spoke is quite clear, and they did not aim to teach astronomy, but only to make it understood by the people. Even those who sentenced Galileo are nowadays known to have repented of having done so, and nothing accredits it so much as the conduct of Italy itself. Throughout Italy, the Copernican and Newtonian system is taught publicly; there is no Religion that refuses to see it published. Fathers Leseur, Jacquier and Boscowick — and even the Academy of Bologna — do not wish otherwise. Can there be more obvious proof than that they do not show the slightest suspicion of heresy (which was the condemnation), and that one the contrary they accept the Copernican system as uniquely true?
>
> Will it be honest — in view of this — to oblige our nation, after explaining the systems and Newtonian philosophy, — to add to each phenomenon dependent on the movement of the earth: but we are not to believe — it as it is against *Holy Scripture*? Is it not an outrage to *Holy Scripture* to try and prove that it is opposed to the most delicate demonstrations of Geometry and Mechanics? Will any learned Catholic be able to understand this without being scandalised? And if there should not be in the whole Kingdom sufficient enlightened persons to understand it, should it be allowed to make so visible a nation that maintains such blindness?
>
> It is not possible that its sovereign, full of wisdom and love, should consent to it. It is necessary that the honour of his subjects should return and it is absolutely necessary that the astronomical systems should be able to be explained without the necessity of refuting them; for as there is no doubt in the exposition, neither should there by any in allowing science to be written without such restrictions.

What has happened between the two key dates (1747 and 1774) in the development of Spanish culture? It was the reign of Ferdinand VI (1746—59), the man who set down the bases for the colossal development of science in Spain in the second half of the eighteenth century. In the first place, his own confessor (he was also an Inquisitor) Fr. Francisco Rávago S. J. (1685—1763)[65] suggested to him the setting up of an astronomical observatory (1752) which was dependent on a large subsidy (paid irregularly, to be sure) and which furnished itself with many pieces of equipment imported from England[66]. This example spread, and we know of several importations of eye glasses, micrometers, etc. and of special note among them was a Herschell telescope brought in by José Mendoza y Rios in 1796.

---

[64] *Ensayo de una biblioteca española de los mejores escritores del reinado de Carlos III — 6* tomos, Madrid, 1785—1789, vol. III, 152—153.

[65] Cf. C. Sommervogel, etc.: *Bibliothèque de la compagnie de Jésus* VI (Brussel, etc., 1890—1930), vol. VI, p. 1496—7.

[66] Cf. J. Simón Diaz, *Historia del Colegio*, p. 122.

The decreed imposition of the teaching of the Newtonian system can be considered as the final point of the whole polemic. This was very well publicised by Hervás y Panduro in his *Viaje estático al mundo planetario* (Madrid, 1775), p. 114–117.

The arguments founded on some texts of *Sacred Scripture* are satisfied.

All these arguments are satisfied with the following considerations:

It would be a rash person who would try to exclude from holy books all metaphors, all comparisons, all the figures of speech accepted among men. Astronomers also say that the sun rises and sets, and they will always say this, without this meaning that they do not know the true state of Nature. If God could speak to men, he would say the same, and Joshua could not do differently. It would be very strange to claim that a general in the army, as was Joshua, should entertain himself giving astronomy lessons while trying to show his army the power and glory of God with a victory; and abandoning the language that his soldiers could understand, should order the earth to stop moving. It would be necessary to explain the reason for such a strange way of speaking and to attempt a very ill-timed or irrelevant dissertation. Thus, even if Joshua had known by divine inspiration something unknown in those days, he would still need to express himself in the manner given in *Scripture*.

We will say the same of other biblical texts in which the holy authors could only speak as one spoke in those days, and just as we speak when we say the rising, setting, movement and inequality of the Sun.

The texts of *Holy Scripture* which seem to be opposed to the movement of the earth should not be understood in their specific, literal sense, but in a general sense as understood by men of that time. There are many texts of *Scripture*, besides those which are quoted against Copernicus, which speak of Astronomy and Physics, and which are obviously not to be taken literally, as when God says: "Tellus fundata super maria" Psalm 23 or when Ecclesiastes says: "Terra in aeternum stat". In the Scriptural texts which mention the movement of the sun, one does not and cannot suspect that the holy writers had the intention to make decisions on questions of Physics, and to express or disprove opinions on this matter.

We have no obligation to believe that through the gift of prophecy, the sacred authors knew of profane matters totally unrelated to the subjects they were writing about, or which did not affect their essence; neither the sacred authors nor the Holy Fathers whose authority one may dispute upon these matters, were acquainted with astronomy. Such a person was St Augustine, one of the great lights of the Church, who denied the antipodes; *De Civit. Dei.* bk 16 ch. 9.

There is no formal decision of the Church against the Copernican system. It is true that the Congregation of the inquisitional Cardinals made a Decree dated March 5th 1616 against Copernicus' works and the works of Zúñiga, Foscarini and another against Galileo, dated July 22nd 1633, condemning him to abjure the error of the Copernican system. But this statement is not classified as a heresy; it is only declared as worthy of suspicion and this does not forbid its justification. It was considered convenient to forbid it, in order to intercept the disadvantages which in those days could result from giving too much freedom to discoverers. But it has always been licit, even in Rome, to admit it as a hypothesis and all those who think this a safer way can do the same.

The works of Jorge Juan and de Bails gave the *coup de grâce* to the last defenders of the geocentric system, and were the open door through which the Copernican system was introduced into the Viceroyalty of la Plata by Pedro A. Cerviño[67] (d. 1816) and into New Granada by José Celestino Mutis[68]. The extraordinary contri-

[67] Cf. G. Furlong, *Matemáticos argentinos...*, p. 157–9.

[68] Cf. A. Federico Gredilla, *Biografía de José Celestino Mutis con relación de su viaje y estudios practicados en el Nuevo Reino de Granada* (Madrid, 1911).

bution of the latter to the scientific development of Hispano-America is reflected in the field of Copernicanism in the defence that was made of this system in 1774 against the Dominican Fathers who denounced it to the Inquisition; and in the report of June 20th 1801 on the pretensions of the Augustinians to go on teaching the geocentric system. From this we extract the following passages:

We will confine ourselves to just what is sufficient to form a true idea not only of the valued and superior reputation which the Copernican System enjoys nowadays among learned men, but also of the Christian liberty with which they teach it and treat it in decrees all the civilised nations of Europe without the exception of Rome itself, theatre of its glories and disgraces according to the fortunes of the times.

Our America is also taking part in these events: it was applauded when introduced into our schools and theatres: and today it is in decay.

So it seems that the novelty of the day is reduced to the examination of whether one may permit in the schools and lecture-theatres of this Capital the teaching and discussion of a system that one had been forbidden to defend positively (because as a hypothesis it has always been allowed), and even with the previous glory of having been used by chosen astronomers for the reform of the Calendar. The solid arguments brought to its defence as a thesis in the year '74 in the Colegio del Rosario, were sufficient to undeceive impartial observers besides that neither the Holy Tribunal of the Faith nor the enlightened superior Government tried to contain this famous philosophical heresy. Before and after that brilliant act in the presence of Viceroyalty, the system was defended as a hypothesis; while the University teachers felt undaunted before the weakness of the spirits startled by an opinion, to them incomprehensible; and it was an intolerable novelty to the people who were in the dark as to what was actually happening in the republic of letters and even less aware of what has happened in literary efforts to obtain the prohibition of the system...

He relates the story of the condemnation initially brought about by the intrigues of the Jesuit Scheiner (1575—1650), Galileo's enemy, and he continues:

As time wipes everything out, and personal complaints are forgotten, we are arriving at the time when, as we said before, the weakest portion of Scholastic philosophers, who want to reach everything through the telescope of their abstract ideas, sustains its party through these means. Without instruments and the continuous watchfulness of observations, one cannot penetrate the organized movements of the true system of the world, which will now and always be the only one which, among wise men, corresponds to the laws of the mechanism of the universe.

[...] from then onwards, the ecclesiastic party was weakened, whilst the freedom to teach the system throughout Italy has prevailed, and within Rome itself the wise Jesuits at the forefront of whom stands the distinguished astronomer Boscowick, the most famous Mathematician the Society has nurtured in its bosom. In his philosophy, in all his separate memoirs, and especially in the commentaries on the charming Latin poem on the Newtonian system by the Benedictine Stay he has poured the most valuable facts for the understanding of Newtonian philosophy, agreeing physically with all the astronomical system of Copernicus. Thus, already it is positively defended by everyone. Thus the innumerable books on these sciences circulate throughout our Spain and America; thus it was tacitly ordered that these doctrines should be taught, with the Supreme Council of the Nation introducing and expressly pointing out the works of Gassendi, Newton, Descartes and Wolff, all Copernican authors; without making exception on this point as would seem normal if the notorious prohibition was in practice.

Whatever may be the progress made by these teachings, we will come across them in some documents which will not admit the least answer. With the motive of the appearance of a harebrained lay Carmelite brother who set about reforming the Calendar and vomited up in his writings as many injuries as his own fury and his protectors dictated to him against the Copernican system, it was ordered that the blundering project be recognised by the Doctors of Mathematics at Salamanca University. As a consequence of such misinformed absurdities, those mathematicians pointed out the error of the project, refuting with well founded argument and wittiness all that the brother had spewed up against the Copernican system. We will transcribe the necessary and apt quotation: "As to the fourth treatise of his project which he wrongly calls "*proofs resulting from what has been said against the Copernican System*", you could have left it out, for to tell the truth, not even your Reverence will understand it [...]. I am in favour of the Copernican System, because my Mother the Catholic Church does not hinder my belief considering it as a hypothesis, because through the constitution I am obliged to explain it in my lectures and because the arguments refuting it — most of them the offspring of misunderstanding of what they oppose — lack all strength [...]". The Roman Inquisition made Galileo abjure the view held by the Copernican System and ordered that no one should follow it: and today even in Rome the public teaching of it is allowed in the view of that venerable Senate, of the Pope, the Apostolic College and many other illustrious and devout ecclesiastics who reside in that metropolis of Christianity. Now, not as a hypothesis — for that was never forbidden — it is taught, discussed and written about in an assertive tone [...]. But here the Father will speak of God and of the Inquisition. What is this? Have patience and listen. The Roman Inquisition did not forbid absolutely to follow the Copernican system: there was a prior exception for the case where one might succeed in making evidence of its truth! And it is certain that the prohibition is conceived in these terms. It so happened that this system became so dominant outside Spain (where already it is beginning to be followed) that almost all modern physicists are Copernicans. And is it not a very rational and wise judgement that, when such learned judges of different interests, nations and religions (of whom the majority respects the authority of *Scripture*, in which lies the only hindrance to the Copernican system) concurred to admit it unanimously, they were undoubtedly motivated by so many strong reasons that for the effect of persuasion one can assess their unanimity as equivalent in some manner, to a complete proof. Cardinal Polignac, in his most learned poem *Anti Lucretius*, defends the Copernican system as really existing, and this work is dedicated to Benedict XIV, of eternal memory, one of the most illustrious and virtuous Pontiffs that every occupied the Chair of St Peter and who knew (there is no comparison) much more than your Reverence, what schould be known concerning the condemnation of this system". With all this Christian freedom and frankness a little work also dedicated to a Prince of the Church, the most emminent Cardinal de Solis was written and published with the necessary licences, in 1766.

In the year '90, the Presbyter, Don Pedro del Río brought out his book on ancient and modern ecclesiastical compotus, in which he advocates the necessity of instructing the Ecclesiastics in those sciences, and, as a result, in the knowledge of the five principal systems of the world; and he gives preference to the one «most well-founded and accredited among learned men, to the one most advocated in civilised nations, and to the one which with greatest facility can explain the phenomena or appearances which we discern in the skies, which without, doubt is the Copernican System». Likewise, the philosophy course entitled *Lugdunense* has just been reprinted in Madrid, with the idea of propagating it throughout the Colleges and Seminaries, so as to gradually improve upon the teaching of our youth, and at the same time to make up for the lack of national courses ordered to be set up in the universities, according to the directions of the Supreme Council which was little or not at all satisfied with the *Goudin*, as the time limit imposed on them suggests. The above-mentioned course, *Lugdunense* which admits the Copernican system as a hypothesis, resolving it with arguments of reason and the authority of Scriptures, has been prefered to all those previously introduced.

With the reprinting of the mathematical compendium of Don Benito Bails, the preface to the first edition has been included, which has passed, unhindered, the scrutiny and careful analysis of the supreme tribunals of the government and of the faith. There the author speaks with complete frankness: "we cannot omit to warn you that this volume (the 3rd) includes a novelty which would perhaps rouse comment among many, and it is that in the principles of astronomy we demonstrate the system of Copernicus, or the opinion of the movement of the earth. Once we hold this as the true theory and it becomes a point of natural philosophy, it does not fit our frankness, to conceal this; and once we can demonstrate it we have the right to ask that before anyone condemns this system, the motives on which we base it should be carefully considered. We know that in other times this opinion was seen as a dangerous novelty, and it was forbidden to follow it; but even in Rome itself nowadays the ban on it is considered so misguided that it is no longer included in the Index of the Expurgatory, and in Spain a posthumous paper was brought out by Jorge-Juan, the thesis of which is to prove the movement of the Earth as admitted by the Copernicans..."

These words allow us to see the degree of freedom which had been attained in Spanish America towards the end of the eighteenth century. And the same occurred within the Peninsula. What importance was held by the Inquisition or censorship – which did exist formally – when in the *Diario de Barcelona* of 1792 the following announcement was published?

If anyone has the works which we name below and wishes to sell them, please send a note – with prices – to the Head Office of this daily paper, whereupon the works will be bought if the prices are not exorbitant. The works are those of John and Daniel Bernouilli, of d'Alembert, of Newton, several works by Euler, the award winning works of the Royal Academy of Science in Paris, the treatise on the shape of the Earth by Clairaut, and other works of this kind[69].

Copernicus had triumphed.

---

[69] Cf. A. Gali, *Rafel d'Amat i de Cortada, baró de Maldà* (Barcelona, 1954), p. 254.

HARRY WOOLF
The John Hopkins University,
Baltimore

SCIENCE FOR THE PEOPLE: COPERNICANISM AND NEWTONIANISM IN THE ALMANACS OF EARLY AMERICA

No learning can long flourish in an inhospitable climate of public opinion, and in the sciences, as in any other branch of intellectual activity, the measure of support which a people give to the learned is often a result of the contact and exchange between the two. The general acceptance of new ideas may well depend upon the frequency and clarity with which they enter the marketplace, whatever else their virtues. In the first two centuries of American life, the ordinary, literate American, for the annual expenditure of a few pennies, could acquire an insight into the practice of contemporary science, and a general understanding of the physical universe in which he lived. Bound up though it was with the factual miscellanea of everyday life, timetables, the weather, the meetings of local courts or religious societies and a varied assortment of calendrical information, the publication which brought him all these things frequently found space for an essay on popular science. In this role and at a time when the printed word was not yet an overwhelming flood but still cherished and savored, the humble almanac takes on special significance.

In 1639, only some two years after the founding of Harvard College, a printing press was installed in the home of its first president, there to strike off as the earliest North American printed matter a broadside called *The Freeman's Oath* and William Pierce's *Almanack calculated for New England*[1]. It was the beginning of a stream which gradually broadened to include psalm books, these of Harvard graduates, the assorted printing needs of a nascent culture and, of course, the regular issue of almanacs.

In their earliest efforts, almanac authors took up the arguments of the new science, particularly in the field of astronomy. Harvard College had come into existence a little less than a century after the death of Copernicus, and only a few years after

[1] Isaiah Thomas, *The History of Printing in America, with a Biography of Printers, and an Account of Newspapers* (Albany, 1847), p. 46.

the Inquisition had publicized the work of Galileo, yet from 1653 to 1717, the *Quaestiones* or theses of Harvard students still disputed the "science" of astrology[2]. This should come as no surprise, for the spread of the heliocentric hypothesis was extremely slow, and Protestants no less than Catholics persevered in a picture of reality which they had acquired from the Aristotelian and Thomistic explanations of the cosmos. In the middle years of the seventeenth century, for example, the astronomical collections at Oxford were for all practical purposes either Ptolemaic or Scholastic, and it was unhealthy to either praise or defend Copernican ideas at the University of Paris before 1686[3]. That aspects of astronomy thus appear to have been almost a politically dangerous, an underground subject even at so late a date is reinforced by the well-know fact that it was not until 1835 that the Church first omitted the works of Copernicus, Kepler and Galileo from the *Index of Prohibited Books*[4].

"About 1670 Copernican astronomy... was getting a little foothold in Harvard"[5], writes one historian, but actually the Copernican system was well established there before Zechariah Brigdon compiled the *New England Almanack of the Coelestial Motions for this Present Year of the Christian Aera 1659*[6]. Brigdon began his almanac with a direct challenge to believers in the old astronomy. Speaking of eclipses for the year, he writes: "Twice shall this Planet whereon we live, and its concomitant the Moon, widdow each other of their Sun-derived luster"[7]. Finally, he concluded his, almanac with a concise but vigorous explanation of the Copernican system, which is the earliest extant essay in popular science published in the English colonies.

Whereas in almost every country the new astronomy had to fight Church and priest, in New England the clergy was its primary propagator, and together with the students and faculty at Harvard (who were for the most part ministers or students preparing for the ministry) they were responsible for most of the better almanacs of the period. Thus, the advance of popular science or at least popular astronomy in America came under the aegis of a "clerical Harvard Corporation, with a clerical President, watched over by clerical Overseers"[8]. Even at this early stage, however, American astronomers did more than write popular essays on the new astronomy, for one of them provided Newton with observations on the path of a comet, that were extremely useful for the *Principia*[9].

Quite often, however, almanac compilers were more eager to demonstrate their poetic abilities or the essential soundness of their general philosophy, than to write

---

[2] S. E. Morison, *Harvard College in the Seventeenth Century* (Cambridge, 1936), p. 214.

[3] *Ibid.*, p. 215.

[4] P. Smith, *A History of Modern Culture* (New York, 1930), Vol. I, p. 58.

[5] T. J. Wertenbaker, *The First Americans* (New York, 1927), p. 124.

[6] S. E. Morison, *The Harvard School of Astronomy in the Seventeenth Century*, "New England Quarterly", Vol. VII (March, 1934), p. 8.

[7] *Ibid.*, p. 9.

[8] S. E. Morison, *Harvard College in the Seventeenth Century* (Cambridge, 1936), p. 217.

[9] *Ibid.*, p. 220.

scientific essays, but the good work which Brigdon had initiated in convincing the American farmer that the earth he plowed was a revolving planet was carried on in appreciable quantities. Samuel Cheever added "A breif Discourse of the Rise and Progress of Astronomy" to his almanac for 1661, and Nathaniel Channcey, in the almanac for 1662, attempted to prove that the planets dance "illipticall Sallyes, Ebbs and flowes" as a result of "Magneticall Charmes" which emanate from the Sun[10]. In the second of his two famous voyages to New England, John Josselyn makes note of an essay in popular astronomy whose title is prescient of the political pamphlets of the later constitutional debates between Britain and the Colonies on the eve of the Revolution. Affixed to the almanac for 1665 and written by Alexander Nowel, "a young studient at *Harvard-College* in the *Massachusetts* Colony"[11], "The Sun's Prerogative Vindicated" is a fine essay in scientific popularization. "Mathematicians", writes Nowel, "have that privilage, above other Philosophers, that their foundations are so founded upon, and proved by demonstration, that reason *volens nolens* must approve them, when they are once viewed by the eye of the intellect..."[12] A statement in support of the heliocentric hypothesis, the essay is outstanding also in its insistence on proof by demonstration and the test of reason. In this sense it represents an important milestone in the gradual, popular acceptance of scientific method in America for Nowel's paper bears all the earmarks of proper procedure: proofs are based on his own observations of stars through "Optick Tubes" and compared with the results of other astronomers such as the "Noble Ticho [Brahe]"[13].

The last three decades of the 17th century bore witness to an ever-increasing number of essays on popular science in the New England almanacs. The impact of all this can be more fully understood perhaps, when we realize that almanacs were the most widely diffused literary form and the only periodical literature in the Colonies during this period. The emphasis throughout was on the physical sciences with items relating to astronom in particular all-prevalent and all-popular. An enumeration of subjects presented and analyzed in these essays reveals the aims and inclinations of writer and readers of the time. For the history of science in early America, it is one of the few available measures of the age and of the limits of its scientific thought.

The almanac for 1674, "compiled by J. S.", contained a note on planetary orbits, indicating that "from Kepler that vigilant and ingenious Mathematician, later Astronomers have received, and are of [the] opinion, that the Planets move in an Ellipsis". John Foster's almanac for 1681 contained remarks "Of Comets, Their Motion,

---

[10] Ibid., p. 217.

[11] J. Josselyn, *An Account of Two Voyages to New England, Made during the years 1638, 1663* (Boston, 1865), p. 40.

[12] Ibid., pp. 40–41.

[13] Ibid., pp. 42–44.

Distance and Magnitude", and as an accompanying essay, Thomas Brattle's[14] *Observations of a Comet seen this last Winter 1680, and how it appeared at Boston.* Cotton Mather's *Boston Ephemeris* for 1683 contained "A Description of the Last Years Comet", Noadiah Russell's *Cambridge Ephemeris* for 1684 dealt with lightning and thunder, and the 1685 edition contained an article, "Concerning a Rainbow" by William Williams. Based on the works of Hooke and Huygens, the *Boston Ephemeris* for 1685 and 1686 presented "A short view of the Discoveries that have been made in the Heavens with, and since the invention of the Telescope", as well as a special discourse specifically designed to awaken an interest in the subject. Written by Nathaniel Mather, this account contained a discussion of sun spots, the craters in the moon and the rings of Saturn. Henry Newman's *Harvard Ephemeris, or Almanack for 1690* quoted "the exquisite observations of the industrious Monsieur Azout", in an essay entitled "A Postscript Exhibiting somewhat Touching the Earth's Motion", and also added a lecture on the tides by John Wallis. Newman's *News from the Stars, An Almanack for 1691* also contained a history of discoveries entitled "Of Telescopes", which was similar to the work of Nathaniel Mather mentioned above[15].

Thus, at the very beginning of the "Newtonian epoch in the American colonies"[16], the popularization of the new astronomy was well under way. Yet at times these young and eager compilers of almanacs wrote in a style that was not too conducive to easy comprehension, for their essays were replete with Latin quotations and a host of allusions designed to demonstrate the learning of the writer rather than facilitate the understanding of the reader. But time was to remedy this fault, and 18th-century almanacs were to be free of much of this academic posturing. Nevertheless, the 17th-century essays were genuine popularizations of a high order and the fact that they were sponsored by the clergy as well as the educated, non-clerical members of society gave them the double sanction of learning and Holy Writ, a factor of major importance in Puritan New England at least.

By the 18th century the American almanac had become the primary medium of communication between ordinary people and the specially educated in the subject matter of science. More than that, its authors were scientifically trained or at least educated enough in the literature of their subject to qualify and criticize, and frequently some of the motivation in producing an almanac, other than economic gain, was the desire to take issue with or support some particular theory. But of course the almanac was more than a magazine of popular science, and no discussion of its role could be considered balanced if it omitted at least some reference to the other vital elements which gave it importance and kept it alive for so long.

---

[14] Thomas Brattle was the American astronomer whose observations were used by Newton in calculations for the *Principia*. F. E. Brasch, *The Newtonian Epoch in the American Colonies (1680–1783)*, "Proceedings of the American Antiquarian Society", N. S., Vol. XLIX (1939), p. 317.

[15] S. E. Morison, *Harvard College in the Seventeenth Century* (Cambridge, 1936), p. 218. Morison discusses these essays in general terms between pages 216 and 220.

[16] Brasch, *op. cit.*, note 14.

No one who would penetrate to the core of early American literature, and would read in it the secret history of the people in whose minds it grew, may by any means turn away, in lofty literary scorn, from the almanac, most despised, most prolific, most indispensable of books, which every man uses, and no man praises; the very quack, clown, pack-horse, and pariah of modern literature; yet the one universal book of modern literature; the supreme and only literary necessity even in households where the Bible and the newspaper are still undesired or unattainable luxuries[17].

Others have also testified to the universal circulation of the almanac, and the list of appreciative commentary is long and respectable. For example, the memoirs of Joseph T. Buckingham, one of the best of early American newspaper editors, contain the following:

We had on our book-shelf a regular file of *Almanacks*, for near... fifty years... These periodicals I read often, and with never-relaxing interest... [A] long poetical account of Braddock's Defeat... accounts of events which led to the Revolutionary War... The Articles of Confederation between the colonies, Petitions to the King, the Declaration of Independence, and many other papers connected with the history and politics of the country, were preserved in these useful annuals, and afforded me ample food for study. But what excited my especial wonder was the calculations of the eclipses, and prognostications concerning the weather[18].

The most important aspects of the legitimate science that continued to be popularized in the pages of the 18th-century almanac were those that dealt with the physical sciences[19]. On occasion there would appear an article discussing a problem in animal husbandry or agriculture, but these were matter-of-fact statements that were more or less common knowledge in an agrarian world. Old wives remedies and a common-sense pharmacopea for everything from colds to epilepsy also appeared frequently, the information being used, like modern newspaper filler to occupy blank spaces on almanac pages. The recommended pharmacopea is quaint and the language of prescription somewhat interesting, but "six Grains of salt of Wormwood" to prevent vomiting, a "Teaspoonful of Pioney Root" to "cure the falling sickness"[20] (epilepsy), do not represent a genuine popularization of science, nor is their subject matter relevant here.

Indeed, in abstracting scientific material from the almanac the unity of thought, the blend of science and superstition so to speak, is destroyed. In minimizing discussion of all essays, save those that deal with the physical sciences, the image of the almanac as it appeared to its contemporaries is further distorted. But history is an artificial memory; as such it foreshortens some events and lengthens others, choosing to emphasize in one age what another has minimized or taken for granted. If the physical sciences interest us here, our concentration upon them is neither rigid nor dogmatic, for the countries of the mind still lay close to one another in the

[17] M. C. Tyler, *A History of American Literature* (N. Y., 1879), Vol. II, p. 120.

[18] J. T. Buckingham, *Personal Memoirs and Recollections of Editorial Life* (Boston, 1852), Vol. I, p. 20.

[19] J. Tobler, *The Pennsylvania Town and Country-man's Almanack, for the Year of Our Lord, 1771, Being the Third after Leap-Year* (Wilmington, 1771).

[20] Ibid., 1772.

18th century, and concern for astronomy could still be based on its alleged importance (via astrology) in the political affairs of men. From another viewpoint entirely, there is Nathaniel Ames' fine substantiation, in the almanac for 1736, of the good scientific doctrine of *ex nihilo, nihil fit*, yet surely not to be limited to the physical sciences.

In the early years of the 18th century, almanacs still bore the inscription "Licensed by Authority" on their title pages[21], but the importance of scientific training in the background of the author had begun to outweigh official sanction, to become a major factor in their saleability. Authors now noted that they were either "Philomaths", "Lovers of the mathematicks", or simply "Students in Physick and Astronomy".

Popular demand for discussion of the weather in the almanacs was enormous and sincere students of science could not resist its pressure, much as astronomers were to maintain a filiation with astrology through a considerable portion of the history of science. But some were sceptical of folk meteorology. Samuel Clough's approach to the "science" of weather prediction is a questioning one. Though he was forced to include such material, he did so reluctantly. "Reader", he writes,

> As for the Weather, I desire a very favorable construction; it is against my mind to put any such thing, in but considering the People have been used to it in the Almanacks of late, I have set it down to gratify those that desire it. I would not have any be so Rul'd by it as to expect such weather always on the days against which it is set... for the very Rules that the foretelling of the weather is ounded upon are not to be trusted... or depended upon...[22]

Yet, by contrast, John Tulley, in an almanac for the same year, gives straightforward weather predictions and adds an essay on "Natural Prognosticks for the judgement of the Weather". A mixture of sense and nonsense, his flowery hand portrays "the resounding of the sea upon the shore, and the murmurs of the Winds in Woods" as sings of an oncoming windstorm, and "the obscuring of the smaller stars... a sign of [a] Tempest"[23].

The prediction of the weather and the explanations of its origins were to occupy major portions of the almanac throughout the century. In their separate approaches to the problem, Clough and Tulley represent, more or less, the main characteristics of 18th-century meteorology, the sceptical, common-sense attitude of Clough on the one hand, and the naive, self-deceived, outlook of Tulley on the other. Experience and local history ruled wherever accuracy obtained, though this can hardly be verified; generalizations were the order of the day, with comments usually reflecting the season of the year. "A Frozen World yields a cold Comfort", wrote Nathan Bowen for December 16 and 17, 1727, after an earlier warning that the weather for

---

[21] S. Clough, *The New England Almanac for the Year of Our Lord, MDCC* (Boston, 1700); Treat, R., *An Almanack of the Coelestial Motions, Aspects, Eclipses, etc., For the Year... 1723* (New London, 1723). Treat's almanac bore the particular notice, "Licensed by his Honour the Governour".

[22] Ibid.

[23] J. Tulley, *An Almanac for the Year of Our Lord 1700* (Boston, 1700).

February 19 would be "Fickle and Onsteady like the Times"[24]. In 1735, Nathaniel Ames, Jr., evidently foresaw the quality of the coming winter and optimistically wrote for January 27:

> The Winter's milder than last year
> Your Hay will last, What need you fear?

Moreover the prescience acquired for the winter did not fail him in the summer and for 6 through 9 July he predicted "very unsettled weather with some strange *Phaenomenon*, and many Remarkable Accidents of both Foreign and Domestick"[25].

And so they went through the cycle of the seasons and down the corridors of time, forecasting the year's weather for the gullible, sometimes with tongue in cheek and at other times with sincere conviction. Many found opportunity in discussing the ever-popular subject to compose poetry or display their wit, frequently combining the two. By and large, however, the roots of weather observation and prediction were grounded in astrology and, to a lesser extent, in the experience of years gone by. In the almanac for 1727, Nathanel Ames could rationalize his forecasts thusly:

> ...[W]hat I have predicted of the Weather... is from the Motions & Configurations of the heavenly Bodies, which belongs to Astrology: Long Experience testifies that the Sun, Moon and Stars have their Influence on our Atmosphere, for it hath been observed for Seventy Years past, That the Quartile & Opposition of Saturn & Jupiter produce Wet Seasons; and none will deny that the Sun affordeth us his benign Rays & kind influence and by his regular Motion causeth Spring, Summer, Autumn & Winter; and if the Moon can cause the daily Ebbing and Flowing of the Tide, and has the vast Ocean subject to ther government, she can certainly change the air which is Thin, and Tenuious[26].

Curiously, in describing the nature of weather analysis, or at least his own scheme for it, Ames' explanation of the four seasons remains ambiguous, being pro-Copernican only if the motion of the sun is taken as apparent. However, the movement of the tides is directly related to the action of the moon. But safely in harmony with New England theology in its accpetance of the new astronomy at large, Ames reminds his readers that "the Stars of Heaven give us such a Noble Idea of the Infinite Power, Wisdom & Glory of God, that they Invite our Thoughts to Soar among the heavenly Glories"[27].

In the 17th century the new astronomy invoked by the groundwork of Copernicus, Kepler and the great synthesis of Newton's *Principia* occupied the minds of scientific scholars everywhere. Attempts at popularization through almanac lite-

---

[24] N. Bowan, *The New-England Diary, or, Almanac, for the Year of Our Lord Christ, 1727* (Boston, 1727).

[25] N. Ames, Jr., *An Astronomical Diary, or an Almanack for the Year of Our Lord Christ, 1733* (Boston, 1733).

[26] N. Ames, *An Astronomical Diary, or, an Almanack for the Year of Our Lord Christ, 1727* (Boston, 1727).

[27] *Ibid.*

rature had already begun in the closing decades of that century, as we have seen, but it was the eighteenth century which was to witness this full flowering. Scarcely an almanac was published, from Massachusetts to Maryland, that failed to attempt an account of some astronomical phenomena. In this category, the prediction of eclipses and the discussion of their significance, a mixture of science and superstition, stands as a major element in popularization[28].

As was usual with the versifying almanacs, techniques "to bring the cause of... Eclipses to... Remembrance", frequently took the following form:

> Th' *Opacous* Moon by day obscures the light
> Of Splendent Sol from men on Earth: by night
> The Sun opposeth her, and o're his face
> Doth cast a Sable Mantle for a space[29].

But Nathaniel Ames' approach to an eclipse is at one and the same time more scientific and more superstitious, for in speaking "Of the Eclipses This Year, 1728" he locates the event among the constellations, but subjects that information to an astrological interpretation:

> The first of these Eclipses (moon) is Celebrated in 6 degrees of Virgo, the Second sign of the earthy Triplicity, which (authors say) portends the scarcity of Fruit and Corn.
>
> The second of these Eclipses, viz: That of the Sun on the 28th of February happens in 20 degrees of Pisces, the House of Jupiter, and Exhaltation of Venus: learned Authors affirm, when Jupiter bears Rule, and is Lord of an Eclipse... [that] he signifys Glory, Fertility... Peace and Plenty; and... especially Ecclesiastical Persons do flourish and live in great Estimation. The Laws are well Executed, and Good; new Customs... Privilages... Honours... are now... conferr'd upon People in general; ...these are the Natural Portends of Jupiter when he bears Rule in an Eclipse[30].

That the good Doctor Ames held favorable views of astrology for all his actual ability at astronomy proper is obvious, yet in the same almanac he expresses his appreciation of the immensity of the universe in an address to the "Kind Reader". "In My last Years Almanack", he writes,

> I gave you a brief hint of the motions & Diameter of the Planets (sic] & their distance from the Sun: but above or beyond all these Planetary Globes is the Firmament or the Region of the Fixed Stars, which are of such Immense Distance from the Sun & this Earth... that a Learned Astronomer writes, "That if a musquet had been shot up at the Mosaick Creation to the nearest fixed star, and continued its swiftest course all the way, it would hardly have arrived there by this time, after the long interval of above Five Thousand Years". These Immensly numerous great & amazing Systems of Globes or Worlds so much surpass this... Earth... that it degenerates into but little more than a point, when... compared with the Regions above, and [the] prodigoius magnitude of the heavenly Space & the Bodies Therein contained...[31]

Here then is good popular science, picturing the size of the universe by means of analogy, in conceptual terms that the layman can grasp. Somewhat unfortunate

---

[28] S. Clough, *The New England Almanack for the Year of our Lord, MDCC* (Boston, 1700).
[29] *Ibid.*
[30] N. Ames, *Ibid.*, 1728.
[31] *Ibid.*

is Ames' failure to identify the "Learned Astronomer" from whom he is quoting. We do know, however, from entries in his diary, that Ames was in communication with Benjamin Franklin[32] and "John Winthrop, Esq., Professor of Harvard Col. F. R. S."[33], men from whom he could certainly have received sound scientific advice.

Like so many of his contemporaries, Ames did his own calculations for the various eclipses, lunations and tables of the planet's positions. Evidently he compared his results with the work of others, and with enough confidence in his own efforts, would enter into heated arguments whenever he found another in error. He found several such, and in an address "To the Legitimate Sons of Urania" warned that:

> All the Ephemeries now Extant among us, and the Tables of that nature containing the Eclipses, Lunations, Planets places, and aspects calculated for the meridien of London, are notoriously false for the first four Months of this Year, and differ from the Truth as far as light from darkness...
>
> And to' my Brethren Almanack makers may be reckoned among that Number... [that have] built upon Colson's[34] Calendar (a rotton foundation)... [and thus] filled the first four months of their Almanacks as full of Errors as there are Day in the same, I am far from laying a foundation for a long Controversy with those of my own fraternity, but... I would [only] have them Re-calculate their Eclipses... They shall find... that they are mistaken: two of 'em make their first Eclipse to be on the 24th Day of February tho' by their own Almanacks the Moon is at her first Quarter at that time, and as for their Eclipse of the Sun... to be on the 19th day of March, it is as far from the Truth as the other, for the Sun at that time is above forty degrees distance from the Dragon's head, which carries the shade of the moon to the north more than three times the Diameter of the Earth, and therefore it is impossible in nature that the Earth should suffer an Eclipse at that time...[35]

Having thus demonstrated his superiority in the world of figures, for he was correct in his calculations, and berated his London contemporary for improper reasoning and false deductions, Ames turns about, and somewhat left-handedly expresses his sympathy for his colleagues and their misfortune in blindly accepting the univerified calculations of the London astronomers.

The explanation and calculation of eclipses were the most extensive parts of astronomy that made their way into the almanacs, but there were other aspects of the subject which, if not as frequently given, were at least as valid. Attempting to give "some Brief Account of the Ebbing and Flowing of the Tides, at the Change and Full of the Moon", Robert Treat, in his almanac for 1723, wrote that,

> ... this Ebbing & Flowing... Proceeds from the Attractive Influence of the Sun and Moon, which together with the Earth and all the heavenly Bodies, have a Power of Gravitation or attraction to their own Centers; whence it follows, that if the Earth were not [so] Disturbed... the Water of the Seas would cleave down continually towards the Center of the Earth without any Tides; But the Sun & Moon having this power of Attraction, do draw and lift up the Water of the Seas towards themselves, and the Moon... being near the Earth, constantly lifts up the Water of the Sea sinto

[32] N. Ames, *Diary of Dr. Nathaniel Ames*, "The Dedham Historical Register" (Dedham: Published by the Dedham Historical Society, 1890), Vol. II, p. 26.

[33] *Ibid.*, Vol. I, pp. 11, 49, 52, 144; Vol. II, p. 97.

[34] Probably Nathaniel Colson, a noted London almanac maker.

[35] N. Ames, *An Astronomical Diary... for... 1729* (Boston, 1729).

a circular heap right under it Self... and... [it] moves along continually Round from East to West with the Moon, and this is called the Moon Tide, and is the greatest; The other... is called a Natural... Tide, being only that position or height the Waters would... abide in if not Disturbed, and this is much less: The Abounding of the Tide at the Change, is because both Sun and Moon with united force attract together. The Tide at the Full, is less than at the Change, because then the Sun and the Moon are separate; and yet bigger at the Quarters, because the Sun is about a Quadrant distant from the Moon and attracts from the place of height to the place of low Water. This is a brief and plain solution[36].

In his almanac for 1731, Ames wrote of "the prodigious Effects of Nature", that year, in "producing terrible Thunder & Lightning... tremendous Earthquakes, great Eclipses of the Luminaries... and strange *Phaenomena* in the Heavens"[37]. The "strange Phaenomena" that caught his attention was the *Aurora Borealis*, which had been unseen in New England until but eleven years previously. He sought to explain the Northern Lights, and he described their emergence from a "concatination of Causes", from "hot and moist Vapours, exhaled from the Earth, and kindled in the Air by Agitation"[38]. Such a striking phenomenon could not be observed without some comment on its meaning for human affairs; and Ames spoke of how the "great God of Nature forewarns a sinful World of approaching Calamities, not only by Prophets, Apostles and Teachers, but... by the Elements and extraordinary Signs in the Heavens"[39]. As was his custom Ames, and others of his school, saw a host of human misfortunes in the wonderful *Borealis.* Wars, famine, deluge, revolutions in kingdoms and states, alterations of laws and customs, and the multiplication of sects and seditions in the Church were but a few of the calamities for man that were read out of the play of light across the darkened skies.

In the almanac for 1732, Ames begged his "Ingenious Reader" to take note of the fact that the consideration of the "Distances... Motions... and Magnetism of the Heavenly Bodies" and the manner in which they "obey the Laws of some Omniscient Contriver" led one to admire the "all disposing Providence", who "not only guides the Rolling Worlds as they Plough the Liquid Aether", but keeps the "Thousand of Atoms that wander up and down... a Sun-Beam... under his cognizance"[40]. In a period when matters of religion were so very widely the subject for debate and conversation, Ames reaffirmed the work of the "Architect of the Universe" and the sublimity of the heavens (and therefore the study of them) as the work of His hands. In this, of course, he demonstrated how much a part of the eighteenth-century mind he was. Yet he foresaw the future destruction of the solar system, saying that "all the Phaenomena of Nature... as with one Voice declare the great catastrophe of

---

[36] R. Treat, *An Almanack of the Coelestial Motions, Aspects, Eclipses, etc. for the Year... 1723* (New London, 1723).

[37] N. Ames, *An Astronomical Diary... for... 1731* (Boston, 1731).

[38] *Ibid.*

[39] *Ibid.*

[40] N. Ames, *ibid.*, 1732.

our System"[41]. The light and heat of the sun would run out some day, and moreover, the eccentric orbit of the earth brought it closer to the sun every year, to result eventually in the scorching of the earth and the end of all life. Comets were also a danger, and the possibility that one of the many which passed through the earth's atmosphere might someday strike the earth was not to be discounted. In this connection, he discussed the "most eminent and remarkable comet that ever appeared to the world"[42], the great comet of 1680. At first, he writes, "this comet in its Aphelion past through so much Cold and Darkness, that its Atmosphere derived a vast Trail of Vapours, and meeting with this Earth at the beginnings of Noah's Flood was the cause of the same". But more importantly he cites Cotton Mather's reference to Isaac Newton's computation of the heat of the 1680 comet.

> Its Heat in its Perihelion was near 2000 times greater than that of red hot Iron. A Globe of red hot Iron of the Dimensions of our Earth... would scarce be cool in 50,000 years. If then this Comet cooled a 100 times as fast as red hot Iron, yet since his heat was 2000 times greater than that of this Earth he will not be cool in a Million of years[43].

Though once removed, Ames has drawn upon Newton to support his popular discourse, but with his penchant for astrology and an inclination to read the divine will into cosmic events, he could not help but wonder at "what Horror & Consternation this Wicked World [will] be in, when they shall behold this vast Comet like a baneful torch, blaze & roll along the unmeasurable Aether, bending its course directly to this Earth with a Commission from Heaven to burn it up"[44].

The almanacs of Nathaniel Ames, father and son, were issued regularly from 1726 to 1775, were widely circulated (as far south as South Carolina) and at their peak had an annual circulation of about 60,000[45]. They were not, of course, the only almanacs to popularize science. In Philadelphia, until 1733, when Benjamin Franklin entered the field with *Poor Richard,* Titan Leeds' *The American Almanac* was the most prominent. As almanacs generally went, *Poor Richard* was one of the best, as an average annual sale of 10,000 copies, from 1733 to 1759 testifies[46], but strangely enough, in the light of Franklin's own interest in science, Richard Saunders never wrote much about the new astronomy, or the tides or even the new field of electricity in which he was so very interested. But this is not to say that *Poor Richard* was altogether devoid of scientific material. Although the first fifteen issues show few references to scientific or rationalistic systems of thought, later numbers demonstrate Franklin's scientific deism and his concern for the popularization of science. In the issue for 1748 he notes that "on the 21st of this month, 1727, died the prince

---

[41] *Ibid.*, 1733.

[42] *Ibid.*

[43] *Ibid.*

[44] *Ibid.*

[45] S. Briggs, *The Essays, Humor, and Poems of Nathaniel Ames, Father and Son, of Dedham, Massachusetts, from Their Almanacs* (Cleveland, 1891), p. 20.

[46] J. Bigelow, (ed.), *Facsimile of Poor Richard's Almanack for 1733* (N. Y., 1894), p. 100.

of astronomers and philosophers, Sir Isaac Newton... who... *Traced the boundless works of God, from laws sublimely simple*"[47]. The same issue also noted the death of John Locke, to whom Franklin referred as "the Newton of the Microcosm". In the 1749 almanac he paid honor to Francis Bacon as "the father of modern experimental philosophy", and in 1751 he wrote an essay on the microscope, an instrument with whose aid it could be seen that "the countless Numbers of... living Creatures, the Profusion of Life every where to be observed is above Measure astonishing, and shews the Maker to be an infinite Being"[48].

One of the best illustrations of Franklin's deism occurs in the almanac for 1753, when he uses the fifty-fourth Psalm as a springboard for discussion of that perpetual question, the conflict between science and religion. He would not permit the Bible to enervate scientific deductions, and wrote that:

> [A]s to the Objections taken... [by] Scripture... which seem to contradict the Theory of the Earth's Motion, it is plain... that Revelation was not given to Mankind to make them Philosophers or deep Reasoners, but to improve them in Virtue and Piety; and... it was therefore proper it should be expressed in a Manner accomodated to common Capacities and popular Opinions in all Points merely speculative... The Truth of the Matter is, that the Demonstrations given by the incomparable Sir Isaac Newton, have established the Doctrine of the Motions of the Earth and other Planets, and the Comets round the Sun, and of the secondary Planets or Satellites round their Primaries, in such a Manner, as leaves no Room for any, but such as do not understand them, to hesitate about it[49].

Thus, like Ames, Franklin saw the hand of God in a system of immutable laws, and the apprehension of His grace through the exaltation of reason. Rather than a series of essays in science as such, he presented to the American colonists a running commentary on the philosophy associated with the new science. The age of enlightenment had opined that the Bible was less important than Nature, and the pages of *Poor Richard* showed that religion could and would flourish without sacred books or traditional authority.

Incredibly curious, Franklin had asked more questions than his friends and his generation could answer. Inquisitive and extremely ingenious, his name is associated with electrical theory, the lightning rod, the open stove, bifocals, navigation, street cleaning and paving, ventilators, preventive medicine, public libraries, fire-fighting and education[50]. But except for a few occasions, *Poor Richard* spoke rather infrequently (in comparison with Nathaniel Ames) of the known Franklin interests in science, and the fame of his almanacs rests primarily on their philosophic content, humor and literary qualities.

---

[47] Quoted by C. E. Jorgenson, *The New Science in Almanacs of Ames and Franklin*, "The New England Quarterly", Vol. VIII (Dec., 1935), p. 559.

[48] *Ibid.*, p. 560.

[49] *Ibid.*, pp. 560–561.

[50] Cf. N. G. Goodman, (ed.), *The Ingenious Dr. Franklin: Selected Scientific Letters of Benjamin Franklin* (Philadelphia, 1931), pp. 1–13.

Moreover, Franklin seems to have geared his almanacs to the squirrel virtues and the early-to-bed, early-to-rise philosophy of a pioneer generation on the make. And as is well known, the pages of *Poor Richard* abound in homely proverbs and wise old saws that articulate the cult of success. Comparing Franklin's Almanacs with those produced by Ames, shows up their deficiencies. Where Ames discussed the significance of the new science and wrote of the latest astronomical achievements, Franklin merely listed forthcoming eclipses or enumerated a series of old wives' remedies for various illnesses. Where Franklin instructed his generation on the best road to wealth, Ames brought them some of the best in English literature, producing for the colonial, perhaps for the first time, fragments from the works of Addison, Pope, Dryden and Milton[51]. In thus comparing the two, it has not been my intention to detract from Franklin the man, but briefly to indicate that there were other almanacs contemporary to *Poor Richard* that were superior in many ways, and that it has been Franklin's great and deserved reputation which has given *Poor Richard* a celebrity and fame beyond all others.

Though the almanacs of Ames and Franklin represent the best of the 18th-century's efforts, there was beneath them a large class of substantial almanac compilation. Among these lesser efforts, there is an extensive if somewhat spotty record of scientific popularization. On occasion Nathan Bowen included such material in his *New England Diary*, and in 1734, he added a history of the calendar. A gem of organization, it divided the subject into eight groups, beginning with the Julian Period, going through the Olympiads and "the Epock of Nabonnassar" to the time of Pope Gregory XIII and the correction of the Julian calendar[52]. In an almanac for 1734, Jacob Taylor briefly discussed stars and constellations from the Pleiades to Sirius, and suggested that only the first part of the night be used for observations. From Dryden's translation of Virgil, he quotes:

> And now the latter Watch of wasting Night,
> And setting Stars to kindly Rest invite.

In addition, he "scattered some fragments of astrology" through the almanac but warned against its use by quoting from Pope's *Essay on Criticism*[53].

> Such labour'd Nothings in so strange a Stile
> Amaze th' Unlearn'd, and make the Learned smile.

In 1751, Great Britain reformed her calendar so as to bring it into harmony with the continental system that had been in existence since Pope Gregory's official change (1582). This reformation corrected the eleven-day lag in the British calendar. The alteration naturally affected the American colonies and of course, was of pri-

---

[51] M. C. Tyler, *op cit.*, Vol. II, p. 123.

[52] N. Bowen, *The New-England Diary, or, Almanack for the Year of Our Lord Christ 1734* (Boston, 1734).

[53] J. Taylor, *An Almanack or Ephemeris... for the Year 1734* (Philadelphia, 1734).

mary concern to the makers of almanacs. One of the first to write an explanatory essay on the shift from "Old Style" to "New Style" reckoning was Job Shepherd, and in his almanac for 1752 he informed his readers "not to be alarmed at such a Deduction of Days, nor regret... the loss of so much Time, but take... for Consolation that... Expenses will... appear lighter... and for those who love their Pillows, [there was now opportunity] to lie down... on the 2nd of this month and not... awake... 'till the 14th in the morning"[54]. With the alteration of the calendar as a result of Lord Chesterfield's bill, almanacs after 1751 advertised that they were made "according to the New Calendar" and Parliament's sanction

Throughout the period under examination, a host of almanacs came and went, to underscore the work of the typical few selected for discussion here. Lesser variations on the theme developed in this paper, it is practically impossible to guess their number or to measure their effectiveness, save to note that they were numerous, that nearly every press in the country produced them, and that they were extensively read for all their brief duration. Daniel Sewall's *New Hampshire Almanac*[55], Nathan Lowe's *The York*, *Cumberland* & *Lincoln Almanack* for Maine[56], James Franklin's *Poor Robin* and Benjamin West's *The New England Almanack* for Rhode Island[57], to name but a few, are examples of those that belong in this category, being competent and of a moderate circulation.

To a large extent the almanac furnishes what is very likely the best index to the tastes and thoughts, the mind so to speak, of the ordinary, eighteenth century American. By exhortations to frugality, temperance, industry, piety, and the "good life", it must have played an effective part in molding his value judgements. In homely phrases that pleased the ear and caught the memory, he acquired a series of proverbs and epigrams to crystallize his thoughts and reduce his philosophy to a formula. "God helps those that help themselves", wrote Poor Richard, and the generations to follow helped themselves to the riches of a continent in the name of the Lord and the free individual. Echoing the agrarian philosophy of his days, Ames wrote in 1765 that "the Farmer only independent lives, he asks but what indulgent Nature gives". Or, what is more indicative of the colonial outlook than Ames' comment that "the most egregious Folly that I can see, is a Man living in Luxury". Proverbs existed to describe every fault and every virtue in the life of man, and it would be quite possible to construct a complete *Weltanschauung* for the 18th-century American from a careful listing of these old saws.

Another aspect of the 18th century almanac worth noting, if only in passing,

---

[54] J. Shepherd, *Poor Job, 1752: An Almanack for the Year of Our Lord 1752* (Newport, 1752).

[55] C. L. Nichols, *Checklist of Maine, New Hampshire and Vermont Almanacs*, "Proceedings of the American Antiquarian Society", N. S., Vol. XXXVIII April, 1928), pp. 87–96.

[56] *Ibid.*, p. 63–64.

[57] H. M. Chapin, *Checklist of Rhode Island Almanacs 1643–1850*, "Proceedings of the American Antiquarian Society", N. S., Vol. XXV (April, 1915), pp. 30–32.

is its direct role in the more familiar phases of colonial history. In his almanacs for 1756, 1758, and 1763, Nathaniel Ames included short accounts of the history and contemporary state of the colonies. Roger Sherman, in his almanac for 1760 gave an account of General Wolfe's reduction of Quebec. From 1759 to 1761 *Father Abraham's Almanack* (Philadelphia) gave brief historical accounts of the military campaigns in Canada, including woodcuts of the plans of various forts and towers[58]. The Seven Years' War found New England rejoicing in the success of British arms, and, as usual, Nathaniel Ames was there to record poetically that emotion.

> Great Alexander, who the World had won,
> Sat down and wept when all his Work was done.
> *Amherst* with Glory triumphs o'er his Foes,
> And rests for want of Countries to oppose.
> *Canada* conquer'd! Can the News be true!
> Inspir'd by Heav'n what cannot *Britons* do.
> The News with Haste to listining Nations tell,
> How *Canada*, like ancient Carthage fell[59].

But by 1766, with the Stamp Act passed, the protests against it filed for posterity by the New York meeting of delegates from nine colonies, and the Virginia Resolves freshly in mind, no almanac-maker could keep silent, and in the issue of his almanac for that year, with his British sympathy fading rapidly, Ames repeats the familiar maxim. "The sole end of government is the happiness of the people", and adds the jingle:

> If each blade, would mind his trade,
> Each lass and lad in homespun clad,
> Then we might cramp the growth of stamp[60].

Towards this growing colonial resentment against Britain, the American almanac was no small contributor. From 1766 to 1775 the almanacs were replete with denunciations of British tyranny and abuses; moreover, suggestions for definite anti-British policy, such as the use of homespun and locally manufactured paper, were directly made[61]. By 1775, Ames had gone far enough in his opposition to British policy to include "The Method of Making Gun-Powder", in his almanac for that year[62]. This was popularization of technology with a vengeance! Yet he was not alone, for fellow almanac-makers like Benjamin West, Abraham Weatherwise, and Edes and Gill were all with him; to ask themselves and their readers, along

---

[58] C. S. Brigham, *An Account of American Almanacs and Their Value for Historical Study*, "Proceedings of the American Antiquarian Society", N. S., Vol. XXXV (October, 1925), p. 204.

[59] N. Ames, *An Astronomical Diary... for... 1761* (Boston, 1760).

[60] *Ibid.*, 1766.

[61] C. N. Greenough, *New England Almanacs, 1766–1775, and the American Revolution*, "Proceedings of the American Antiquarian Society", N. S., Vol. XLV (October, 1935), p. 299.

[62] N. Ames, *Ibid.*, 1775. These instructions were copied and repeated in other almanacs.

with Ames in 1769, as to: "Who would sell his Birth Right for a Mess of Soup, or risque his Constitution for a Sip of Tea?"[63]

Finally, it is important to note that the faithful reader of New England almanacs from 1766 to 1775, even if he read nothing else, would have been prepared to act intelligently in the events before and after 1775. Almanacs for those years would have given him a fairly accurate record of recent public events of great importance. He would have possessed the texts of various charters and public documents, and been familiar with some of the verses of the better English poets, especially in praise of liberty. He would have been introduced to the idea of economic resistence and shown several schemes for reducing his dependency on the mother country. He would have been told who his best friends were in Britain — Wilkes, Chatham, Mrs. Macaulay — and also known something about John Dickinson. Finally, he would have been shown the possibility of armed resistance, and of course, even how to manufacture his own gunpowder!

Seen thus the almanac was obviously more than a journal for popular exegesis of science, but the latter remains one central and concluding theme. In a very real sense, the scientific concerns of the 18th-century American encompassed and expressed a good part of the American mind. Believing in the ultimate triumph of reason and sharing that faith of the Enlightenment with their European contemporaries, intellectual Americans very early found the study of nature and the pursuit of the new science congenial to their tastes. To this interest their new environment gave constant nourishment. By strange and wonderful flora and fauna, the discovery of new resources, and the geographic exploration of unknown regions, the American imagination was stimulated and expanded. To plant a civilization successfully on a new continent called for the maximum use of natural ingenuity and conditioned a pragmatic approach to the problems of life. The physical world was immediate and demanding, and it was thus no accident that early American science, striving to meet new social and economic needs, should confine itself largely to astronomy, navigation, botany, cartography and mechanics, although, as I have shown, the almanac dealt primarily in only the first of the sciences.

In the last years of its history, those of the 18th century, the American almanac gave popular expression to all of these scientific inclinations. That it was nurtured in an environment highly favorable to its growth during this period goes without saying, but in its own way it also contributed to the maintenance of that very ambience favorable to its continued existence. The question of the relationship between science and society is often one of communication. A new theorem in mathematics revolutionary as it may be, can be uttered without disturbing the calm of an uncomprehending public opinion. To a lesser extent in various other sciences new discoveries may clash with the armies of professionals among them but precipitate little or no combat from an indifferent public. But the almanac brought home to the whole

---

[63] Ibid., 1769.

of literate America the new world that emerged from the Copernican hypothesis and the great Newtonian synthesis. This it managed to do broadly and effectively without raising too much oppositional fever.

Yet heat there was as these new ideas came into contact with the world picture created by the faith of the fathers. Because it sold in the marketplace, the temperature produced by the friction between the old and the new had to be reduced. Consciously or not, the almanac authors made their compromises. Sometimes it was astrology, sometimes it was deism, sometimes it was both that were blended together in the final version of an almanac. Thus did Franklin's *Poor Richard*, Ames' *Astronomical Diaries* and all their contemporaries sell, and sell well, and in their modest way help to narrow the gap between the spiritual, invisible world of the 17th century and the allegedly comprehensible world of the 18th.

JOLAN ZEMPLÉN
Technical University of Budapest

THE RECEPTION OF COPERNICANISM IN HUNGARY

A contribution to the History of Natural Philosophy and Physics in the XVIIth and XVIIIth centuries

INTRODUCTION

Intention. This study does not set out to discuss the historyforming importance of the appearance of Copernicus's theory. It is well known to every historian of science that the Copernician ideas constituted a serious blow to the Aristotelian-Scholastic concept of the structure of the universe. At the same time, however, it is also well known that the book *De Revolutionibus Orbium Caelestium*, published in 1543 by the already 70-year-old Copernicus became common knowledge only much later. The objective reasons as well as the subjective motives of this fact are more or less well known[1]; it may also be regarded as common knowledge that the ecclesiastical-Aristotelian reaction took notice of the dangers involved in the new doctrine only at about 1600 in the time of the burning of Giordano Bruno at the stake, though it was essentially the work of Kepler and Galileo which initiated the official campaign against the followers of Copernicus. Giordano Bruno's atomistic panteism was considered at least as great a sin as his adherence to the Copernican theory[2].

It is a special point of interest that first the Protestant churches—Melanchton, Calvin and Luther—objected to the "lunatic throwing everything into confusion"[3].

By way of introduction I should like to stress the point that I do not wish to deal with the dissemination of the Copernican theory from the astronomical point of view; it is rather my intention to discuss the problems as they present themselves

---

[1] D. Stimson, *The gradual Acceptance of the Copernican Theory of the Universe*, Hanover — New Hampshire, 1917.

[2] L. Mátrai, *A Contribution to the Acceptance of the Copernican Theory of the Universe* (in Hungarian), Hungarian Academy of Sciences. Communications of the History and Social Sciences Class, 1952, 233.

[3] Stimson, *op. cit.*, pp. 18 and 21.

in connection with natural philosophy and physics in an area which from the scientific-cultural point of view was rather remote not only in space but also by fundamentals from the focal points of European development, mainly represented by the rapidly spreading bourgeoisie. Outstanding development places of were Firenze, the Netherlands, England and later France and Germany.

The area discussed in this treatise is old Hungary which includes to-day's Slovakia and smaller parts of Rumania (Transylvania) and Yugoslavia. The conclusions drawn can, however, partly be applied to the old Austria (including Bohemia of the XVI–XVIII centuries) which though nearer to the West European development had little to contribute to the evolution of the history of physics.

Another question which may arise refers to the general conclusions which may follow on investigation of the circumstances of the propagation of the Copernican ideas within a single country.

As will be seen from the following investigation, the conclusions can be summarized as follows:

1. First, of course one finds exactly the same conditions as everywhere else. In other words, one can differentiate approximately theree groups: the adherents, who belonged to the most progressive and courageous natural philosophers and scientists; the antagonists, recruited mainly from the Jesuits, and who opposed chiefly for religious reasons; and those who hesitated. This last-named group, led by Descartes and Tycho Brahe, was the most numerous, influenced by religious, psychological reasons or simply not daring to agree with Copernicus.

2. The Hungarian situation was precisely such, but the time shift was considerable. This was partly due to the fact that the various phases of European development reached this central European area afflicted by Turkish occupation and wars of liberation against the Habsburg empire at a later period; this, however, does not explain everything, since for instance Rheticus lived and died in Kassa (Košice) (although unfortunately no data on his influence there exist. It might be perhaps stated that observation of the spread of Copernicus's teaching in Hungary in the XVII–XVIII centuries reminds of a slow motion picture).

3. One must, however, consider that on a world scale developments connected with teachings of Copernicus were quite different from those which became usual concerning more recent theories. Planck writes one of his papers: "Some highly important scientific theory does not develop by gradual conversion of its opponents [...] but rather by those opponents dying off while the rising generation becomes from the outset accustomed to the new ideas"[4].

Planck is certainly right as regards either quantum theory, relativity or the older kinetic-atomistic theory of gases; but his statement does not apply to Copernicus's theory. Copernicus's work was first published in 1543, Galileo's two most impor-

---

[4] Max Planck, *Physikalische Abhandlungen und Vorträge.* III, Braunschweig, 1958, p. 208.

tant books in 1632 and 1638 and Newton's *Principia* in 1686. But when were the works of Copernicus and Galileo expurged from the Index? In 1757. Look at it how you will, more than two hundred years had to pass before Copernicus became officially acknowledged.

In the centuries between this last date and 1543, quite diverging situations are found in the various European countries. Turn where one will one finds that in this connection the Planck quotation does not apply, because the acceptance of Copernicanism everywhere took longer than one generation.

Taking all this into consideration, one feels that the analogy comparing the spread of Copernican ideas in Central Europe to a slow motion picture is justified. Everything which can be observed in connection with our problem in Europe can be found also in Central Europe, but the single historical phases may here be considerably better observed.

4. Another consequence of this slowing down consists in that data referring to the history of natural philosophy and physics are rather difficult to detect in the universal history. Such by-product is, for instance, the spread in Hungary of Carthesianism, which is here inseparable from Copernicanism.

After these preliminary remarks, let us try to survey Hungarian scientific literature of the XVII century, and also in the XVIII century, since consequent on the time lag already referred to Copernicanism constituted a problem in Hungary still in the XVIII century, even in the second half of it.

Copernicus's theory and physics. Perhaps it is not superfluous to point out an important circumstance. The acceptance of Copernicanism as an astronomical problem depended mainly upon the perfection of measurements and measuring instruments. This explains why the work of Tycho Brahe — who by the way was an opponent of Copernicus's ideas — played so important a role in the stabilization of the new theory. However, full understanding of the Copernican cosmic system called for an entirely new physics. While this had to be realized by Kepler and Galilei, Newton was able to build his celestial mechanics almost without difficulty on the cosmic system constructed by Copernicus and Kepler. For this reason, the Copernican theory does not mean simply the substitution of the heliocentric world system into the Ptolemaic geocentric system, but also complete rejection of Aristotelian physics, replacing it by an entirely new conception of the world. Though with Copernicus the celestial bodies retain the circular rotation postulated by Aristotle, with the new theory the "heavy" earth is also a part of this rotation. Even the kinematical view was new, but it really became convincing together with the new dynamics. The Galileo-Newton type of dynamics, however, was created only after Copernicus. For the great majority of the scientists, educated on scholasticism and inadequately, trained n mathematics, to comprehend such dynamics was far from easy. This explains — as will be seen below — why one finds so frequently in various publications (still in the XVIII century) the statement that the problem of the world system is a "private affair" of mathematicians. All this applies of course to a much higher degree to the Carpathian basin where the population lived at that time in relative social-economic backwardness.

Hungary in the XVII and XVIII centuries. Though Copernicus's work was published in 1543, it has already been shown in the introduction that its influence began to make itself felt only later in the XVII century, by way of a dispute lasting for approximately two hundred years. The author wishes to give an account of the Hungarian development of these discussions. It seems to be the more reasonable to begin this treatise with the XVII century, because relatively little is known of the development of sciences in Hungary up to that time, selected as a starting point[5].

It is, however, unquestionable that though the Copernican discussions were by far not concluded in Hungary in the XVII century, the character of the dispute changed by shifting to a Newton versus Descartes controversy, while the Copernican problem appeared to be only a function of this new main issue. On the other hand anti-Copernican ideas still lived in part in some backward brains[6]. This motivates the arrangement of the material into two periods of which the first lasted from 1600 till 1750, and the second was the latter half of the XVIII century.

These two great time units, however, may be in essence divided into sub-periods of approximately equal lenght. Prior to a more detailed discussion of these sub--periods it appears to be necessary to summarize briefly the economic, social and cultural situation in which lived and worked those Hungarian scientists who were in a position to join in the discussions on the Copernican problem.

After a flourishing economic and cultural life in the XV century under the rule of King Mathias Hunyadi, the whole country was almost entirely destroyed during the 150 years of Turkish occupation following the lost battle of Mohács in 1526. The central parts of the country were occupied by the Turks, the Western areas were ruled by the Habsburgs, while in the East was established a more or less independent country — the principality of Transylvania. The ruling princes of Transylvania from time to time conducted wars of independence against the Habsburg dynasty but only with temporary success. The leader of the last war of independence was prince Ferenc Rákóczi II, but his war, too, was lost in 1711. After this time, nothing stood in the way of the Habsburg policy of oppression and Germanization. None of this, of course, favoured scientific progress. Let us add that parts of the feudal ruling class were—in order to defend their privileges—in alliance with the Habsburgs, whereas a smaller part—mainly Protestants—constituted an opposition.

The well-to-do ruling class was exclusively interested in politics, and hence scientists were not recruited from it. Great masses of the people lived in poverty and ignorance. The burgesses, on the other hand, who were in considerable numbers and whose counterparts in West Europe were pioneers of the up-to-date sciences, could not develop.

---

[5] Zemplén I, pp. 21–29.

[6] Zemplén II.

It follows logically from this distressing picture that the Hungarian intelligentsia of the XVII–XVIII centuries were mainly recruited from among theologians, Catholics (primarily Jesuits) as well as Protestants (followers of Luther and Calvin, and also Unitarians).

Before the battle of Mohács, teaching was carried on in Hungary in monastic and parochial schools, a large number of which were destroyed. To replace them new schools were founded by Protestants, some of which later became flourishing institutions of higher learning (Sárospatak, Debrecen, Eperjes, Késmárk (Kežmarok), Löcse, Nagyenyed, Kolozsvár). The cream of the Hungarian intelligentsia came from the students and professors of these schools.

The rapid spread of the Reformation, of course, had its reactions. Cardinal Peter Pázmány (1570–1673) wanted at all costs to recover the lost souls, mainly by means of the Jesuits whose schools soon became considerable competitors of the Protestant schools, and—as will be seen later—if not by the standard of education, at least by advantages recurring to their pupils. The first University founded by Pázmány in 1635 in Nagyszombat (Trnava), was directed by the Jesuits until the dissolution of their order in Hungary in 1773. The new professors of what became a state university, however, were initially former members of the Jesuit Order, and the University's first significant professor of physics Anyos Jedlik in the XIX century when the university already worked in Pest, was still a Benedictine monk.

At the end of the XVII century also, the Piarists appeared and founded many excellent schools in the country.

Towards the end of the XVIII century, beside the university, many royal academies were founded with professors who were also mainly churchmen.

Summing up, it may be said that in Hungary there were to be found no scientists who were economically independent (as for instance Boyle) or who were supported by rich patrons (as was e. g. Galileo). Only theologians could become professors of the university or of the royal academies. The famous physician of Debrecen, István Hatvani, was also a Protestant clergyman. In general, only an ecclesiastical career could enable student of peasant or poorer gentry origin to continue their studies if they wished to do so.

According to the educational policy of the Jesuits, every member of the Order spent only from one to three years in teaching; thus Jesuits professors rarely had the opportunity to devote themselves entirely to a single subject. On the other hand, if a scholar obtained by his outstanding talents a professorship at any of the Protestant colleges, his ambition was even so to obtain a living in a parish, which gave him greater security: in those confused times, the salaries of college professors quite frequently went unpaid for years.

From all this, two things follow. First, one seeks in vain among Hungarian scientists working in the XVII and XVIII centuries for men comparable to Galileo or Newton, men who created anything original. What one finds is, having regard to the prevailing conditions, a rich literature of natural philosophy, which yields ample opportunities to followt he Hungarian development of the Copernican problem[7]. Another consequence, however, is that in analyzing this literature it always must be born in mind that its authors were, practically without exceptions, theologians and this, irrespective of their religious tenets determined from the onset their attitude.

The division of the material. Within the framework of the time division indicated, the grouping of the material follows from the facts examined. In the first period, between 1600 and 1750 one may find each and all of the three types already mentioned: violent anti-Copernicans, hesitant (dubious) teachers, and—this is only a small number—Copernicans. The representatives of the three types belonged to various religious groupings and were the adherents of the most manifold concepts of natural-philosophy[8]. Perhaps the Jesuits constituted an exception: they were unanimously devoted to the peripatetic scholastic philosophy and were anti-copernicans.

However, this no longer applies to the second period—that is, for the second half of the XVIII century—though general acceptance of Copernicanism was in some instances, with Protestants as well as with Jesuits, still hindered by religious--theological considerations.

I. THE MAIN TRENDS OF HUNGARIAN NATURAL PHILOSOPHY AND THE COPERNICAN PROBLEM

Approximately from 1600 to 1750 A. D.

1. Scholasticism and the Jesuits

A survey of Hungarian scientific literature, printed publications as well as manuscripts, reveals text books, commentaries, lecture notes and dissertations, all closely connected with higher education. At the University of Nagyszombat (Trnava) superintended by the Jesuits — only anti-Copernican attitudes can be found, though the number of written documents is small and the standard rather low: at this university, Aristotelian physics were taught, and the dissertations were prepared on the basis of Aristotle's theories. Nevertheless one may venture to say that an open anti-Copernican attitude is preferable to absolute ignorance.

The lectures of Cardinal Peter Pázmány, the counter-reformer already referred to and founder of the University of Nagyszombat, delivered most interesting lectures at the University of Graz. Though he never taught in Nagyszombat, it may safely be assumed that the manuscript lectures in philosophy from the years 1598/99

[7] J. Zemplén, *Kopernik i Węgry* [Copernicus and Hungary], „Kwartalnik Historii Nauki Techniki", 1962, 7, p. 259—284.

[8] J. Zemplén, *Copernicus and the Development of Physics in Hungary*, in: *Actes du XIe Congrès International d'Histoire des Sciences*, 3, Warszawa, 1968, pp. 60—67.

and 1599/600, left in full to posterity, served as a manual to the frequently changing professors throughout the XVII century. These lectures were printed in the XIX and XX centuries. Two rather comprehensixe chapters deal with Aristotle's physics and astronomy[9].

As to their content, at first glance these chapters do not deviate too much from the usual Aristotelian commentaries taught from the Middle Ages practically at every University, and at the Jesuit universities still in the XVIII century.

Nevertheless, it seems to be worthwhile to discuss in some detail from the point of view of our subject those lectures which cardinal Pázmány wrote down in the form of disputes as comments on Aristotle's book about the Heavens (*De coelo*). The highly educated cardinal felt at the end of the XVI century that he had to deal with the new concepts. Though Pázmány naturally rejects Copernicus's views, it is rather interesting that he actually was acquainted with the Copernican literature of his age. His knowledge of the material (of course up to the year 1600) can be compared with the best monographs of the XX century[10].

He reviews Copernicus's theory with remarkable knowledge of the subject, discusses the arguments in favour of the new concept, treats of the Spaniard Didacus de Sunica, Patritius, and Fracastaro, and refutes their arguments according to Clavius.

Even so, the reader should not conclude that the author thinks or wishes to assert that Pázmány agreed in the depth of his soul with Copernicus. This could be inconceivable in a faithful Jesuit, leader of the Hungarian counter-reformation. In his later years he showed no interest in these problems.

Undoubtedly, Pázmány's standard was not reached in the Jesuit dissertations[11] left over from the XVII century which dealt more or less with questions of physics; what is more, Copernicus's theory made no appearance in these works.

One exception must be mentioned — the Jesuit polyhistor Márton Szentiványi (1633—1705). He is the only author who published a larger book entitled "Some Interesting and Various Scientific Problems"[12] (see fig. 1), which has been reprinted several times.

Szentiványi is an interesting though not too attractive figure in the history of Hungarian Science. Of aristocratic birth he was an absolute and faithful adherent

---

[9] *Petri Cardinalis Pazmany Archiepiscopi Strigoniensis et Primatis Regni Hungariae Physica quam codice propria auctoris manuscripto in Bibliotheca Universitatis Budapestiensis asservato*; prepared for the press by István Bognár, Budapest, 1895; *Tractatus in libros Aristotelis de Coelo, de Generatione et Corruptione in Libros Aristotelorum*, Budapest, 1907.

[10] Cf. D. Stimson, *op. cit.*, or Thomas S. Kuhn, *The Copernican Revolution*, Cambridge, 1957.

[11] Some of these are: Miklós Szekhelyi, *Exercitatio Philosophica...*, Posoniensi, 1638 and András Mokchai, *Triplex Philosophia...*, Tyrnaviae, 1640.

[12] Márton Szentiványi, *Curiosiora et Selectiora variarum scientiarum miscellanea in tres artes divisa, Decadis Pars I—III*, Tyrnaviae, 1689, RMK II, 1652. Further parts and editions ibidem 1691, 1697, 1702, 1709; RMK II, 1700, 1906, 2132, 2133, 2384.

1. The title page of the book by Szentiványi

of the counter-reformation and the Habsburg dynasty and as — among other things — head of the censor's office a deadly enemy of progressive ideas.

In spite of this his works are not uninteresting, though from the point of view of their value they may be regarded as mixed and contradictory. It is undoubtedly to his credit that in his work, three comprehensive volumes, he practically summarized encyclopaedically the entire knowledge of his age. His book deals with sciences — physics, astronomy, biology; it contains sections of history, literature, geography, philosophy and so on. Its weakest part is the physics which is entirely treated according to Aristotle. The astronomy is somewhat better. Szentiványi was the first in Hungary to undertake astronomical observations, at Nagyszombat (the observatory was built only in the XVIII century), giving a correct description (confirming the Ptolemaic theory) of the firmament (celestial heavens), since quite obviously one can carry out correct astronomical measurements and observations by any of the world systems. In spite of his artificial conception of the universe, Tycho de Brahe was correctly considered the greatest observing astronomer of his age.

Consequently, one need not be surprised that Szentiványi — though a good astronomer — not only rejected the Copernican concept of the universe, but assumed a somewhat anti-scientific point of view on this question, which he regarded as purely fictive. In his chapter *De rebus falsae et dubiae existentiae*[13], however, he went as far as to repudiate some of Aristotle's propositions which also he considered outdated. Thus no primum mobile existed, the movement in continuous directions of the stars was also erroneous, perhaps there was even no "fire" below the moon. Similarly, no movement of the earth as imagined by Copernicus and no precession of the earth's axis existed just as there was no Salamandre living in fire or any fabulous bird called the Phoenix... (fig. 2).

From a study of the works of Jesuit authors of the XVII century one may obviously deduce that the scientific standard was not very high, and perhaps it is no exaggeration to suggest that the situation must have been quite similar in the Austrian Hereditary Provinces.

The turn of the century brought no qualitative change at Nagyszombat. Quantitatively, the number and extent of the dissertations increased[14], but—with the exception of a few works—their standard approximately reflects the value of Szentiványi's publication. Perhaps one may observe this much change — that backwardness in knowledge of the new physics was already more apparent. The publications entitled "Physics" are educational, mostly, with a single exception, chemical and meteorological dissertations. This exception was a book on mechanics, printed after the 40's of the century, in which the basic problem of this age, the Descartes-Newton dispute, was already discussed[15].

---

[13] Szentiványi, *op. cit. Dec. Sec. Pars I*, pp. 301—344, 1961 edition.

[14] Zemplén II, pp. 145—158.

[15] Mihály Lipsicz, *Statica...*, Cassaviae, 1740.

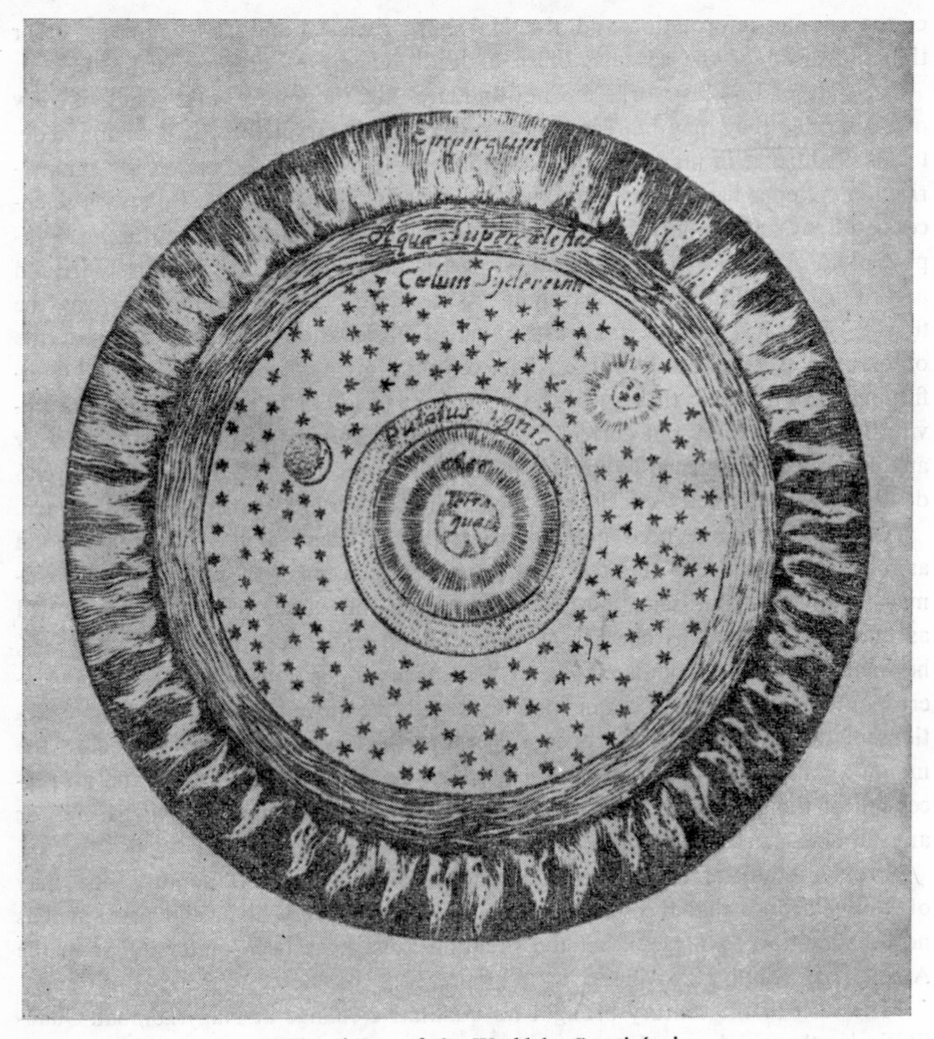

2. The picture of the World by Szentiványi

A study of dissertations dealing definitively with astronomy only confirm our opinion that the Copernican problem could not be solved without a thorough knowledge and acceptance of the new physics. One of these dissertations is Professor Peter Mayr's (1709–1753) work of 203 pages prepared for the inauguration of a young nobleman, Xaver Ferenc Klobusitzky[16].

The astronomy, written in the form of questions and answers, shows a marked scholastic influence. The book consits of four chapters. The first deals with what

---

[16] Péter Mayr, *Systema Mundi Coelestis per quaesita et responsa synopsis propositum*, Cassoviae, 1726.

is common to every star and planet; the second especially with the planets; the third with the starry heavens; and the fourth with the Earth.

A clear picture of the value of this work can be obtained from a consideration of only a few parts: thus for instance, the world was created by God with the help of angels, the stars made of water are moved by the angels. Otherwise the work is based on Aristotelian physics, with many of Aristotle's astronomical views. This indicates that in Jesuit schools of higher education science stood still not only between 1600 and 1689 (Pázmány's lectures and Szentiványi's book), but even in 1726 and, what is more, neither in the XVI nor in the XVII century can be found such a strange mixture of physics, supersit on and religion[17].

It appears that the students examined in Kassa in 1737 knew no more. On the contrary, it seems that Mayr's book served as a model for later years, since the dissertation of 536 pages entitled *Cosmographia* prepared in honour of János Gruber by professor Kristóf Akai (1706—1766) took over much from the Mayr's book in arrangement as well as in its contents. Beside positively astronomical chapters, it also contains the whole "physics"[18] (fig. 3).

Also here a few citations seem to be quite sufficient to characterize the author's knowledge of physics as well his attitude. The "folly of those philosphers" who think that the world consists of atoms had already been disproved (by the *Bible* and the fathers of the Church). God created the world on the 20 March. — The earth was the centre of the world. — The ebb and flow was caused by the Moon since by its increase it attracted more humidity. — The strength of the stars was in Hungary greater because there were many metals there, and so on[19]. It should be added that this book was still reprinted in 1739, 1741 and in 1749.

As already indicated, there were also other publications at this time (in 1740) at Kassa (Ko°ice) and Nagyszombat, though the general impression of the Jesuit scholasticism can in principle be summed up by stating that neither adherents nor worthy opposers of Copernicus are found in this period till 1755 (see later).

---

[17] One must remark that the Jesuit "dissertations" of Nagyszombat represented a special literary form, which differed somewhat from the disputes prepared at the Dutch and German universities to be considered later, and in connection with which it is difficult to decide the authorship between the "praesidents" and "respondents" participating in the dispute except when "author responds" both occur on the title page. The Jesuit dissertations were written practically without exception by professors and not by the candidate, and not necessarily by the president. Further there was usually not only one doctorand, but a whole group, the name of the most distinguished member of the group being printed in capital letters (one may assume that the dissertation was printed at his expense). The author's name frequently does not occur at all. In these cases, the authorship can usually but not always be established either by comparing biographical data, or other printed works published or by contemporaneous quotations (as example see the title page of the fig. 3. Here also the author's name appears, the names of other doctorands can be found on the second title page).

[18] Kristóf Akai, *Cosmographia seu Philosophica Mundi descriptio*, Cassoviae, 1741.

[19] *Ibid.*, pp. 3, 5, 11, 64 and 124.

D. O. M. A.

QUOD

BONUM FELIX FAUSTUM,

FORTUNATUMQUE SIT

Huic Almæ Epiſcopali Univerſitati Caſſovienſi Societatis JESU, Senatui Philoſophico, totíque Reipublicæ Chriſtianæ

SUB

Admodum Reverendo, ac Clariſſimo Patre

GEORGIO ARVAI

è SOCIETATE JESU,

AA. LL. & Philoſophiæ, ac SS. Theol. Doctore, ejusdémque Profeſſore Ordinario, nec non Inclytæ Facult. Philoſ. DECANO SPECTABILI.

Perilluſtres, Reverendi, Prænobiles, Nobiles, ac tam Virtute, quàm Eruditione Conſpicui AA. LL. & Philoſophiæ

CANDIDATI

In Aula Univerſit. A. M.DCC.XLI. Menſe Aprili, Die Horâ 8. matutinâ.

Primâ AA. LL. & Philoſophiæ Laureâ Condecorati ſunt.

PROMOTORE

R. P. CHRISTOPHORO AKAI

è Soc. JESU, AA. LL. & Philoſ. Doctore, ejusdémque in Phyſicis Profeſſore Ordinario,

3. The title page of the dissertation of Akai

2. General charakterization of the Hungarian natural philosophy

The appealing critisicm in the works of the Jesuit authors is far from implying any denominational prejudice. After all, Copernicus, Galileo, Gassendi and such were all faithful Catholics. The mainly denominational deviations in Hungary can easily be deduced from the characteristic class structure of the country in the period under investigation. The Jesuits represented the interests of the Roman church, and thus the interests of the Habsburgs, of the high clergy and partly of the aristoctrats; consequently, they opposed every innovation in the same way as the Pope and the Inquisitors were antagonistic to the bourgeoisie. The consequences of this attitude naturally became manifest also in the scientific field. The Protestant clergy and professors — who formed practically the whole Protestant intelligentsia — were linked with the lesser gentry and the petty bourgeoisie by close class relations. Consequently, they were more qualified to accept the new science. Since, however, their class structure in politics became almost unanimously apparent in religion and religious disputes, which were in the centuries under investigation the main field of ideological controversy, a much more complicated situation is found within the sphere of science.

This follows partly from the fact that science is at the same time base and superstructure, but also from the fact that Aristotle was a common source for both parties. The science of the XVII century — disregarding the greatest, really revolutionary minds — could be regarded as a sudden revolutionary change only by an anachronistic, narrow-minded historian. Enlightenments are not known even if the new discovery occurred in a really revolutionary way, as for instance at the turn of the XIX and XX centuries. The new and old ideas still co-existed simultaneously even in such cases. The reason for this was that the prevailing ideology (world outlook) was only one element of science, whereas the steadily increasing stock of knowledge, which the scientist must first learn by long, onerous work enabling him to criticize, select or develop it according to his ideology and class relations, was as important as the first element. Protestants and Catholics, progressives and conservatives could acquire in principle the same body of knowledge and could select from it only by rejecting or retaining its individual teachings. Of course, it is quite natural that the selection could have been more or less unanimous only in the case of the conservatives.

With the Hungarian Protestant scientists the foreign university which taught them was of decisive importance. The Lutherans came mainly from the upper northern Hungarian territories (today parts of Slovakia), or from German language areas in South-Transylvania (today Rumania). They studied in Wittenberg, more rarely in Jena or Halle (mainly in the XVIII century). The Calvinists of Sárospatak and Debrecen, and the Transylvanian Unitarians studied principally in Holland, but they also frequented the Swiss Universities (Basel), and rather rarely the universities in England. As a result, the Lutherans were, apart from a few exceptions, anti-Copernicans, or rank at most among those who abstained from any open expression of opinion. The Calvinists brought home from Holland the most various theological or ecclesiastical political ideas (cocceianism, presbyterianism, independentism, the social manifestation of these two latter ternds of ideas being Puritanism). It would be of no great interest from the standpoint of our subject to go into the details of all these ideas, since in the XVII century (and at the beginning of the XVIII) the Hungarian Calvinist scientists and physicists without exception accepted the Carthesian philosophy and physics. In addition to the scholastic survival from the past and the philosophically most up-to-date Cartesianism, also further effects can be detected in the Hungarian literature of natural philosophy. Above all, Johann Henrik Alsted (1558–1638) and Johannes Amos Comenius or Komensky (1592–1670) exerted a considerable influence on Hungarian authors. Partly because of their personal Hungarian connections, and partly because of their theological attitude their ideas appealed to Hungarian Protestant philosophers.

Alsted was Professor of Philosophy at Herborn, and when in the course of Thirty Years War this German town was devastated, the Prince of Transylvania, Gábor Bethlen, invited him to Gyulafehérvár to become head of the university college founded by the Prince himself. Alsted spent there 9 years and maintained a lively correspondence with numerous Hungarian scientists in many other parts of the country.

Alsted's best known works were his *Encyclopaedia* published in two editions in 27 and 7 volumes[20], and his *Physica Harmonica*[21], less known, but having a very great influence in Hungary.

Like Alsted, Comenius also fled to Hungary from the troubles of the Thirty Years War. He spent four years from 1650 to 1654 in this country and reformed the educational system of the College of Sárospatak. It is not intended here, however, to discuss either Comenius the pedagogue or the important influence he exerted in Hungary; instead, his relatively less known *Physics*[22] and its influence in Hungary as regards the Copernicus question should be dealt with in some detail.

The trend of physical ideas represented by Alsted and Comenius can be characterized by two important features. One of them is their clearly declared anti-Aristotelianism. It is interesting, however, to observe that with both this is rather an intention than an actual reality. They could have considered themselves much more deeply rooted in scholasticism. Nevertheless, and this appears to be important, both reject Aristotle as an authority; in his place, however, another authority emerges: Revelation — i. e., the *Bible*.

The only source of knowledge in Alsted's *Physics* is the Revelation. In this way this author created, in principle, a Protestant scholasticism which is in many ways

[20] J. H. Alsted, *Cursus Philosophici Encyclopaedia Libris XXVII*, Herborn, 1620, and J. H. Alstedii *Encyclopaedia Septem Tomis Distincta*, Herborn, 1630.

[21] J. H. Alsted, *Physica Harmonica Quattuor Libellis Methodica Proponens*, Herborn, 1616.

[22] J. A. Comenius, *Physica ad Lumen Divinum Reformata Synopsis*, Lipsiae, 1633.

identical with the Aristotelian concept in so far as that is consistent with the *Scripture*.

Alsted exerted influence on the younger Comenius, who, however, went further than his master, as was to be expected from the creator of up-to-date pedagogy. Also according to Comenius, one source of knowledge is the holy (*Scripture*; but according to his teaching commonsense (*mens sana*) and the experience also play an important part in acquiring knowledge.

It is not for us to go into a deeper analysis of the works of Alsted and Comenius. Nevertheless, from what it has been said, it is quite obvious that both were anti-Copernicans. It should be added that Comenius was in many respects a follower of Francis Bacon, himself—as is well known—a keen anti-Copernican. Incidentally, Bacon exerted a very great influence on Hungarian natural philosophy.

Perhaps one further characteristic of the two physical trends may be stressed. Both arose out of a belief in the Revelation, closely linked with certain mystic trends of the Renaissance.

Now, our statement that the Hungarian Protestant natural philosophy showed more shades and more nuances perhaps sounds more convincing. The more so since the many contradictory trends in which scholasticism, rationalism and mysticism became mixed up with Bacon's empiricism created manifold possibilities of choice, and manifold indeed were the choices.

It is of further interest that—as can be shown by a more detailed analysis—all these natural philosophies are far from being unanimously anti-Copernican. Among their adherents can be found hesitating scientists (in numbers by far the largest) as well as (though smaller in number) followers of Copernicus.

The survey of Hungarian trends in natural philosophy is by far not exhausted by the above discussion. Alongside scholasticism, Cartesianism, mention might be made of Protestant scholasticism, Bacon's empiricism, mysticism and also atomism (Sennert, Sperling) or other mixed trends. This, however, has been avoided since only those ideas are discussed which seem to be important from the point of view of the Copernican question. Only essentially pronounced anti-Copernican are dealt with, and it is highly interesting to see that the results are not unanimous.

Study of the works of the various non-Jesuit scientists, however, clearly shows that the influence of this or that teaching is properly reflected, and that in some cases pupils accepted standpoints more or less independent from the idea developed by the master. Consequently, it seems to be more purposeful to group the works to be discussed according to the various philosophical ideas and not to the authors, belief or disbelief in Copernicus.

3. Scholasticism and the Lutherans

One thing may be pointed out at once. The relation of the Protestants to scholasticism seems to be profoundly different from that of the Jesuits. Alsted and Comenius in principle adhere to the philosophy of Aristotle simply by their loyalty to the *Scripture*, without any special religious semblance.

Since—as has already been pointed out—many Hungarians studied in Wittenberg, the dissertations and books of the students learning in that famous German university, best show this effect. Much decisive criticism of Aristotle can be found, but similarly to the Jesuit works systems of the universe are either not considered or the authors try to avoid expressing an opinion. Consequently, while they take a decisive stand against Aristotle in the question of the substantiality of atoms, or in general in the question of matter and form they can scarcely free themselves, when the motion of the Earth is discussed, from the influence of Luther or Melanchton.

The works of many Hungarian scholars studying in Wittenberg in the XVII century show considerable differences. One finds scholars who appear only as the authors of dissertations on physics or astronomy, and neither their later scientific work, nor their biography is known. In many instances the nationality of the author is disclosed only by the title page of the dissertation. In other cases, the situation is more favourable, not only the author's biography but his whole work is well known.

Thus for instance many details of the life and work of the Hungarian scientist Jakab Schnitzler, are well known. Schnitzler (1634—1674) was one of those Hungarian scientists (e. g. János Apáczai-Csere, Istvan Hatvani) who did not accept a foreign university chair offered but came home to help to improve the backward national cultural situation. Schnitzler was in a rather stormy period first Professor of Philosophy and later chaplain in Nagyszeben. In Wittenberg, where he studied, as well as in Nagyszeben, he wrote a considerable number of philosophical, theological and geographical papers, and studies on architecture; he also presided several times at sessions for the defence of inaugural dissertations. He was very active in the field of astronomy, set up an astronomical telescope on the market, and taught the elements of astronomy.

Even so, his studies in Wittenberg influenced his whole life, he never reconciled his views with Copernicus, and added to the existing arguments against Copernicus's teachings new arguments of his own.

Schnitzler was no mystic philosopher; on the contrary he was a sober and highly practical man. Yet to explain phenomena he frequently turned to the supernatural. A quite characteristic example of this is his book *De Terra*[23], in which he rejected Copernican notions as expounded mainly by Galilei and maintaining that the holy *Scripture* is popular, its "vulgar" style addresses the common people and is consequently easily intelligible; instead, Schnitzler affirms that the miracles as described in the *Bible* are actually realistic. According to Schnitzler there exist both natural (*naturalis*) and supernatural (*supernaturalis*) ones. Thus, for instance the comets and the new stars. It is true — he says — that in philosophy one does not like to turn to the supernatural, but "cum haec naturalibus nonnisi per naturam et causas

---

[23] Jakob Schnitzler, *Disputatio Physica de Terra*, Wittenberg, 1658. RMK III 2045.

secundas agat, ubi tamen cessant naturae leges et nulla causarum secundarum intercidit, id ipsum tutissime fieri potest".

Schnitzler's whole life work shows this duality: the curiosity of the scientist goes hand in hand with the cautiousness of the devout theologian. Of course, to go into a detailed discussion of his voluminous works, would lead us too far[24]. So let us be stisfied with a few typical matters concerning the Copernican question, which, however, are characteristic not only of Schnitzler but also of a considerable number of scientists similar to him who lived in XVII century not only in Hungary, but in Central Europe (and occasionally here and there also in West Europe).

For instance, not in a scientific work but in a sermon to his parishioners on earthquakes one reads the following sentences: "No one should believe that this (the earthquake) refers to the motion of the whole Earth — that is, this does not refer to the idea which holds that the Earth revolves around its axis like a ball while the sun and the other stars of the sky remain immobile, as was once taught by the scientist Copernicus and with him (and through his work) many others in clear contradiction to the holy *Scripture* and the express word of God"[25].

This is a regularly recurrent thought in his arguments against Copernicus: nothing is impossible for God. He answers Kepler's witty analogy that the cook who revolves the kitchen fire around the spike ewer must be mad by saying: *non videndum esse, quid fieri possit, sed quid acta fiat in Natura*[26].

Nevertheless, as can be seen from his counterarguments, Schnitzler's erudition must have been similar to Pázmány's though the standard of his scientific arguments is not very high, and only rarely surpasses that of Melanchton's: the discharged cannon ball would miss the target, always an east wind should blow, a stone falling from the peak of a tower could not fall to its foot, everything would fall off the rapidly revolving earth, and so on. He invented also two new counter arguments: as a result of the violent friction with the air the revolving Earth must catch fire; it is well known from architecture that the base must be solid, yet the Earth is the base of the world...[27] According to this author, that Cartesian argument that our senses easily cheat us does not hold, and in general the physics of Descartes-Regius is thoroughly incorrect since it leads to atheism. For this reason, Aristotle is not always right, even great men can fall into error. Following Sperling[28], he makes a few modest corrections in Aristotle's physics, though he opposes Maestlin.

---

[24] Zemplén I, pp. 139–144.

[25] J. Schnitzler, *Bericht aus Gottes Wort und der Natur von der Erdbewegungen Ursprung und Bedeutung*, Szeben, 1681. RMK II 1495.

[26] J. Schnitzler, *De Terra and Tractatio Astronomica de Globo Coelesti*, Wittenberg, 1661. EMK III 2152.

[27] Idem, and *Tractatio Geographica de Globo Terrestri*, Wittenberg, 1662. RMK. III 2180.

[28] Johann Sperling, professor of philosophy in Wittenberg, pupil of the atomist physician Daniel Sennert (1572–1637). Both were well known representatives of the antiperipatetic school exerting a great influence also on Hungarian scientific life, though their criticism of Aristotle is somewhat modest.

Finally, he explains quite characteristically how the undoubtedly well educated and excellent astronomer Copernicus (and together with him Galileo and Kepler whose astronomical works he not only knew quite well, but frequently quoted them and highly estimated their astronomical discoveries) could believe in such an erroneous theory:

"One can sincerely doubt whether Copernicus accepted physically the motion of the earth, since he treated this question mathematically..."[29]

This subterfuge one finds in the dissertations published in Wittenberg quite frequently, more so than the allusion to the supernatural, in which—by the way—Schnitzler stands rather alone among the rationalistic thinkers.

Thus, for instance, János György Graff defended his thesis *De Coelo* in 1648 under the patronage of Sperling. In his dissertation, he disproves many Aristotelian theses: the material of the sky is neither more noble than that of the Earth, there is no fire below the Moon and so on. On the other hand: the distance between the Earth and Sky is always the same and does not change with time, since *the Earth is exactly in the centre* and is the starting point of every exact calculation[30]. This, however, is mentioned only occasionally. For example, the statement that in the work of Mihály Unger the standpoint of the author can be ascertained only by his occasional statement that the construction of the sky as given by him can be deduced only from the motion of the Sun[31].

Somewhat earlier than *De Coelo*, another publication is to be found which is decidedly anti-Copernican. In his disputation called "geographical", György Sebastiani writes that the subject of geography is the shape, dimension, position and immobility of the Earth[32].

In connection with these two latter features, the author says that the Earth is exactly in the centre of the Universe. Concerning the problem of mobility, a dispute existed between the "physicists" and the "mathematicians" (though it does not become clear which party is meant by the former). Today—he writes—many scholars attack the thesis of the immobility of the Earth; though this question had already been buried for centuries until Copernicus, an excellent astronomer, revived it two hundred years (!) ago, thereby achieving a great reputation. He had many followers who thought that it was easier for the small Earth to revolve round its axis in 24 hours than for the sun together with the whole firmament: "The authority of the most famous and most ancient authors, however, is against this..." Then follow the well known arguments based on physics and on the *Bible*.

In a thesis[33] on earth-quakes written in 1674 by Kristóf Mazar (d. 1709) the fol-

---

[29] Schnitzler, *De Terra*.

[30] János, György Graff, *Disputatio Physica de Coelo...*, Wittenberg, 1648.

[31] Mihály Unger, *De Aequatore et Zodiaco...*, Wittenberg, 1662. RMK 2188.

[32] György Sebastiani, *Disputatio Geographica de Affectionibus Terrae*, Wittenberg, 1659. RMK III 2100.

[33] Kristóf Mazar, *Disputatio meteorologica de Terra Motu...*, Wittenberg, 1674.

lowing lines can be read: The earthquakes are caused according to Athanasius Kircher by light *effluvia spirituosa* buried in the depths of the earth because, according to Aristotle, the "natural place" for the light bodies is the edge of the Universe which they try to reach by uniform motion; motion is not the nature of the Earth "if only one does not with Copernicus by a principle of his own want to assert that the earth-globe has a universsal motion". This opinion, however, was already condemned by the cardinals under Pope Paul V, and Galileo, an adherant of this theory, was forced to renounce it on oath. The author adds, however, with the already eminent caution, that this does not belong to the essence of the question.

Two dissertations on comets by György Gastitzius are also characterized by this cautiousness[34].

It is characteristic of the XVII—XVIII century that various authors prefer to discuss rare phenomena. The rarer a phenomenon the most likely was its discussion. They wrote not only on earthquakes and volcanos, but on phenomena rarely observable on the sky, such as the halo, double sun, draco volcans and so on, and of course also on the comets.

Though the comets are really rare phenomena, they have a double meaning for the Copernican theory. Superstitious beliefs together with believing in astrology regarded the appearance of a comet as a premonition of calamity, as an evil—wars, famine, epidemies. Any opposition to this belief may be regarded as a progressive step toward the up-to-date science, though it still can be made consistent with the acceptance of the peripatetic-scholastic construction of the firmamentum.

If, however, one also knows that the comet cannot move "below the Moon", but races in Aristotle's eternally unchangeable firmamentum over immeasurable distances, this is already a new, though not final step in breaking the Aristotelian world-system.

Two such works can be found in the Hungarian literature of natural philosophy, which exemplify the fact that progressive thinking, anti-Aristotelian and not superstitious, does not necessarily mean the acceptance of Copernicus's system of the universe. The author of one of these two works is the famous Hungarian humanist and physician, András Dudith (1533—1589), whose dissertations on comets is his only astronomical work[35]. Dudith rejects every superstition, but he also confesses

---

[34] György Gassitzius, *Hypothese de Cometis, quas Praeside... Dn. Michaelo Waltero... propugnavit. G. G. Hungarus*, Wittenberg, 1679. RMK III 3010, and *Exercitatio Academica de Comentarum Natura et loco, quam... Praeses G. Hung...*, Wittenberg, 1679. RMK III 3011. The thesis first quoted was defended by Gassitzius on 10th September; three days later he alrealy presided over an examen. Thus one may assume that in his dissertations his *own ideas* were developed though he also quotes the opinion of the presiding Michael Walter.

[35] *Andreae Duditi viri clarissimi de Comentarum Significatione Commentariolus, in quo non minus eleganter quam docte et vere mathematicarum quorundam in ea re vanitas refutatur*, Basel, 1579. RMK III 679. For Dudith's biography and complete literary work, see e. g. Costil Pierre, *André Dudith Humanist Hongrois 1533—1589. Sa vie, son oeuvre et ses manuscrits grecs*, Paris, 1935.

that physically as well as mathematically only little is known of the essence of comets. It is of course quite understandable that such a statement could be made in 1579.

The influence of Dudith can be seen almost a hundred years later in György Csipkes' (1628—1678) book published in 1665. Csipkes was physician and professor in Debrecen and wrote his book in Hungarian[36], a very rare event in this period, showing the author's educational and pedagogical intentions. In spite of the 100 years, however, Komáromi did not get further than his master. He too, would only be happy to know more about the comets though he is sure that whichever of the many contradictory theories about the comets proved to be right nothing can be predicted from their appearance.

The progressive character of the viewpoints of Dudith and Komáromi are stressed by the fact, for example, the already discussed Schnitzler used the comet which appeared in 1680 to admonish his parishioners in a magnificent sermon designed to convert. It is true that according to Tycho the comets are of physical origin, nevertheless they are created by God to warn men of the punishment to be expected[37].

After this interposition let us turn back to Wittenberg to see what was the opinion of comets held by György Gassitzius in 1679.

He also complains of insufficient theories, because up to now — he says — nobody could throw any light on this problem "nisi peripaticus fuerit", and he feels he cannot accept this already outdated theory. Indeed he is—on the other hand—no Cartesian, does not sympathize with the ideas of Kepler, Galileo and Gassendi. The scientists are so modest that though "they bestowed on us many physical hypotheses, they are still presented as mathematical theories". Finally, on the basis of the works of Sethus Vardus[38] and his chairman Michael Walter, he comes to the conclusion that the comets rise to very great altitudes, which is why they can rarely be seen, and move on elliptic orbits.

If, however, he accepts an elliptic orbit for the comet he necessarily must attribute to the other heavenly bodies some other motion. Then in turn one has to face the following "paradox": 1. The Earth moves. 2. A new world-system must be constructed. 3. The Universe is infinite. The name of Copernicus is not mentioned. The author only remarks thet these suppositions of more up to date authors (recentiores) constitute no part of his subject, and is not sure whether the elliptical orbit of the comet can be proved by deciding this problem only. Thus, highly characteristically of the methods of Wittenberg, he does not decide at all.

These few examples from Wittenberg show clearly that the scholastic structure, though already strongly shaken, still does not collapse, and the geocentric concept at which the first was aimed already in 1543 still kept its ground even a hundred years later.

---

[36] György (Komároni) Csipkés, *On the Judicaria Astrology and on the comets* (in Hungarian), Debrecen, 1665. RMK I 1023.

[37] Jakab Schnitzler, *Comet-Stern predigt...*, Szeben, 1681. RMK II 1494.

[38] Vardus, Sethus or Seth Ward (1617—1689), professor of astronomy in Oxford.

4. The title page of the book of Bayer

It is true, that except for Schnitzler, whose work was rather extensive, the examples presented (and many others similar could have been quoted) are from authors who did not remain scientists. Thus these works are somewhat characteristic of Wittenberg thinking.

Nevertheless, Wittenberg did turn out scientists who carried out their scientific work on a larger scale, thus proving that in addition to the more or less mediocre professors, they not only knew others also, but were capable of reproducing their ideas on natural philosophy in a synthesis which proved their independent manner of thinking.

In this connection, only János Bayer of Eperjes is briefly mentioned for works reflecting the influence of Comenius and Francis Bacon[39] (fig. 4).

However, Bayer's two natural philosophical, methodological works[40] are of great interest for the history of Hungarian physics. His attitude in the Copernican question can be seen in advance.

Though Bayer propagates quite vigorously the efficiency of the inductive method, the importance of science, the indestructibility of matter, yet the Earth stands in the centre of his universe, immobile and around this centre revolve the Ptolemaic spheres. There is only one thing he admits when discussing the division and characterization of sciences: he acknowledges that it is the task of astronomy to set up "artificial system of the universe"[41].

With Bayer the XVII century trend in which scholasticism and the *Bible* played a decisive role was approximately exhausted; the new ideas present themselves only in very pale contours. Let us pass over to another line of thought, to be summed up as Cartesianism. The connections between the philosophy of Descartes and the universe of Copernicus survived in Hungary into the XVII century and extended deep into the XVIII century.

4. Cartesianism in Hungarian natural philosophy and the Copernicanian question

The authors discussed in the previous section were without exception Lutherans. The authors of the various treatises to be dealt with in this section, however, were followers of Calvin, who — as already pointed out — went from the College of De-

[39] János Bayer, *Ostium vel atrium Naturae...*, Cassoviae, 1662. RMK II, and *Filum Labirinthi vel lux mentium universalis...*, Cassoviae, 1663, RMK II 984.

[40] Thus e. g. János Kvacsala, writes in his paper — *A Half-Century of the History of Philosophy in Hungary*, (*Budapesti Szemle 175*, 1891, p. 186) — that it is "the best product of Hungarian XVII century natural philosophy"; Morhof (*Danielis Georgi Morhofi in Tres Tomos Literarium Philosophicum et Brassicum Opus Posthumatum...*, Lübeck, 1747, II 102) regards him as even more zealous and dilligent than Comenius. According to J. F. Budde, (*Introductio ad historiam philosophiae hebraeorum*, Halle, 1710, 256) he was an original thinker who deviated from the accustomed way of philosophy.

[41] Bayer, *Lux Mentium*, p. 27.

brecen, Sárospatak and Transylvania (Gyulafehérvár, Nagyenyed, Marosvásárhely, Kolozsvár) to complete their studies in Dutch and Swiss universities.

Also in this section brief disuptations as well as voluminous text books, discussing and practically entirely accpeting physics of Descartes will be met with.

It appears to be superfluous to deal within the scope of this treatise with the physics of Descartes, his vortex theory as treated in these works, since it is well known that he distinguishes three types of variously fine matter. Light is made of the finest of these; less fine is the ether, consisting of extremely small spheres filling up the entire universe, whereas the particles of the rougher material are everywhere cornered and edged so that the ether can permeate everything. At the creation God gave the ether a well determined *mv* impulse, which in the whole universe since its creation is constant. The heavenly bodies move in the universe by friction with vortices consisting of the closed ether chains. Also, the Earth drifts in such a vortex that it stands unmoved in the centre of the vortex. It follows from Descartes' concept of matter and the vortex theory that Carthesian physics can be made to concur without any serious concessions with atomism as well as the Copernican universe.

It will be of interest to see how the Hungarian authors receive the Copernicus theory. We shall thus investigate the reactions in this case of the various Hungarian scientists.

The first to write about and taught Cartesian philosophy and physics was János Apáczai Csere, who had a tragic life and died young (1625–1659).

Apáczai studied as a Calvinist student in the College of Gyulafehérvár (whose first rector was Alsted), whence he went to Holland. He studied almost five years in Utrecht and came home with a Dutch wife to Transylvania, where the young philosopher infected with the philosophy of Descartes and Puritanism was received with suspicion. He experienced many difficulties and for some time was even prohibied to teach.

In spite of all his difficulties, Apáczai succeeded in finishing his great work which he began still in his years in Utrecht. His aim was to write in *Hungarian Encyclopaedia* of sciences[42].

To write in the native tongue meant in the XVII century (Stevin, Galileo) a progressive attitude. By constrast with the already cited György Komáromi Csipkes, however, Apáczai did not want to write a popular book; he set out to show that even the deepest scientific ideas can be treated in the Hungarian language.

Of course, it is beyond the scope of this paper to deal with, in addition to Apáczai's mathematics, physics and astronomy, the gramatical, logical or economical chapters of his book; only two remarks need be made:

---

[42] Janos Apáczai Csere, *Hungarian Encyclopedia or a proper arrangement of every true and useful Wisdom...* (in Hungarian), Utrecht, 1653, RMK III 876. This publication is, however, missing, and the quotations are from a later publication published in 1655, RMK III 1941.

One refers to the quite natural fact that one of Apáczai's models was Alsted's work, and the other, the work of Gisbert Voetius (1588–1676)[43], a famous anti-Cartesian philosopher, who always showed great goodwill towards Hungarian students and thus also towards Apáczai. Even so, neither Alsted's religious mystycism nor Voetius's pedantic orthodoxy are reflected in Apápczai's work which represents the most up-to-date Dutch natural philosophy of that time, though with a certain independence.

The second remark concerns the fact that Apáczai was not only the first Hungarian follower of the philosophy of Descartes, but also Copernicus's first disciple.

In his work Apáczai deals with the astronomy and the structure of the universe in the Part VI, entitled *On the haevenly things*. Here he discusses the Carthesian vortices and Descartes's three types of matter. With him the break with peripatetic philosophy is complete, though according to his work the vortices are so constructed that "the sun is in the very centre of our firmament where the sky with all the animals in it revolves at a terrific speed".

The further Cartesian physical dissertations and book of this period are either non-commital as to the Copernicanian question or their independence is of a lesser degree. One may, however, state in general that — as has been already pointed out — the vortices of Descartes can easily be reconciled with Copernican ideas, though this is not always explicitly stated. Even so, one feels justified in thinking that among the Hungarian trends in natural philosophy the most progressive was undoubtedly Cartesianism.

In connection with Apáczai, the position of the Utrecht school of Cartesianism has already been briefly mentioned. Let us now investigate two dissertations, both from Leyden. The author of one of them is János Köpeczi[44], who became professor in Sárospatak and later physician in Transylvania, he presented his thesis in 1666. Like the work of Gassitzius in Wittenberg, this work also deals with the comets, though with a much more determined attitude: the author is in every question not a disciple of Descartes but of Aristoteles. Matter and its motion is entirely sufficient to explain the universe. In connection with the motion of the comets, the motion of the planets is also treated of course, but without mention of the Earth.

The situation is quite similar with the other Leyden dissertations dealing with the system of the universe. One author is an interesting polyhistor, Samuel Köle-

[43] G. Voetius succeeded in obtaining a decision passed by the senate in Utrecht forbidding the teaching of Descartes', and what is more even his name was not allowed to be mentioned at the university lectures (this explains why in the works of many Hungarian authors the name of Descartes never appears). *R. Descartes et le cartésianisme hollandais*: *Etudes et Documents*, by several authors, Amsterdam 1960, including C. Louise Thijssen Schoute, *Le cartésianisme aux Pays-Bas* (p. 183–260) and Vrijer, *J. Henricus Regius Een* "*Cartesiansch*" *Hoogleeraar an de utrechtsche Hoogschool*, Amsterdam, 1952.

[44] János Köpeczi, *Disputatio Philosophica de Cometis*... (president Jean de Raey, Cartesian philosopher), Leyden, 1666, RMK III 2342. The dates of the birth and death of J. Köpeczi are unknown. This is usually the case also with other authors where such dates are not mentioned.

séri (1663–1732)[45] (fig. 5). Köleséri first studied philosophy and theology in Leyden; later, he became a certificated physician in Basle. As a municipal physician he also dealt with mining, metallurgy and wrote much on these subjects. He was, perhaps, the first Hungarian member of the Royal Society.

The dissertation in question is an inaugural thesis which the candidate defended "sine praeside", in those times considered a great distinction. Nevertheless, it is known that Köleséri was a disciple of the great Burcher de Volder, Cartesian professor, one of the first to teach experimental physics, since in the same year he also defended, this time under de Volder, an optical work written in the Cartesian manner[46].

One interesting feature of his dissertation on the system of the universe is that in this work he mentioned Descartes by name. He treats his subject in a faithful Cartesian manner. Though he mentions Galileo as well as Copernicus, he can say no more that everything which seems to be acceptable in their ideas is already included in the theory of vortices.

Apáczai's imposing experiment had a more profound impact on only a few of his fellow citizens, since both dissertations are characteristic mainly of the Dutch universities. However, in 1678 the first entirely Cartesian physics was published in Debrecen (it will be seen later that something similar had already happened in Sárospatak, though since that is not a purely Cartesian book it will be discussed later). The author was Márton Szilágyi Tönkö (in Latin Silvanus) (1642–1700) professor in Debrecen[47].

From the standpoint of the history of Hungarian physics, this book indicates that in one of the oldest and most important colleges of the country the Cartesian teaching had become accepted. It is of interest to note, however, that also in this book the name of Descartes is cited only one or twice.

What does this mean as regards the Copernican question? The answer is again somewhat uncertain. As concerning the position of the Earth the author points out the following "paradoxon". The Earth stands still — this must be so since the *Holy Scripture* says so — in spite of this however, it may move *per accidens* with a hidden motion, which man does not observe, because he notices only the motion of the other planets[48]. Yet in describing in detail the motion of the planets he contradicts this statement, since he says that the planets revolve round the sun, and the Earth is a planet[49]. Thus again we see that Cartesian physics is in general attractive, but in 1678 Copernicus seems to be more highly valued. Galileo and Newton are still not sufficiently known, though Silvanus sometimes quotes them.

As already indicated, Sylvanus's philosophy in Debrecen marked the beginning

---

[45] Sámuel Köleséri, *Disputatio philosophica Inauguralis de systemate mundi...*, Leyden, 1681.

[46] Sámuel Köleséri, *Dissertatio mathematico physica de lumine...*, Leyden, 1681.

[47] *Martini Sylvani in Illustri Schola Debrecina Rectoris Philosophiaeque et L. Professoris Philosophiae...*, Heidelberg, 1678, RMK III 2899.

[48] Ibid., pp. 168–169.

[49] Ibid., p. 208.

5. Samuel Köleséri, a member of the Royal Society

of a more extensive Cartesian period. This is proved for example, by a thesis defended in Debrecen in 1702. The author and defender was István Csapó[50], the president Mihály Vári (d. 1723), probably successor to Sylvanus. (fig. 6). The thesis deals with the planets in the manner of orthodox Cartesianism but without any concessions to Copernicus such as his professor Sylvanus had made.

In the course of investigating the spread of Cartesian philosophy in Hungary, we have thus reached the XVIII century. The evaluation must be somewhat different at this period, since not only are we after Copernicus and Galileo, but also Newton's *Principia* was already published. The new physics was created, the lack of which might have been an acceptable excuse to anyone who contradicted the Copernican ideas or was at least hesitating in their acceptance. Consequently, Cartesian philosophy can no longer be regarded as progressive in the XVIII century, but represented a force which was rather pulling back and thus retrograde, together with the scholasticism which still lived in Nagyszombat, though the contents of Cartesianism were more up to date than the scholastic teaching.

Consequently, just as in the Jesuits colleges scholastic philosophy was still taught, in the Calvinistic colleges Cartesianism within the frame of which the Copernican question developed either in a positive or in a negative direction, became inflexible.

The author of the first printed book of this period was Mihály Szathmáry (1681–1744) professor first in Marosvásárhely (Transylvania) and later in Sárospatak[51]. His text book on physics[52] was published in 1719.

The concept of Szathmáry's book is purely Cartesian. The Copernican problem is even more facilely solved that in the works of his predecessors. After reviewing the systems of Ptolemy, Copernicus and Tycho Brahe, he states that while the systems of Ptolemy and Ticho de Brahe result, because of their complexities, in absurdity, "Descartes follows Copernicus's theory"[53].

The change of time is perhaps best indicated by the fact that in the text books of physics the various possibilities of describing the universe — by the systems of

---

[50] István Csapó, *Dissertatio Physico-Astronomica de Planetis...*, Debrecen, 1702, RMK II 2102.

[51] The lot of the Protestant colleges in Hungary at the end of the XVII and the beginning of the XVIII century was somewhat turbulent. Thus, e. g., the famous college of Sárospatak was expelled from the town by the Jesuits in 1671. The students and professors had to take to the road. Finally, the Prince of Transylvania settled them in Gyulafehérvár, whence the school of Alsted and Apáczai had to flee from the Tartar invaders to Nagyenyed. Expelled again, this time by the imperial mercenaries, they finally established themselves in Marosvásárhely. Though the college was later reopened in Sárospatak, the school in Marosvásárhely also remained as a flourishing educational centre of Transylvania. For these and similar reasons the biography and affiliation of the professors presents a somewhat difficult problem.

[52] Mihaly Szathmáry (paksi), *Physica contracta juxta Principia Neotericorum in usus ill. coll. s. p. Maros-vásárhelyiensis...*, Claudiopoli, 1719.

[53] Ibid., pp. 106, 108 and 113.

DISSERTATIO
PHYSICO-ASTRONOMICA,
De Planetis;
Quam
Auspice DEO;
Sub Præsidio CELEBERRIMI ac CLARISSIMI VIRI,
D. MICHAËLIS VÁRI,
A. L. M. & sanioris Philosophiæ Doctoris, ejusdemque ut & Sacrorum Bibliorum in Schola DEBRECZENIENSI Professoris:
STEPHANUS CSAPO
Nobilis JAURIENSIS, Auctor defendet,
Ad diem Octobris.

DEBRECZINI,
Per GEORGIUM VINCZE. *Anno* 17

6. The title page of the dissertation of Csapó

Ptolemy, Copernicus, or Tycho system — Cartesianism is described either as an independent system or, as in Szathmáry's book, as a system depending on some other theory.

This line, which may be called orthodox Cartesianism, continues though because of the more and more sparse material[54] it is increasingly difficult to trace it. The handwritten lecture notes found in Debrecen and Kolozsvár, however, stress our opinion that in the first half of the XVIII century mainly this Cartesian physics was taught.

A more detailed discussion of these notes would lead the reader too far; thus a few examples need be mentioned. In Debrecen several notes on physics were left to posteriority, which contain the lectures of Márton Szilágyi, of whom only the name is known[55] (fig. 7). In these Cartesian lectures, all the systems of the universe are reviewed, and the conclusion arrived at that Copernicus's theory must be regarded as correct.

There are also a few other data available from which it becomes quite clear that in Debrecen Cartesian physics were in part taught till 1749, and on the other hand that the professors teaching astronomy were undoubtedly Copernicans.

One of the professors who began the reform of teaching was György Maróthy who died at an early age (1715–1744); the other was his successor Samuel Szilágyi. According to the notes of one of their pupils both taught the most up to date astronomy[56].

Unfortunately, only very few written records of György Maróthy remained, but his recently published correspondence[57] makes some interesting contributions to the knowledge of European propagation' of Copernicanism. In the course of his study trip abroad, Maróthy reached the Dutch town of Groningen where he became friendly with professor Daniel Gerdes. Maróthy had no money to publish his thesis and thus he undertook to defend Gerdes's thesis in a public dispute on condition that he might complete them with his own 24 theses. In the first five of these theses he defended Copernicus's theories. Maróthy relates all this in one of his letters and writes that in the course of the dispute lasting 7 hours he succeeded in routing his conservatice opponents, the audience took his part *nisi de terrae an solis motu agebatur*[58].

These examples may be multiplied, however, without yielding any new data

---

[54] For the reasons mentioned in note 51, the professors of the Hungarian Protestant colleges, including also the professors of the Lutheran schools of North Hungary (Slovakia), were scarcely able to publish books in the XVIII century.

[55] Library in Debrecen, MS. R. 303, a volume of 710 pages.

[56] Manuscript in Hungarian by the minister Ferenc Ujfalusi, in Library of Debrecen No. 0.380. His Debrecen studies were written in 1767 in the form of memoirs.

[57] Study tour abroad of *György Maröthy* by Imre Lengyel, and Béla Tóth. *Book and Library* VIII: 1, Debrecen 1970. 29.

[58] Ibid., p. 106: Letter to Jacob Cristoph Beck (1711–1785), Swiss theologian.

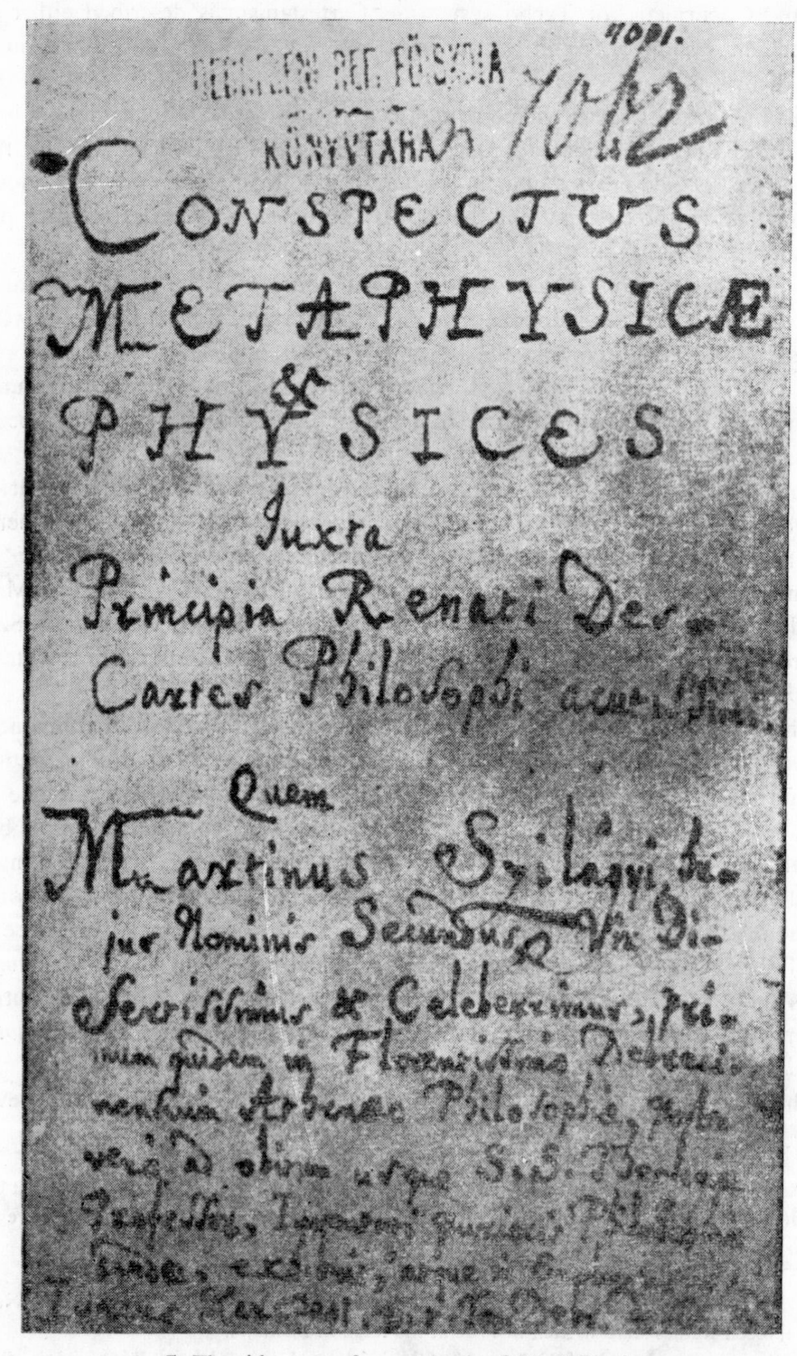

CONSPECTUS
METAPHYSICÆ
&
PHYSICES
Iuxta
Principia Renati Des-
Cartes Philosophi acutissimi.

Quem
Martinus Szilágyi, hu-
jus Nominis Secundus, Vir Di-
sertissimus & Celeberrimus, primum quidem in Florentissima Debreci-
nensium Athenæo Philosophiæ, postea
vero ad obitum usque S.S. Theologiæ
Professor,

7. The title page of a notebook of M. Szilágyi

in this line. Neither were more documents Cartesian science mentioned, since in them the Cartesian theory was mixed up with others. The authors of these papers might be called eclectics.

5. Eclectic trends in natural philosophy and the Copernicus problem in Hungary

Before beginning to review the principally Cartesian natural philosophy reference must be made to Copernicus's first Hungarian disciple known to us, who stands quite alone in the history of the Hungarian science, and cannot be ranged with either of the various trends of thought.

Dávid Frölich (1600–1648) lived in North-Hungary (today Slovakia) in a small town called Késmárk, in Szepesség (Spiš), where was a college which was an important centre of the cultural life of the XVII and XVIII centuries. Frölich is also unique since he is the only one of the scientists so far mentioned who was no theologian and not even a professor. He differs from the others also in that he did not select as places for studies abroad exclusively Wittenberg or one of the Dutch universities. He spent his longest period of study in Frankfurt-on-Oder, though during his European study trip he travelled practically all over the continent. Coming home, he lived in his native town as a private scientist.

Finally, another distinction: Frölich did not approach the Copernican problem from physics. If one may use in the case of this polyhistor the word "profession" in its good sense, Frölich expressed his opinion on this question from the point of view of geography. His text book on geography[59] published in 1639 attracted attention even abroad[60].

The geographical sciences already had a tradition in Hungary. János Honterus of Brassó (Brașov) was in the XVI century well known throughout Europe for his *Cosmographia*; with him, naturally, the Copernicus problem still did not arise[61].

In addition to geography, Dávid Frölich spent a good deal of time in preparing calendars. The calendar played an important role in the educational literature of the XVI and XVII centuries. Beside the Calendar in the proper sense of the word which also included a *Prognosticon* – i. e., astrological prodictions for the coming year—the calendar contained manifold historical, geographical, agronomical, medical, and rather frequently also physical information. Thus the calendar partly

---

[59] Dávid Frölich, *Medulla Geographiae Practicae* (the full title practically outlines the whole contents of the book), Bárfta, 1639.

[60] E. g. Antoine de Croilly French ambassador to Hungary wrote in a copy that the book is known also in France. L. Lipták, S., *Geschichte des evang. District Lyceum A. B. in Kesmark*, Kesmark, 1933, 46.

[61] Johannes Honterus's books were published in several editions between 1542 and 1564 in Hungary as well as abroad. Some of the copies can be found in Hungary, mostly with the title *Rudimenta Cosmographiae*; Brassó, 1542, RMK II 28; Brassó, 1570, RMK II 123; Zürich, 1548, RMK III 123 etc.

satisfied those needs which today are met by the daily journals, educational periodicals, broadcasting and television.

Dávid Frölich was a master of this genre. During thirty years of his short life he published calendars with various titles in Hungarian, Latin and German, several publications a year with various titles[62]. For this activity he obtained the title "imperial and royal mathematician" among other distinctions. He was rather proud of this title, and like to use it.

Of course Dávid Frölich was not the only editor of calendars in the XVII century. Among many unknown authors (and translators), Jakab Schnitzler and Márton Szentiványi published similar works, but Frölich was unparelled in this field, because among other reasons, he used popular education to raise the Copernican problem.

His interest in this field is quite clear also from the sentecne in his geography "Copernicus the excellent mathematician was born in Thoronium on Vistula"[63].

After a great variety of historical, geographical, natural history, "physical" and astronomical subject matter the treatment of which reflects a large scale between real science and superstition (magicians, witches, astrology) he rather unexpectedly raises the question: "Does the earth revolve with us every day in an immobile sky which is always at rest"[64].

This, says Frölich, is a profound question requiring careful consideration. Many scientists and untrained (ungelehrte) brains try to think it over. He himself answers positively, but in order that the prejudiced heads (Einseitigen) should better understand it, he is going to explain it more closely.

What follows are arguments and counterarguments which today may be regarded as classical. If Frölich had had contributed only this to the propagation of Copernican ideas in Hungary, his achievement would in 1640 have been creditable, for Galileo's Discorsi was scarcely known. He knew the Dialogue quite well, since he also quoted from it.

Frölich, however, did not mediate only on the empty pages of the calendar, but already *in 1632*—that is before Galileo's trial, he wrote a book on the revolution of the Earth[65]. This indicates that during his study trip abroad he must have

---

[62] The full list of David Frölich's calendars would lead too far. Many of them listed in RMK though not all of them can be found. See. J. Zemplén I, 154—155, which contains the most important ones; the copies in question are indicated in this work.

[63] Frölich, *Medulla Geographiae...*, 400.

[64] David Frölich, *Schreibkalender*, 1840.

[65] Because of its importance the full title of the work is quoted: *Anatome revolutionibus mundanae non solum bissextilis post Christum natum MDCXXXII, verum etiam annorum effluxorum et subsequentis saeculi ex infallabilibus Astronomiae principiis ad vivum quasi representans. Huis annexum est De antipodibus et Telluris Quotidiana circum versione edita Devide Frölicho Astron. Pract* (at this time the title "royal and imperial mathematician" did not exist) RMK II 478. It should be noted that one reads that no copies exist any more and consequently the title is incorrect. However, the author has found one copy in the Teleki Library in Marosvásárhely (today Tirgu Mures, Rumania).

become acquainted with the Copernican theory, and by contrast with many of the contemporaries did not allow either the burning of Giordano Bruno or the interdict against Galileo to deter him from acceptance of this theory. Religious considerations apparently did not much influence his works, since he expressed his opinion quite openly even in 1640 after Galileo had been sentenced.

In spite of all this, the work in question does not unequivocally present Copernicus's system. The first part deals with the preparation of calendars, the calculation of holidays. Frölich's other works are also of this type. After this follow 16 unnumbered pages discussing the theory on the revolution of the Earth. Reviewing the various reasons for the lunar eclipse, he writes as follows: "It is by evidence of divine providence that one sees the universe, which contains all bodies seen to remain in the same position, though the earth, our dwelling place, revolves with a rapid and uniform motion from west to east (according to the reformed astronomy which for the sake of practice I follow in this year) around his centre, its axes and poles fixed, which generally causes the Earth to be regarded as a gyroscope fastened to a nail. From this originate day and night. The motion itself, however, cannot be observed, only the lights surrounding the earth. We get the impression that they (the lights) rise and set".

It is worth noticing that even Frölich was cautious enough to use the expression "for the sake of practice". Even so, there is no doubt that he firmly believed in the revolving of the earth around its axis. With regard to the revolving around the sun he is somewhat uncertain: he speaks of the "Sun's road" or about Earth's revolving around the sun "according to Copernicus..."

It seems to be of interest that Frölich mixes up all this with a mystic, mythological philosophy: all is penetrated by a spirit of the Universe (mundi spiritus). This is doubtless the influence of Alsted, who by the way was much respected by Frölich. The spirit of the universe ensures, according to him, the *harmonia universalis*. The motion of the Earth and the planets is inevitable because only so can they fulfill their function. All this is rather confused; he becomes clear and specific only when he writes about the revolving of the Earth. He says that the Earth's revolving which in the preceding century was detected by Copernicus and Kepler should be sufficiently evident to all. Yet there are people in Pannonia who make fun of it, and say with the English poet:

Stare negas terram, nobis miracula narras
Haec cum scribebas, in rate forsan eras.

Everything that is now old was once new and became outdated. It is also true that the teaching of the Earth's revolving is not new (he enumerates the old adherents of the heliocentric theory). After this follow 14 counter-arguments and their refutation; for the most part these are discussed in the calendar quoted. The 14 counter-arguments are of a philosophical nature, and after those follow counter-arguments based on the *Bible*, though these are as easily disposed of as the previous ones: the *Bible* was dictated by the holy Spirit so that it might be descriptive and easily comprehensible by the common people (Galileo). If Joshuah had really stopped the Earth, it would be a miracle as great as the miracle of stopping the Sun. In general, to accept the *Bible* word for word may fall into grave mathematical as well as physical errors.

Thus the reader has so far encountered two lines of definitive acceptance of Copernicanism, one of which was the philosophy of Descartes linked with a mystical conception of natural philosophy. The works to be discussed also show strong Cartesian influence though the first elements of dynamics already appear in them. Even so, uncertainty still appears.

Like Frölich, János Pósaházi (1628/32–1686), professor in Sárospatak, author of the first Hungarian *Philosophia Naturalis*[66] (fig. 8) is also a contradictory personality. In his physics, which he calls "Physiology", he tries to comprise between the mechanics of Galileo and Descartes – i.e., between the most up to date trends of his age. Nevertheless, he was also influenced by the religiosity of Comenius and by Bacon's empiricism[67]. In his ecclesiastical views, on the other hand, after he and his school had to flee to Transylvania, Pósaházi was the upholder of the most conservative opinions.

What is to be expected from him in the Copernicus problem? He was too great a scientist to reject it, and too fanatical in religious questions to accept it definitively. Thus there remained only the possibility of a cautious evasion.

In reviewing the motion of the planets, Pósaházi says that the Copernicans believe that the Earth also is a planet revolving together with the moon in one day and round the sun within one year: "I leave it to the judgement of the wise—writes Pösaházi, whether this is true. I myself do not consider it as impossible"[68]. The first part of this sentence can be found in Sperling's physics[69] (much quoted by Pósaházi who himself studied in Utrecht and Franeker). The second part, however – the cautious, hesitating approval – is the author's own. Pósaházi is more "theological" than Apáczai, but undoubtedly much more of a "physicist" than his contemporaries.

Some further parts of Pósaházi's work also show the cautiousness of the theologian coupled with the behavious of the physicist. Gravitation for example is (following Descartes) the result of the pressure of the ether; however, this holds even if the Earth revolves around its axis in 24 hours. "Though this hypothesis is absurd, there are still many who adhere to it"[70].

---

[66] János Pósaházi, *Philosophia Naturalis Sive Introductio in theatrum naturae...*, Sárospatak 1667.

[67] Zemplén I, p. 275–287. With full literature.

[68] Pósaházi, *op. cit.*, p. 179.

[69] Sperling, *op. cit.*, p. 465.

[70] Pósaházi, *op. cit.*, p. 216.

PHILOSOPHIA
NATURALIS.
Sive:
INTRODUCTIO IN
THEATRUM NATURÆ.

Authore JOH. POSAHAZI.
Art. L. M. In Illuſtri Schola
Saros-Patakina publico
Profeſſore.

PATAKINI,
Impenſis, JOHANNIS ROSNIAI,
M. DC. LXVII.

8. The title page of the *Philosophia Naturalis* by F. Pósaházi

In reviewing the various theories, he points out the difficulties and contradictions which arise from accepting the views of either Ptolemy, Tycho or Descartes, while the Copernicans easily refute every counter-argument. Otherwise, when discussing the vortex theory and speaking of the centre he always adds that it is either the Earth or the Sun. Thus, for example, the apparently no-uniform motion of the planets was solved by Ptolemy by introducing the epicicles — with the ether theory this could eventually be explained by supposing that the velocity of either changes along the orbit of the planet", ...but if the hypothesis of the motion of the Earth is accepted, these phenomena can apparently be explained"[71].

Pósaházi's book and his scientific attitude exerted a great influence on his contemporaries as well as on his successors. His political discussions have long been forgotten, while in the teaching of physics in Transylvania and Sárospatak his basic tenets can still be found. Of course, positive characteristics can also be found: the beginning of the dynamical way of thinking, which with Newton's mechanics was gradually completed by his successors. Now let us rather examine the survival of the negative part of Pósaházi's teaching concerning the Copernicus problem.

Mihály Szathmáry's book already reviewed in fact reflects Pósaházi's influence. Since, however, he follows in the copernicanian question rather the lines of Apaczai and Köleseri, this author was ranked among the Cartesian philosophers.

Pósaházi — as we already know — did not return to his school to Sárospatak, but died in Transylvania. His successor in the school, working at that time u Gyulafehervar, was Sámuel Kaposi (1660—1713) who was also one of the characteristic polyhistors of the time. His works are extant, unfortunately, only in manuscript, including a complete physics entitled *Physiologia* which follows Pósaházi practically word for word[72]. Expressing himself more briefly, he is of the same opinion as his master in the Copernican question. This is proved by demonstrating the various systems of the universe, side by side, as can be seen in fig. 9.

The connecting links between Gyulaféhervár, Marosvásárhely and Sárospatak have already been mentioned[73]. In this way was manifested in 1723 the Pósaházi-Kaposi influence in Sárospatak by way of the physical[74] and geographical manuscripts of Csécsi János the younger (1689—1769)[75]. Csécsi was apparently a follower of the Cartesian ine of thought, and was also a good experimental physicist; even so, in his geographical manuscript appears the following sentence: "Either it (the Earth) is the centre of the universe or somewhere else about which physicists still

[71] Ibid., p. 182—184.

[72] Sámuel Kaposi, the exact title of the note is: *De Physiologiae Natura et Partibus*, 1708. Bound together with other notes in the Bolyai Library, Marosvásárhely.

[73] See note 51.

[74] Janos Csécsi, junior, *Compendium Physicum*, copied by Andras Balogh, MS. 164, Library of the Reformed College, Sárospatak.

[75] János Csécsi, junior, *Introductio in universalem Geographiam...*, Sárospatak 1738.

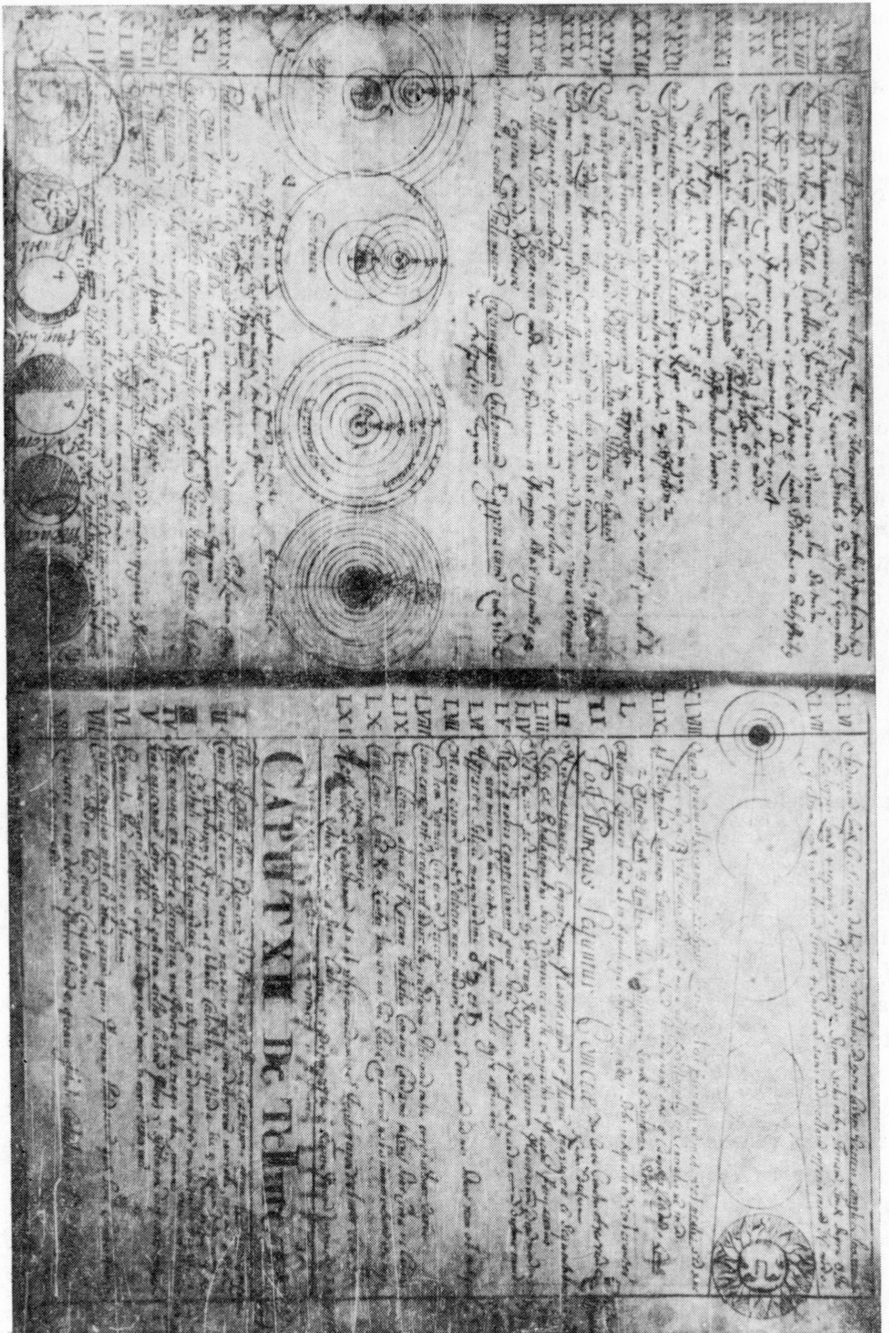

9. World systems in a manuscript of S. Kaposi (early XVIII century)

dispute; geography leaves it on its place; ... though it is also said that the Earth revolves around its axis".

Still in the year 1738 such views can be found. In 1736 in Transylvania a most interesting scientific experiment was carried out. István Töke (dates of his birth and death unknown) taught philosophy in Nagyenyed from 1725, and his book published in 1736 shows that he tried to reconcile the Cartesian speculations with experimental physics[76]. Of course, one knows in advance the result of this experiment, though even so it is interesting. From our point of view it is interesting that this author declares it as self evident that the sun is the centre of every turbulent motion of the firmament and that the Earth revolves round the sun.

In discussing the eclectical physical trends from the point of view of Copernicus's theory only one question remained untouched Among Hungarian trends in natural philosophy up to the turn of the XVII–XVIII centuries, the influence of Christian Wolf played an important role. Those who because of their education in Wittenberg did not come in touch with Cartesianism learn through Wolf the new physics and become followers of Newton. One finds, of course, here also various shades and mixtures of thoughts. One thing, however, is quite obvious: the Copernicus problem was really never a problem to the "Wolfians", since Wolf accepted the greater part of Newton's mechanics, which automatically meant acceptance of Copernicus (the elliptical orbit of the planets constitutes no major difference).

For this reason, we need mention only briefly that Wolf's only representative in the first half of the XVIII century in Transylvania was Sámuel Nádudvari (?–1754), professor in Marosvásárhely, also a polyhistor. Many manuscripts of his are extant including works dealing with physics and astronomy[77]. His contemporaneous biographer stresses Nádudvari's great erudition but also disapproves of his revolutionary ideas—e. g., that he taught astronomy entirely according to Copernicus and Newton, and what is more he preached his revolutionary ideas even from the pulpit[78].

Samuel Nádudvari, as a follower of Wolf was an exception in Cartesian Transylvania. In upper Hungary, however, whence the Lutherans of the XVIII century went mainly to Halle to study and took up the ideas of pietism, Descartes physics had practically no followers.

After Bayer, no book was printed, though according to manuscripts and lecture notes physics was here taught according to Wolf. The Copernican problem already appears only by reviewing the various systems of the universe.

---

[76] István Töke, *Institutiones Philosophiae Naturalis Dogmatico-experimentalis, quibus veritates physicae luculentis observationibus et experimentalis illustratae ac confirmatae nexu scientifico methodice proponuntur...*, Cibini 1736.

[77] A detailed enumeration in J. Koncz, *History of the Lutheran and Calvinists college in Marosvásárhely*, 1889, p. 182 (in Hungarian).

[78] Péter Bod, *St. Policarpus of Smirna... history of the Transylvanian bishops*, Nagyenyed, 1766, pp. 175 and 202 (in Hungarian).

Thus, for example, in Vilmos Frigyes Beer's (1691–1767) physics lectures delivered in 1720, one can observe that this professor reviewed first the astronomical discoveries of Galileo and Huyghens, and then explained the advantages of the Copernican system as compared with the Descartes vortex theory[79]. In further examples, similar statements and teachings can be found.

* *
*

Our review covers approximately 200 years from the publication of *De Revolutionibus*, and one may rightly say that the new concept of the universe and with it the new physics was able only slowly to invade the economically and socially backward Carpathian Basin.

One might perhaps think that there is not much more to be said, and that in the second half of the XVIII century the question as to whether the earth revolves or not simply could not arise. But this was not so: the problem was still alive and lived until it had conquered also the brains of theologians and influenced what they had to say.

II. THE QUESTION OF COPERNICANISM IN THE SECOND HALF OF THE XVII CENTURY IN HUNGARY

1. The Survival of Anti-Copernicanism

Though it occasionally occurred, outright anti-Copernicanism can be regarded as a rather exceptional phenomenon in the second half of the XVIII century.

The author of the first geography written in Hungarian, Pál Bertalanffy (1706–1763), a Jesuit professor, in 1767 firmly took the side of the geocentric theory, and writes that the followers of Copernicus cannot properly explain the teachings of the *Scripture*; "I should rather be ready to believe that Copernicus's brains revolved than accept that the earth revolves round the sun like a planet"[80].

There is still a later fact. Bernat Sartori (1735–1801), a Franciscan monk, published in 1772 a complete scholastic philosophy[81], in which he took over word by word Bertalanffi's (mostly erroneous) data and opinion concerning Copernicus.

---

[79] F. W. Beer, *Theorema Physices ex Johanni Chr. Gottschedii Philosophia excerpta*... (the author has not succeeded establishing the identity of Gottsched). In the archives of the former Lutheran Lyceum in Késmárk, No. 17.

[80] Pál Bertalanffy, *A concise two-fold review of the world* (in Hungarian), Tyrnaviae, 1767, p. 9.

[81] Bernát Sartori, *Philosophy in Hungarian*... (in Hungarian), Eger, 1772.

2. Change in Nagyszombat (Trnava)

In 1757, also the Jesuit Bertalanffi may be considered as an exception. From the fifties of the XVIII century new, voluminous text books on the new physics were published almost suddenly in Nagyszombat. In connection with a few, characteristic phenomena may be observed, which are outstanding in the history of sciences.

As already indicated, in 1737, and even later, books such as Akai's *Cosmographia* were published. In 1753, Maria-Theresa issued an order in connection with the university reform in Nagyszombat making it become compulsory for every professor to write text books and to cease the dictation of lecture notes which up to that time was quite customary[82]. Following this order and beginning with 1755 voluminous physics text books were published in quick succession: works by András Adány (1715—1795)[83], András Jaszlinszky (1715—1783)[84] (fig. 10), Antal Reviczky (1723—1781)[85] and Mihály Klaus (1719—1792)[86].

This is certainly a surprising phenomenon. Practically without any antecedents, serious university text books are found. This can only be explained by assuming that the manuscripts must have been already prepared, and the authors only awaited the royal order and a loosening of the rigid Jesuit *Ratio Studiorum* to come out into open with their books[87].

Much more surprising, however, is the contents of the books. After the publication of Newton's principia, Cartesianism was in principle out-dated, at most a few years can be accepted. Yet these physics books were — though the authors well knew the Newtonian physics — essentially Cartesian books. A rather curious paradoxon of the history of physics here took place practically before our eyes. The philosophy and physics of Descartes regarded by the churches as atheistic (or at least as a danger leading to atheism) was first brought into the country and disseminated by Calvinist theologians between the years 1653 (Apáczai) and 1736 (Istvan Tőke). When, however, Newton's physics throughout the country were already accepted, the Cartesianism appears in the Jesuit Nagyszombat.

---

[82] See e. g. E. Fináczy, *History of Hungarian Public Education in the age of Maria Theresa*; Budapest, 1899, I. 308; Further J. Szentpétery, *History of the Faculty of Philosophy 1635—1935*, Budapest, 1935, p. 76, and T. Pauler, *History of the Hungarian Royal University Budapest*, Budapest 1880, 171 (in Hungarian).

[83] *Philosophiae Naturalis Pars Prima physica generalis in usum discipulorum a* R. P. Andrea Adány... *anno 1755*, Tyrnaviae, pp. 640.

[84] *Institutiones Physicae Pars Prima* a R. P. Andrea Jaszlinszky... *Tyrnaviae 1756 et... pars altera 1756*. A volume of 782 pages.

[85] *Elementa Philosophiae Naturalis Pars Prima ab* Antonio Reviczky... *Tyrnaviae 1757 et... pars altera Anno 1758*. A volume of 551 pages.

[86] *Naturalis Philosophiae seu Physicae Tractatio Prior... Viennae et Pragae 1756... et altera Viennae 1756*. A volume of 993 pages.

[87] Csaba Csapodi, *On the boundary of two worlds. A chapter on the past of the Hungarian Enlightenment* (in Hungarian), Századok, 1945—46, 85—137, and Zemplén II, 221—240.

INSTITUTIONES
PHYSICÆ
PARS PRIMA,
SEU
PHYSICA GENERALIS
IN USUM
DISCIPULORUM CONCINNATA
A R. P. ANDREA JASZLINSZKY
E SOCIETATE JESU
PHILOSOPHIÆ DOCTORE,
EJUSDEM IN UNIVERSITATE
TYRNAVIENSI
PROFESSORE PUBLICO
ORDINARIO.

TYRNAVIÆ,

TYPIS ACADEMICIS SOCIETATIS JESU,
ANNO MDCCLVI.

10. The title page of the textbook by Jaszlinszky

One of the antecedents of this phenomena is that Ferenc B. Kéri, professor (1702–1768) who in Nagyszombat was engaged also in astronomy[88], was Rector of the university in 1753, at the time of the university reform. He himself wrote books on mechanics in which he tried to reconcile the Carthesian and Newtonian theories[89]. With Kéri, as in general with physicists of the XVIII century, the Newtonian concept of force and the problem of action in distance caused serious difficulties. Kéri did not succeed in finding a satisfactory solution. Nevertheless, he may be regarded as the spiritual ancestor of the above text books of physics; he exerted a very great influence, and is frequently quoted.

However, this "philosophical" explanation does not seem to solve the problem. The late Cartesianism of the Jesuits in Nagyszombat (including Kéri himself) had deeper ideological roots. This is shown by Kéri's standpoint on the Copernican problem.

It has been respectedly said that the acceptance of Copernicanism is in close relation with the dissemination of the new physics. Whoever accepts the Newtonian dynamics and the law of general gravitation must necessarily be Copernican. Descartes, Bacon or a follower of any other discipline of natural philosophy may still disaprove of, accept, or even "correct" the Descartes vortex theory. An empiricist may be a Copernican, though Bacon did not accept it, but with Newton's acceptance there is no longer freedom of choice.

And precisely this was what the authors of the first text books wanted to avoid, though in many instances they did not state it openly.

It is still characteristic of these books that they consider it important to review the various systems of the universe, and that astronomy is treated in what was called "special" or "particular" physics and not within the framework of mechanics; consequently, they still did not accept the new celestial mechanics.

In rewiewing the systems of the universe they go so far as to state that only Copernicus's system is rational though they do not say it decidedly. We have stil not come so far as that. Nevertheless the dispute lost its edge. The author knew thel consequences for the reader should he get into trouble by any open statement[90].

3. Cartesianism disappears

As far as is ascertainable, and disregarding the short intermezzo discussed in the previous section, Cartesianism had disappeared earlier in the other Hungarian schools. It is known for instance that István Hatvani (1718–1786), famous physi-

[88] In a detailed review of Kéri B. Ferenc's, *Studies in physics and his biography*, see Zemplén II, 214–221.

[89] Keri B. Ferenc's works, *anonymous* dissertations similar to those discussed in the 2nd section. These mechanical topics are treated in the following works: *Dissertatio Physica de Corpore generatim deque opposito eidem vacuo*, Tyrnaviae, 1752.; *Diss. Phys. De motu corporum*, ibid. 1735; *Diss. Phys. De Causis motuum in corporibus*, ibid., 1754.

[90] M. Klaus *op. cit.*, II, p. 32.

cian and professor of physics, certainly taught in Debrecen from 1749 Newton's physics, though no book of his on this subject does exist. His book on philosophy however, amply proves that he was y disciple of Newton and Copernicus[91].

After István Töke's book, only one printed work was published in Transylvania, a translation from German to Latin in 1774[92]. Though the data (manuscript material) are scarce it can be assumed with great probability that in Transylvania Cartesianism was followed much earlier by Copernicanism and Newton's physics. Nevertheless, also in Nagyszombat Cartesianism soon disappeared.

Thus, for instance, one finds in Nagyszombat an anonymous thesis promising astronomy according to its title but treating actually the mechanics[93]. The dissertation consists of two parts: the first contains theses including physical ones. Thesis no 37 reads as follows: "Copernicus's system is better than that of Tycho". In the text itself the author fully and decidely accepts Newton's mechanics.

This latter fact is important over and above its already discussed significance, because the authors in Nagyszombat still make a detour, though already a smaller one, to react Newton. Pál Makó (1724–1793)[94], Antal Radics (1726–1773)[95] and finally János Horváth (1732–1799)[96], author of a text book on physics well known also abroad, hurl themselves after Descartes into the most extreme Newtonianism while they build up their physics based on interesting but rather idealistic mechanics and forcecurve of Roger Boscovich (1711–1787)[97]. Though these authors are still all Jesuits (the Jesuit order was dissolved in Hungary by the Emperor Joseph II), they taught later as priests in the university, first transferred to Buda (1777) and then to Pest (1780), Boscovich type dynamics which can in no way be reconciled with any anti-Copernican attitude. This becomes apparent from the fact that the authors referred to treat the motion of the planets *and* the Earth within the framework

---

[91] István Hatvani, *Introductio ad Principia Philosophiae Solidioris...*, Debrecini, 1757.

[92] The original book: Johann Gottlob Krüger, *Naturlehre*, Halle, 1740—1749. Translated by István Kováts, professor in Nagyenyed. Title of the translation: *Elementa Philosophiae Naturalis...*, Kolozsvár, 1744.

[93] Full title: *Astronomia physicae juxta Newtoni Principia Breviarium Methodo scholastica ad usum juventus*, Nagyszombat, 1760. This dissertation was prepared for István Ranics' (student in Pécs) doctorate, but the authorship of no professor lecturing in that time in Nagyszombat was able to be established (as e. g. in the case of Kéri, B. Ferenc).

[94] Pál Mákó, *Compendaria Physicae Institutio...*, *Pars I. Phys. Experimentalis*, Viennae, 1762; Pars II, ibidem 1763.

[95] Radica devoted three works to a review of Boscovich's theory: *Introductio in Philosophiam Naturalem P. Rogerii Boscovich accomodata...*, Buda (S. a.); *Principia Boschovichii singulari tractata illustrata*, Buda, 1765, and *Institutiones Physices...* Buda, 1766.

[96] Bapt. János Horváth wrote more text books on physics, which were republished several times; the Boscovich-theory is dealt with only in the first publications in some detail, e. g. *Institutiones Physicae Generalis...*, Tyrnaviae, 1767, and *Inst. Phys.*, *ibidem* 1770.

[97] Boscovich's works attained several editions in his time. The most up to date critical edition is a Latin-English bilingual work used also by the present author, *Rogeris Boscovich: Theoria Philosophiae Naturalis, Venice 1763*, Chicago—London, 1922, introduced by J. M. Child.

of mechanics, and no further mention is made of the various systems of the universe; at most they polemise with Descartes, but even with him on other questions.

In the later editions of Horváth Janos's text books, also the Boscovich curve disappears together with Cartesianism, and gradually the most up to date physics became developed at the university in Pest[98].

Our review concerning the second half of the XVIII century, however, still can not be concluded. There is a field in which one still meets with some hesitation in relation to the Copernican theory — popular science literature.

4. The Copernican problem appears again in popular literature

In general, one encounters in popular literature only well developed, already proved subjects, and as a result this literature always shows a certain time lag as compared with the actual state of science.

Moreover, in Hungary two further factors influenced disadvantageously the standard of the popular literature.

One of them is that this literature had to be written in *Hungarian* (Apáczai's attempt to write in Hungarian had no followers for a long time); consequently, the authors struggled to use proper scientific expressions and this — naturally — to some degree influenced also the standard.

Another reason is that the authors — still clergymen — prefered to translate, and to translate older, mainly religious books from English, German and French literature.

In essence, the development of the 200 years of Hungarian scientific literature written in Latin as so far followed, was repeated in concentrated form over 30 years, so that all the typical attitudes concerning the Copernican problem can be found again.

The open anti-Copernicanism in Bertalanffi's and Sartori's works has already heen noted.

Let us now examine the work of a Jesuit author of much greater erudition. We refer to János Molnár's book published in 1777[99]. Even the title of this book shows that it was prepared according to Newton's physics. The author calls the Cartesian ether a "poetic fiction" and yet he is extremely cautious when writing about the Earth's motion. Thus for example, the Earth can be called only with a certain caution a planet, and "... for the sake of simplicity, according to Copernicus and Newton the Earth is considered as a planet", "though — he says — there are philosophers who think that this is not entirely certain"[100].

---

[98] E. g. J. Horváth, *Elementa physicae opus novis elaboratum curis et a prioribus editionibus diversum*, Buda, 1790.

[99] Bapt. János Molnár, *The first principles of Physics, on the nature according to Newton's pupils* (in Hungarian), Pozsony and Kassa, 1777.

[100] Idem, pp. 117—120 and 127—128.

As in the XVII century, one finds hesitancies, also among the Protestants. It is interesting to observe that Derham's extremely popular book, which was also translated and served as a model to Protestant authors[101], did not find any problem in the Copernican theory: God's wisdom ensures that with the Earth rotating at high speed gravitation should maintain equilibrium against centrifugal force[102].

Perhaps a few further typical examples: Benjamin Szönyi translated Charles Rollin's religious natural history work and added some parts written by himself in verse. In these, he uses alternatively the motion of the Sun and Earth, though sometimes he speaks expressly of the Earth's revolving and orbit[103].

The Calvinist minister György Horváth, meditating in his "scientific prayer book on the system of the universe acknowledges that the theories of Descartes and Copernicus are more rational than Tychos's. Yet he prefers the latter because it can be better reconciled with the *Holy Scripture*, but: "I do not dispute with any of them... leaving the Copernican system to the philosophers I take my stand with Tycho in what is shown by the *apparentia* and taught by the *Holy Scripture*"[104].

However, as one approaches the end of the century such examples become more rare. Adam Pálóczi Horváth, son of the lastnamed author praises Newton, who solved the problems concerning the Copernican theory, with eloquent distinction[105].

Finally the best and the most complete popular book on physics, published in 1808 by Márton Varga, by finally accepting Newton closes not only the Copernican question, but also the 100 years old dispute between Descartes and Newton[106].

SUMMARY

Some 200 years of the history of the Hungarian natural philosophy and physics are reviewed. To sum up once more the oft repeated conclusion can be drawn: in order to accpet Copernican theory, it was not only necessary for the new physics to emerge but this new physics had to be understood and become common knowledge. The Hungarian economic and social situation in only a few instances up to the end of the XVIII century enabled the third step to be taken in the development of the new physics. Nevertheless, Hungarian scientists, too, participated in the European ideological and scientific disputes, and it is perhaps no overstatement to say that even their errors contributed to put into motion the Earth which for so long stood still in Hungary.

---

[101] The title of the work in question by William Derham (1657—1735) minister of Upmister is: *Physics-Theology or a demonstration of the being and attributes of his works creation*, London, 1713. The Hungarian translation, which is quoted, was published in Vienna in 1793. The translator was G. Segesvári.

[102] Idem, p. 48, 53.

[103] Charles Rollin (1661—1741) was professor of theology of he Collège de France. The title of B. Szönyi's translation is: *Our Children's Physics* (in Hungarian), Pozsony, 1774.

[104] György Horváth, *The School of Nature and Grace*, (in Hungarian), Györ, 1775, p. 216.

[105] Adám Horváth, *The shortest summer night...* (in Hungarian), Pozsony, 1791.

[106] Márton Varga, *The Science of Nature Beautiful*, Nagyvarad, 1808 (in Hungarian).

SOME ABBREVIATIONS

RMK —Regi Magyar Kőnyvtar (Old Hungarian Bibliotheque). This is an extensive Bibliography in four volumes, which contains all works published from 1472 to 1711 in Hungarian (Vol. I.), in Latin or any other language (Vol. II), or works written by Hungarians and published abroad (Vol. III in two parts), ed. by Károly Szabó, Budapest, 1896 and 1898. This work will be published in a new edition. Up to now one volume (1473—1600) has been published, through it was not possible to quote from this book.

Zemplén I — M. J. Zemplén, *The History of Physics in Hungary up to 1711*, Budapest, 1961 (in Hungarian), 317 pp.

Zemplén II — M. J. Zemplén, *The History of Physics in Hungary in the XVIII Century* (in Hungarian), 495 pp.

MTA — Hungarian Academy of Sciences.

TABLE DES AUTEURS ET DES PERSONNAGES HISTORIQUES

TABLE DE MATIÈRES